Advanced Condition Monitoring and Fault Diagnosis of Electric Machines

Muhammad Irfan
Najran University, Saudi Arabia

A volume in the Advances in Computer and Electrical Engineering (ACEE) Book Series

IGI Global
DISSEMINATOR OF KNOWLEDGE

Published in the United States of America by
IGI Global
Engineering Science Reference (an imprint of IGI Global)
701 E. Chocolate Avenue
Hershey PA, USA 17033
Tel: 717-533-8845
Fax: 717-533-8661
E-mail: cust@igi-global.com
Web site: http://www.igi-global.com

Library of Congress Cataloging-in-Publication Data

Names: Irfan, Muhammad, editor.
Title: Advanced condition monitoring and fault diagnosis of electric machines / Muhammad Irfan, editor.
Description: Hershey, PA : Engineering Science Reference, [2019] | Includes bibliographical references.
Identifiers: LCCN 2018011682| ISBN 9781522569893 (h/c) | ISBN 9781522569909 (eISBN)
Subjects: LCSH: Electric machinery--Monitoring. | Fault location (Engineering)
Classification: LCC TK2313 .A38 2019 | DDC 621.31/042--dc23 LC record available at https://lccn.loc.gov/2018011682

This book is published in the IGI Global book series Advances in Computer and Electrical Engineering (ACEE) (ISSN: 2327-039X; eISSN: 2327-0403)

British Cataloguing in Publication Data
A Cataloguing in Publication record for this book is available from the British Library.

For electronic access to this publication, please contact: eresources@igi-global.com.

Advances in Computer and Electrical Engineering (ACEE) Book Series

ISSN:2327-039X
EISSN:2327-0403

Editor-in-Chief: Srikanta Patnaik, SOA University, India

MISSION

The fields of computer engineering and electrical engineering encompass a broad range of interdisciplinary topics allowing for expansive research developments across multiple fields. Research in these areas continues to develop and become increasingly important as computer and electrical systems have become an integral part of everyday life.

The **Advances in Computer and Electrical Engineering (ACEE) Book Series** aims to publish research on diverse topics pertaining to computer engineering and electrical engineering. **ACEE** encourages scholarly discourse on the latest applications, tools, and methodologies being implemented in the field for the design and development of computer and electrical systems.

COVERAGE

- Sensor Technologies
- Digital Electronics
- Computer science
- Microprocessor Design
- Electrical Power Conversion
- Algorithms
- Qualitative Methods
- Power Electronics
- Circuit Analysis
- Computer Hardware

IGI Global is currently accepting manuscripts for publication within this series. To submit a proposal for a volume in this series, please contact our Acquisition Editors at Acquisitions@igi-global.com or visit: http://www.igi-global.com/publish/.

The Advances in Computer and Electrical Engineering (ACEE) Book Series (ISSN 2327-039X) is published by IGI Global, 701 E. Chocolate Avenue, Hershey, PA 17033-1240, USA, www.igi-global.com. This series is composed of titles available for purchase individually; each title is edited to be contextually exclusive from any other title within the series. For pricing and ordering information please visit http://www.igi-global.com/book-series/advances-computer-electrical-engineering/73675. Postmaster: Send all address changes to above address.

Titles in this Series

For a list of additional titles in this series, please visit:
https://www.igi-global.com/book-series/advances-computer-electrical-engineering/73675

Applications of Security, Mobile, Analytic, and Cloud (SMAC) Technologies ...
P. Karthikeyan (Thiagarajar College of Engineering, India) and M. Thangavel (Thiagarajar College of Engineering, India)
Engineering Science Reference • ©2018 • 300pp • H/C (ISBN: 9781522540441) • US $215.00

Smart Grid Analytics for Sustainability and Urbanization
Zbigniew H. Gontar (SGH Warsaw School of Economics, Poland)
Engineering Science Reference • ©2018 • 306pp • H/C (ISBN: 9781522539964) • US $215.00

Vehicular Cloud Computing for Traffic Management and Systems
Jyoti Grover (Manipal University Jaipur, India) P. Vinod (SCMS School of Engineering and Technology, India) and Chhagan Lal (Manipal University Jaipur, India)
Engineering Science Reference • ©2018 • 281pp • H/C (ISBN: 9781522539810) • US $235.00

Multidisciplinary Approaches to Service-Oriented Engineering
Mehdi Khosrow-Pour, D.B.A. (Information Resources Management Association, USA)
Engineering Science Reference • ©2018 • 389pp • H/C (ISBN: 9781522559511) • US $245.00

Cyber-Physical Systems for Next-Generation Networks
Joel J. P. C. Rodrigues (National Institute of Telecommunications (Inatel), Brazil & Instituto de Telecomunicações, Portugal & University of Fortaleza (UNIFOR), Brazil) and Amjad Gawanmeh (Khalifa University, UAE)
Engineering Science Reference • ©2018 • 293pp • H/C (ISBN: 9781522555100) • US $210.00

Augmented Reality for Enhanced Learning Environments
Gerardo Reyes Ruiz (Universidad Autónoma del Estado de México, Mexico) and Marisol Hernández Hernández (Universidad Autónoma del Estado de México, Mexico)
Information Science Reference • ©2018 • 322pp • H/C (ISBN: 9781522552437) • US $195.00

For an entire list of titles in this series, please visit:
https://www.igi-global.com/book-series/advances-computer-electrical-engineering/73675

IGI Global
DISSEMINATOR OF KNOWLEDGE

701 East Chocolate Avenue, Hershey, PA 17033, USA
Tel: 717-533-8845 x100 • Fax: 717-533-8661
E-Mail: cust@igi-global.com • www.igi-global.com

Table of Contents

Preface xv

Chapter 1
Fault Diagnosis of Induction Motors Using Motor Current Signature Analysis: A Review 1
Muthiah Geethanjali, Thiagarajar College of Engineering, India
Hemavathi Ramadoss, Thiagarajar College of Engineering, India

Chapter 2
Fault Detection and Isolation in Power Systems Using Unknown Input Observer 38
Hassan Haes Alhelou, Tishreen University, Syria

Chapter 3
Diagnosis of the Broken Rotor Bars Faults by Root-MUSIC Method 59
Ahmed Hamida Boudinar, University of Sciences and Technology of Oran "Mohamed Boudiaf", Algeria
Azeddine Bendiabdellah, University of Sciences and Technology of Oran "Mohamed Boudiaf", Algeria
Noureddine Benouzza, University of Sciences and Technology of Oran "Mohamed Boudiaf", Algeria

Chapter 4
Advanced Fault Diagnosis Monitoring Scheme in Asynchronous Motor Using Soft Computing Method 89
K. Vinoth Kumar, Karunya Institute of Technology and Sciences, India
Ramya K. C., Sri Krishna College of Engineering and Technology, India
Muhammad Irfan, Najran University, Saudi Arabia

Chapter 5
A Mobile Augmented Reality System for Enhancing Electrical Machine Supervision 98
Jesús Gimeno, University of Valencia, Spain
Sergio Casas, University of Valencia, Spain
Cristina Portalés, University of Valencia, Spain

Chapter 6
Invasive Methods to Diagnose Stator Winding and Bearing Defects of an Induction Motors 122
Muhammad Aman Sheikh, Universiti Teknologi PETRONAS, Malaysia
Nordin B. Saad, Universiti Teknologi PETRONAS, Malaysia
Nursyarizal Mohd Nor, Universiti Teknologi PETRONAS, Malaysia
Sheikh Tahir Bakhsh, King Abdulaziz University, Saudi Arabia
Muhammad Irfan, Najran University, Saudi Arabia

Chapter 7
Types of Faults and Condition Monitoring Methods of Induction Motors 131
Maheswari M., Malla Reddy Engineering College, India
Gunasekharan S, Malla Reddy Engineering College, India

Chapter 8
Vibration Monitoring of Rotating Electrical Machines: Vibration Monitoring 163
Achintya Choudhury, Sir Padampat Singhania University, India

Chapter 9
A State of Art Critique Review of Role of Optimization in NDTS for Effective Condition Monitoring 189
Ashish Khaira, Samrat Ashok Technological Institute, India
Ravi K. Dwivedi, Maulana Azad National Institute of Technology, India

Chapter 10
Induction Machine Rotor and Stator Faults Detection by Applying the N-F Network 205
Souad Saadi Laribi, Ibn Khaldoun University, Algeria

Chapter 11
A State-of-the-Art Critique Review of the Role of Decision-Making Techniques in NDT for Condition Monitoring 222
Ashish Khaira, Samrat Ashok Technological Institute, India
Ravi K. Dwivedi, Maulana Azad National Institute of Technology, India

Chapter 12
Identifying Parameters for Nonlinear Degradation-Based Prognostics: A Case Study With Exponential Degradation on Induction Motors 237
*Feng Yang, Institute of High Performance Computing (A*STAR), Singapore*
*Mohamed Salahuddin, Institute of High Performance Computing (A*STAR), Singapore*

Chapter 13
Sensor and Its Application 261
Saifur Rahman, Najran University, Saudi Arabia
Abdullah S. Alwadie, Najran University, Saudi Arabia
S. Hasan Saeed, Integral University, India
Faizan A. Khan, Integral University, India

Compilation of References 275

About the Contributors 299

Index 306

Detailed Table of Contents

Preface xv

Chapter 1
Fault Diagnosis of Induction Motors Using Motor Current Signature Analysis: A Review 1
Muthiah Geethanjali, Thiagarajar College of Engineering, India
Hemavathi Ramadoss, Thiagarajar College of Engineering, India

Induction motors are termed as horses of modern industry because they are playing a vital role in industries. They are simple, efficient, robust, rugged, and highly reliable. The feasibility of mishap in induction motors is less, but they are prone to faults, which are left unobserved most of the time. Hence, more attention has been paid to detection and diagnosis of incipient faults to prevent damage spreading and increase the lifetime of the motor. To detect and diagnose the faults, online condition monitoring of the machine has been utilized in a wide manner. At present, focus is made on optimization procedures for fault diagnosis in induction motors to obtain a quick assessment at industry level. This chapter discloses an overview of various types of possible faults in induction motors. In addition, the conventional (invasive) and innovative techniques (noninvasive), especially motor current signature analysis (MCSA), techniques for fault detection and diagnosis in induction machines are covered with a focus on future research.

Chapter 2
Fault Detection and Isolation in Power Systems Using Unknown Input Observer 38
Hassan Haes Alhelou, Tishreen University, Syria

Unknown input observers (UIO) find application in many cases for successful fault detection and isolation (FDI). In this chapter, a scenario where the unknown input observer is applied to load frequency control loops in interconnected power systems is analyzed. A UIO was chosen because load demand is not always constant and it can be considered to introduce an unknown disturbance to the system. Mathematical formulations on how to detect and isolate sensor faults are presented which are then implemented in MATLAB Simulink for simulations. Based on this historical survey on the application of UIO, a thesis on UIO application in FDI in distributed generation is done.

Chapter 3
Diagnosis of the Broken Rotor Bars Faults by Root-MUSIC Method................59
Ahmed Hamida Boudinar, University of Sciences and Technology of Oran "Mohamed Boudiaf", Algeria
Azeddine Bendiabdellah, University of Sciences and Technology of Oran "Mohamed Boudiaf", Algeria
Noureddine Benouzza, University of Sciences and Technology of Oran "Mohamed Boudiaf", Algeria

This chapter describes a new diagnosis approach, the Root-MUSIC (RM) method, for identification of the progressive cracking in the rotor of induction motors. This method used initially in the domain of RADAR for localizing mobile targets is being applied to the domain of induction motors diagnosis. This approach has several advantages compared to the conventional power spectral density estimation (PSD) by periodogram technique. Indeed, the main advantage of this approach is its very good frequency resolution for a very short acquisition time, something impossible to achieve with the conventional method. However, in order to reduce the computation time which is the main drawback of the RM method, this method will be applied to only a specified frequency band: one that carries information about the sought fault.

Chapter 4
Advanced Fault Diagnosis Monitoring Scheme in Asynchronous Motor Using Soft Computing Method..89
K. Vinoth Kumar, Karunya Institute of Technology and Sciences, India
Ramya K. C., Sri Krishna College of Engineering and Technology, India
Muhammad Irfan, Najran University, Saudi Arabia

This chapter deals with the implementation of a PC-based monitoring and fault identification scheme for a three-phase induction motor using artificial neural networks (ANNs). To accomplish the task, a hardware system is designed and built to acquire three phase voltages and currents from a 3.3KW squirrel-cage, three-phase induction motor. A software program is written to read the voltages and currents, which are first used to train a feed-forward neural network structure. The trained network is placed in a Lab VIEW-based program formula node that monitors the voltages and currents online and displays the fault conditions and turns the motor. The complete system is successfully tested in real time by creating different faults on the motor.

Chapter 5
A Mobile Augmented Reality System for Enhancing Electrical Machine Supervision 98
Jesús Gimeno, University of Valencia, Spain
Sergio Casas, University of Valencia, Spain
Cristina Portalés, University of Valencia, Spain

Electrical machines are used almost everywhere, and our daily life depends on them. For this reason, it is important to articulate mechanisms to control, supervise, and perform proper maintenance of these machines, especially those used in critical industrial process. The SCADA protocol is one of the technologies that eases the operation and supervision of electrical machines. However, the absence of a spatial connection between the SCADA signals and the machines being supervised suggests the use of augmented reality (AR) to fill this void. This chapter describes SIMARA: A Mobile AR application based on a dual computer-vision system (QR-codes and fiducial markers). SIMARA provides a robust client application for the integration of AR and SCADA signals by means of virtual panels shown on top of real SCADA machines. An authoring tool is also provided in order for users to customize the application to their particular needs, allowing to create, by means of web services, customized virtual panels, and links between SCADA signals and the virtual information shown in the AR application.

Chapter 6
Invasive Methods to Diagnose Stator Winding and Bearing Defects of an Induction Motors 122
Muhammad Aman Sheikh, Universiti Teknologi PETRONAS, Malaysia
Nordin B. Saad, Universiti Teknologi PETRONAS, Malaysia
Nursyarizal Mohd Nor, Universiti Teknologi PETRONAS, Malaysia
Sheikh Tahir Bakhsh, King Abdulaziz University, Saudi Arabia
Muhammad Irfan, Najran University, Saudi Arabia

Condition monitoring in an induction motor must concentrate on the root causes of the failure modes that exhibit a slow failure sequence. According to published surveys, two-fifths of the faults are due to bearing failures. Inter-turn short circuit faults in stator windings are approximately responsible for one-third of the motor faults. In the last few decades, various methods and alternative techniques have been proposed and implemented to diagnose induction motor faults. In an induction motor, stator winding and bearing faults account the largest percentage of motor failure. Due to the fact these faults can lead the motor to catastrophic failure that are expensive in term of maintenance cost, wastage raw material, and unplanned shutdown. Thus, to diagnose the state of motor and overcome existing problem, the chapter provides detailed invasive methods which are proposed and are currently in practice. Moreover, the chapter also highlights the limitation, scope, and the challenges of existing invasive condition monitoring techniques.

Chapter 7
Types of Faults and Condition Monitoring Methods of Induction Motors........131
Maheswari M., Malla Reddy Engineering College, India
Gunasekharan S, Malla Reddy Engineering College, India

Induction motors are the electromechanical devices used to convert electrical energy into mechanical energy and work under the principle of mutual inductance. They have stator and rotor as two major parts. They run at constant speed when the supply voltage and frequency are constant and are suited for constant speed drives. They have rugged construction but working environment causes different faults. As per IEEE and EPRI study on induction motor faults, bearing fault and stator faults are 46% and 36%, respectively. The broad categories of the fault are stator winding fault, broken rotor fault, rotor mass unbalance fault, bowed rotor faults, single phasing fault, bearing fault, and crawling. Unbalanced stator voltage and current, oscillations in torque, drop in efficiency and torque, overheating and unwarranted vibration are the effects of these faults. Undetected faults cause complete failure of motor and it is costly in terms of lost production time, maintenance cost, and wasted raw materials.

Chapter 8
Vibration Monitoring of Rotating Electrical Machines: Vibration Monitoring 163
Achintya Choudhury, Sir Padampat Singhania University, India

Vibration monitoring is applicable to all rotating machines for defect detection and diagnosis. Measurement and analysis of vibration have also been applied to rotating electrical machines with the objective of fault detection and predictive maintenance. The sources of vibration generation in electrical rotating machines, both electrical and mechanical, have been identified in this chapter. The vibratory characteristics associated with these defects have also been discussed in detail. Analyses of vibratory signatures in time domain, frequency domain, and time frequency domain have been dealt with, and different features and indicators associated with each domain have been described. The details of vibration measurement schemes such as transducers, different signal conditioning elements, as well as characteristics of recording and display devices and their applicability to electrical machines have also been included in the chapter.

Chapter 9
A State of Art Critique Review of Role of Optimization in NDTS for Effective Condition Monitoring.. 189
Ashish Khaira, Samrat Ashok Technological Institute, India
Ravi K. Dwivedi, Maulana Azad National Institute of Technology, India

Nondestructive testing (NDT) techniques play a pivotal role in a condition-based monitoring system. Generally, using an optimization technique for optimizing the available solution gives added advantage. By keeping this fact in mind, this chapter is an attempt to provide a detailed review of research work from 2000-2018 that covered the role of optimization for effective condition monitoring using NDTs. It was observed from literature that little work found related to NDT process optimization. The researcher and practitioner can use the outcome of this work as a beacon and improve it further.

Chapter 10
Induction Machine Rotor and Stator Faults Detection by Applying the N-F Network..205
Souad Saadi Laribi, Ibn Khaldoun University, Algeria

This chapter focuses on the monitoring and diagnosis of induction machine faults, particularly the broken rotor bars. The design of a system for monitoring, detecting, and locating incipient faults for different loads of the machine is achieved by the use of advanced intelligent techniques based on ANFIS-based neuro-fuzzy network. The knowledge base is based on indicators derived from the stator current spectral analysis of the machine which in addition has to detect and assess the number of faulty bars.

Chapter 11
A State-of-the-Art Critique Review of the Role of Decision-Making Techniques in NDT for Condition Monitoring 222
Ashish Khaira, Samrat Ashok Technological Institute, India
Ravi K. Dwivedi, Maulana Azad National Institute of Technology, India

Nondestructive testing (NDT) is a vital tool in maintenance. Each NDT technique has some benefits and hindrances; therefore, the selection is crucial. Generally, the selection of a technique relies on operating personnel experience, and very few research papers shows uses of the decision-making (DM) approach. It was highlighted by various researchers that if a proper DM approach is used, it will save time and increase fault detection reliability. By keeping this fact in mind, this chapter is an attempt to provide a detailed review of research work from the year 2000-2018 that covered the role of DM techniques while making combinations of NDT for effective condition monitoring. It observed from the literature that very few researchers effectively utilized the power of DM tool. The researcher can use the outcome of this work as a beacon and improve it further.

Chapter 12
Identifying Parameters for Nonlinear Degradation-Based Prognostics: A Case Study With Exponential Degradation on Induction Motors 237
*Feng Yang, Institute of High Performance Computing (A*STAR), Singapore*
*Mohamed Salahuddin, Institute of High Performance Computing (A*STAR), Singapore*

Prognostics and health management (PHM) methodologies are increasingly playing active roles in improving the availability, reliability, efficiency, productivity, and safety of systems in many industries. In predicting the remaining useful life (RUL), this chapter introduces a prognostics framework with health index (HI) formulation, with specific emphasis on incorporating and validating nonlinear HI degradations. The key issue to the success of this framework is how to identify appropriate parameters in describing the behavior of the nonlinear HI degradations. Using exponential HI degradation as an example in predicting the RULs of induction motors, this chapter discusses three different explorations in verifying the existence of good parameter values as well as identifying the appropriate parameters automatically. Comprehensive experiments were carried out with degradation process (DP) data from eight induction motors, and it was discovered that good parameters can be automatically determined with the proposed parameter identification method.

Chapter 13
Sensor and Its Application...261
Saifur Rahman, Najran University, Saudi Arabia
Abdullah S. Alwadie, Najran University, Saudi Arabia
S. Hasan Saeed, Integral University, India
Faizan A. Khan, Integral University, India

Electronic nose systems are used to deliver a pattern response to a listed odor, and pattern recognition software is used to perform odor recognition and discrimination by using a series of sensors. The method of electronic noses generally includes time taking measurements in a non-standard test and error process. The sensory panel problem can be solved by electronic nose. For this purpose, a sensor model is used to design sensor array. The generated signal of these sensor array is used further to classify a mixture of two gases using principle component analysis (PCA)-based classification analysis. During classification, the efficiency of PCA classification has been checked over the different signal preprocessing technique. Continuous real monitoring of odor is done at specific sites in the field over hours, days, weeks, or even months. An electronic machine can also avoid many other troubles linked with the employ of human panels. Each and every variability, adaptation (becoming minimum sensitive during extended exposure), and revelation to hazardous compounds all come to mind.

Compilation of References 275

About the Contributors 299

Index 306

Preface

INTRODUCTION

Induction motors are the industry workhorse due to the fact that they are rugged, reliable and economical. Induction motors play a pivotal role in industry for conversion of electrical energy into mechanical energy. Induction motors make up 95% of the prime movers and utilize up to 40 to 50% of all the electric energy generated. Induction motors have been applied in various applications, such as, in the petroleum industry, chemical processing plants, nuclear power plants, paper mills, cooling water systems and the mining industry. Induction motors have also been used in more general situations such as with compressors, pumps, crushers, fans, lifts, air conditioners, machine tools, tractions, robotics and etc.

Due to enormous electric energy consumptions, reliability of induction motor operation in harsh industrial environment has been a major requirement in many industrial applications. It is especially important where an unexpected breakdown might result in the interruption of critical services such as military operations, transportation, aviation and medical applications. An unexpected breakdown of a motor might result in costly maintenance or loss of life in applications where continuous process is needed and where down time is not tolerable. Although induction motors are very dependable with low failure rate and require only basic maintenance, still as with other kinds of motors, they will breakdown and fail after some time. The unexpected breakdowns of the motors cause a great deal of unacceptable production loss. This is unacceptable in applications that are vital for the industry. Consequently, detecting initial failures and replacing damaged parts according to schedule will prevent the problems of unexpected breakdowns of the machines. The prevention of unscheduled downtime for electrical drive systems has been the goal of every industry for a long time as this would help in reducing the costs associated with maintenance. The 50% of operating cost of manufacturing and processing plants is related to maintenance. Therefore, this is a major area of concern in industries. The researchers are constantly looking for new techniques that can lower the cost of maintenance and minimize the possibility of unexpected breakdowns.

METHODS OF CONDITION MONITORING

In general, three methods for the machine maintenance used in industrial areas are classified as scheduled replacement, scheduled maintenance and condition based maintenance. In the scheduled replacement, replacement of machine parts is conducted on a regular basis. It is very simple and easy method but proves to be very expensive. On the other hand, in the scheduled maintenance, the checking and/or overhauling of the equipment has to be done on a regular basis. It is the method typically used in industries as it is less expensive than the schedule replacement technique. In condition based maintenance, the condition of a machine is determined by taking measurements using sensors. By this method, the time periods between maintenance can be increased and machines can be monitored continuously so that maintenance can be scheduled on a needed basis. Condition monitoring (CM) is a technique that serves the condition based maintenance (CBM). It is an effective type of predictive maintenance (PM).

Normally, the condition monitoring maintenance process would be monitoring the specific parameters like vibration, overheating, over current of equipment for early sign of coming failures and to forecast the need of maintenance before rigorous failures and to estimate the machines health. It could be achieved by visual inspection or through sophisticated intelligent diagnosis system. It embraces the life mechanism of the machine parts, the different data acquisition methods and exploitation of the data to forecast the trends. CM is mainly appropriate for continuous process plants where breakdowns can be very costly. Before CM, time based maintenance (TBM) was the technique served for predictive maintenance. In TBM method, maintenance was performed on predefined running hours of equipment, without the information of current conditions of machine so most of the times it causes wastage of manpower, time and money. As maintenance was performed off-line, it causes many unnecessary shutdowns. However unpredicted breakdowns may still occur in the intervals.

The methods of condition monitoring are categorized into two primary classifications, namely the off-line tests and the on-line tests. The off-line tests are performed by isolating the machine from main AC power supply. Although this is a comprehensive approach, sometimes causes unnecessary shut-downs on machines. Alternatively, the sensors which are installed on the machine are used to detect faults for on-line condition monitoring and fault diagnosis in an induction motors during the operation of the machine. The on-line tests cause fewer disturbances than off-line tests but the results produced from on-line testing are more complicated and their interpretation is difficult than the off-line tests. Over the past two decades, there has been an abundance of research work done in the on-line condition monitoring techniques for diagnosing problems in induction motors. For the detection of the various faults usually affecting machines, several different techniques have been not only proposed but also used successfully. However, a good understanding of the mechanical and electrical properties of the machine in healthy and in faulty conditions dramatically influences the accuracy and reliability of the on-line condition monitoring methods. On-line condition monitoring techniques can be classified into two categories: firstly the classical method and secondly the digital method. In classical method, electromechanical devices are used to protect the motors. The electromechanical devices are expensive, less efficient, having very slow response and not reliable as some of the devices have even shorter life than the motor itself. The digital method is the latest method for the condition monitoring and involves integrated circuits, microcontrollers, microprocessors and programmable logic controllers.

ORGANIZATION OF THE BOOK

The book is organized into 13 chapters. A brief description of each of the chapters follows:

Chapter 1 presents a comprehensive review on induction motor fault diagnosis through current signature analysis techniques. This study discloses an overview of various types of possible faults in induction motor. In addition, the conventional (invasive) and innovative techniques (noninvasive) especially Motor Current Signature Analysis (MCSA) technique for fault detection and diagnosis in induction machine are wrapped with focus on future research.

Chapter 2 presents a scenario where the unknown input observer is applied to load frequency control loops in interconnected power systems is analyzed. An UIO was chosen because load demand is not always constant and it can be considered to introduce an unknown disturbance to the system. Mathematical formulations on how to detect and isolate sensor faults are presented which are then implemented in matlab Simulink for simulations. Based on this historical survey on the application of UIO a thesis on UIO application in FDI in distributed generation will be done.

Chapter 3 describes a new diagnosis approach, the Root-MUSIC (RM) method, for identification of the progressive cracking in the rotor of induction motors. This method used initially in the domain of RADAR for localizing mobile targets, is being applied to the domain of induction motors diagnosis. This approach has several advantages compared to the conventional Power Spectral Density estimation (PSD) by Periodogram technique. Indeed, the main advantage of this approach is its very good frequency resolution for a very short acquisition time, something impossible to achieve with the conventional method. However, in order to reduce the computation time which is the main drawback of the RM method, this method will be applied to only a specified frequency band; one that carries information about the sought fault.

Chapter 4 deals with the implementation of a PC based monitoring and fault identification scheme for a three-phase induction motor using artificial neural networks (ANNs). To accomplish the task, a hardware system is designed and built to acquire three phase voltages and currents from a 3.3KW squirrel-cage, three-phase induction motor. A software program is written to read the voltages and currents, which are first used to train a feed-forward neural network structure. The trained network is placed in a Lab VIEW based program formula node that monitors the voltages and currents online and displays the fault conditions and turns the motor. The complete system is successfully tested in real time by creating different faults on the motor.

Chapter 5 describes SIMARA: a Mobile AR application based on a dual computer-vision system (QR-codes and fiducial markers). SIMARA provides a robust client application for the integration of AR and SCADA signals by means of virtual panels shown on top of real SCADA machines. An authoring tool is also provided in order for users to customize the application to their particular needs, allowing to create, by means of web services, customized virtual panels and links between SCADA signals and the virtual information shown in the AR application.

Chapter 6 highlight the limitation, scope, and the challenges of existing invasive condition monitoring techniques.

Chapter 7 addresses the Types of Faults and Condition Monitoring Methods of Induction Motor.

Chapter 8 The sources of vibration generation in electrical rotating machines, both electrical and mechanical, have been identified in this chapter. The vibratory characteristics associated with these defects have also been discussed in details. Analyses of vibratory signatures in time domain, frequency domain and time frequency domain have been dealt and different features and indicators associated with each domain have been described. The details of vibration measurement schemes such as transducers, different signal conditioning elements as well as characteristics of recording and display devices and their applicability to electrical machines have also been included in the chapter.

Chapter 9 presents an attempt to provide a detailed review of research work from year 2000-2018, that covered the role of optimization for effective condition monitoring using NDTs. It was observed from literature that very less work found related to NDT process optimization. The researcher and practitioner can use the outcome of this work as a beacon and improve it further.

Chapter 10 focuses on the monitoring and diagnosis of induction machine faults, particularly the broken rotor bars. The design of a system for monitoring, detecting and locating incipient faults for different loads of the machine is achieved by the use of advanced intelligent techniques based on ANFIS-based neuro-fuzzy network.

The knowledge base is based on indicators derived from the stator current spectral analysis of the machine which in addition has to detect and assess the number of faulty bars.

Chapter 11 provides a detailed review of research work from the year 2000-2018, that covered the role of DM techniques while making combinations of NDT for effective condition monitoring. It observed from the literature that very few researchers effectively utilized the power of DM tool. The researcher can use the outcome of this work as a beacon and improve it further.

Chapter 12 introduces a prognostics framework with Health Index (HI) formulation, with specific emphasis on incorporating and validating nonlinear HI degradations. The key issue to the success of this framework is how to identify appropriate parameters in describing the behaviour of the nonlinear HI degradations. Using exponential HI degradation as an example, in predicting the RULs of induction motors, this chapter discusses three different explorations in verifying the existence of good parameter values as well as identifying the appropriate parameters automatically. Comprehensive experiments were carried out with Degradation Process (DP) data from eight induction motors, and it was discovered that good parameter can be automatically determined with the proposed parameter identification method.

Chapter 13 provides a sensor model to design sensor array. The generated signal of these sensor array are used further to classify mixture of two gases using Principle Component Analysis (PCA) based classification analysis. During classification the efficiency of PCA classification have been checked over the different signal preprocessing technique. Continuous real monitoring of odor is done at specific sites in the field over hours, days, weeks or even months. An electronic machine can also avoid many other troubles linked with the employ of human panels. Each and every variability, adaptation (becoming minimum sensitive during extended exposure), and revelation to hazardous compounds all come to mind.

Muhammad Irfan
Najran University, Saudi Arabia

Chapter 1
Fault Diagnosis of Induction Motors Using Motor Current Signature Analysis:
A Review

Muthiah Geethanjali
Thiagarajar College of Engineering, India

Hemavathi Ramadoss
Thiagarajar College of Engineering, India

ABSTRACT

Induction motors are termed as horses of modern industry because they are playing a vital role in industries. They are simple, efficient, robust, rugged, and highly reliable. The feasibility of mishap in induction motors is less, but they are prone to faults, which are left unobserved most of the time. Hence, more attention has been paid to detection and diagnosis of incipient faults to prevent damage spreading and increase the lifetime of the motor. To detect and diagnose the faults, online condition monitoring of the machine has been utilized in a wide manner. At present, focus is made on optimization procedures for fault diagnosis in induction motors to obtain a quick assessment at industry level. This chapter discloses an overview of various types of possible faults in induction motors. In addition, the conventional (invasive) and innovative techniques (noninvasive), especially motor current signature analysis (MCSA), techniques for fault detection and diagnosis in induction machines are covered with a focus on future research.

DOI: 10.4018/978-1-5225-6989-3.ch001

INTRODUCTION

From domestic applications to many complex industrial applications, where the transformation from electrical energy to mechanical energy is required it would be possible with the help of electric Induction Motors (IM). This motor is an Asynchronous Motor falls under the type of Alternating Current Motor, which operates, on electromagnetic induction principle to supply power to the rotor. They are the spine of the industries for their low cost and maintenance (Bhowmik, Pradhan & Prakash, 2013; Merizalde, Hernandez-Callejo & Duque-Perez, 2017; Bayar & Terz, 2016).However, this machine undergo for failure due to electrical or mechanical defect. This is because of heavy-duty cycles, unfortunate working environment, improper installation, excitation and manufacturing factors etc (Bhowmik, Pradhan & Prakash, 2013).The factors involving in failure of induction motor are likely to be environmental, operations, equipment, human and electrical power (Merizalde, Hernandez-Callejo & Duque-Perez, 2017). Under these factors, so many faults lie which lead to the mishap of induction motors have covered in section 2. When the induction motor is subjected to failure it may lead to unwanted interruption in industrial plant, safety measurement and product excellence with results in revenue losses (Messaoudi & Sbita, 2010). Hence, the field of fault detection and diagnosis in Induction motor is attaining more importance due to the elevation of demands for reliability and proficiency.

However, the maintenance techniques available such as default type, proactive technique, preventive maintenance, discard technique and online-offline type there is always an anxiety for the decision maker on choosing the type of maintenance that results with more optimal. Henceforth, it is necessary for the decision maker to have a concern on suggestions from the manufacturer, their own experience on the plant and the data regarding the condition of plant offline or online (Sonje & Monje, 2011). As a whole, considerable amount of research has been carried out on condition monitoring and fault diagnosis techniques for induction motor in last many years (Bhowmik, Pradhan & Prakash, 2013; Mehala & Dahia, 2007; Kucuker & Bayrak, 2015). The summary of the techniques are given below:

- Fast Fourier Transform based fault diagnosis.
- Fault detection and diagnosis with use of vibration signal.
- IR recognition, measurement of temperature, RF (Radio Frequency) emission observing.
- A detection technique based on harmonic analysis of machine speed fluctuation measurement.
- An approach based on state and parameter estimation.
- Fault detection based on flux analysis or airgap torque approach.

- Measurement of acoustic noise radiation.
- Analysis of airgap magnetic field.
- Fuzzy logic, artificial intelligence, model and neural network-based approaches.
- Space vector angular fluctuation (SVAF) based detection.
- Motor Current Signature Analysis (MCSA).
- Instantaneous Power Signature Analysis (IPSA).

From the above techniques, it is inferred that MCSA technique is better than other methods because of its non-intrusive nature. Unlike other approaches, the machine details obtained from MCSA technique is reliable because this approach doesn't necessitate any access to the machine (Prakasam & Ramesh, 2016; Granda et al., 2017). The spectral analysis of the induction motor supply current is utilized in this technique to locate the incipient fault of the machine. This MCSA approach is the most consistent one as it evaluate the entire health condition of a rotor system. This study is provided with a comprehensive review of the faults may occur in induction motor, conventional protection schemes for the induction motor and the non-invasive protection schemes especially MCSA technique for the protection of induction motor. In addition, more attention is paid on three-phase induction motor fault diagnosis to bring out the effectiveness of the MCSA approach.

THREE PHASE INDUCTION MOTOR: CONCEPTS AND SIGNIFICANCE

The achievement of making Dynamo to work as a motor aided as precedent for the invention of first Induction AC motor in 1888, by Nicola Tesla. To overcome the low performance and insufficient starting torque in this type of machine, Chares Brown and Michael Dolivo-Dobrolowsky developed the present wound rotor and squirrel cage induction motor type (Bhowmik, Pradhan & Prakash, 2013). After undergone for many improvisations by the inventors, induction motors occupied a principle position in industrial applications until now. The simplicity and reliability of the induction motor have made it as a valuable invention. In addition, IM has occupied a superior role in research area for new design, construction and maintenance techniques (Bhowmik, Pradhan & Prakash, 2013).

The IM is comprised of two main parts called stator and rotor. The working of induction motor in based rotating magnetic field. In addition, most of the industrial applications use three-phase induction motor, which is smaller, economical and has less starting current. They are integrated in commercially available equipment and industrial processes. In spite of the availability of monitoring tools and protection techniques, still most of the industries have the challenge of unanticipated system failure and reduced motor lifetime. Hence, a deep knowledge on the abnormal behavior of the induction motor during faulty conditions and the diagnosis techniques on these conditions are the challenging areas for the researchers (Mehala & Dahia, 2007).

Possible Faults in Three Phase Induction Motors

Several types of faults can occur due to so many factors lead to the deterioration of the machine and the factors are listed in Figure 1. In addition, mechanical/electrical failure can occur in IM due to the factors given in Figure 2.

Apart from the factors mentioned above, the induction motors are subjected to some other stresses (Figure 3), which leads to intense effects on stator winding, rotor, bearings and shaft.

Due to the factors discussed above, unexpected failures can occur in IM which tend to major problems. From the studies conducted by IEEE, EPRI and Allianz it is incurred that the following are the major faults can occur in IM (Granda et al., 2017). These faults are broad in manner and the comparative study on the faults is given in Figure 4.

- Stator faults
- Rotor faults
- Bearing faults
- Mechanical Disparities (Other faults)

Figure 1. List of factors for faults in induction motor

Environmental	Operations	Equipment	Human	Electrical Power
• Temperature • Moisture • Rust • Ventillation • Pollution • Strange Objects	• Vibration • Overload • Excessive Starts • Alignment • Resonance of the system • shaft currents • Stator rotor friction • Partial Discharge	• Aging • Quality • Design defects • Manufacturing defects	• Bad selection of Motor • Bad usage • Lack of maintenance • Repairs • Poor quality parts • Lubrication	• Transients due to Short circuit, fluctuations, Resonance, Transfers, reconnections, capacitors, Insulation, Drives • Voltage drop • Voltage low • Voltage unbalance • Harmonics • Defective Electrical Installation

Figure 2. Factors for electrical and mechanical failure

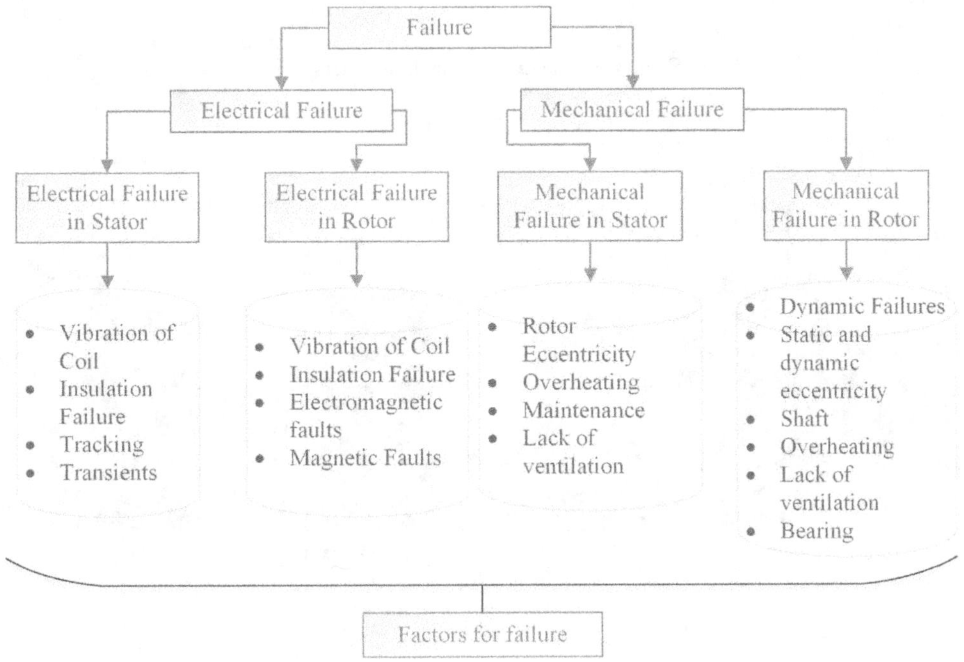

Figure 3. Stresses cause effects on induction motor

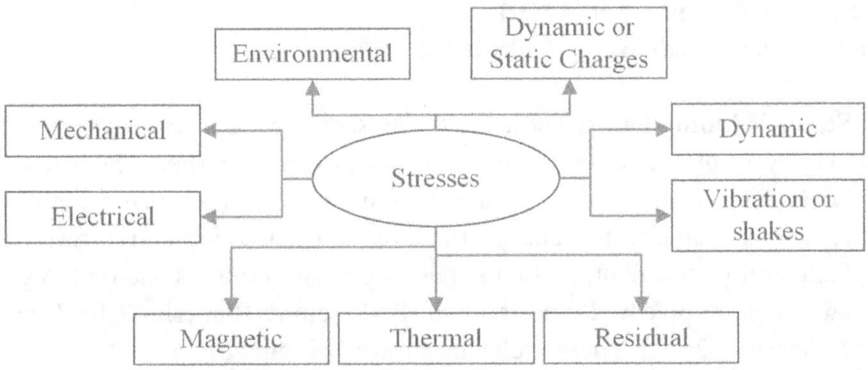

There may be different types of faults can occur in stator and rotor. On the whole faults in IM can be listed as follows for identification: a) stator winding fault b) broken bar fault c) airgap eccentricity d) load torque influence e) bearing fault f) unbalanced supply voltage g) single phasing fault h) rotor mass unbalance faults i) crawling etc (Messaoudi & Sbita, 2010; Karmakar et al., 2016; Bhardwaj & Agarwal, 2012). As well, there is a chance for the occurrence of combination of faults in IM.

Figure 4. Comparison on occurrence of major faults in induction motor by IEEE, EPRI and Allianz

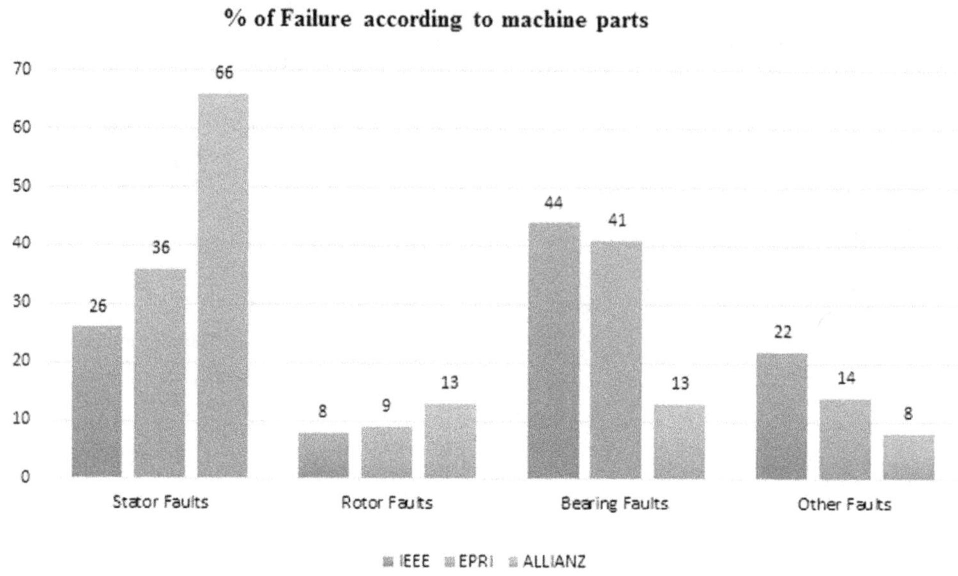

In that case, it is tedious to locate the initial problem which leads to the incidents of oscillations in torque, stator and rotor voltage imbalance, overheating and lessen the efficiency (Karmakar et al., 2016).

In this study, the following faults in IM are discussed.

1. **Stator Winding Fault:** These are the most critical faults to diagnose and this may occur due to several stress types as discussed earlier. The stator short circuit faults results in flow of large amount of circulating current in the turns. Hence, excessive heat is generated as a consequence which tends to turn-to-turn fault, coil-to-coil fault, phase to phase, open circuit faults and coil to ground fault that are potentially destructive (Prakasam & Ramesh, 2016; Bhardwaj & Agarwal, 2012). These faults are shown in Figure 5.
2. **Broken Bar Fault:** In squirrel cage induction motor, the bars are get partially cracked or completely broken due to the following reasons:
    a. Fatigue of the rotor bar metal
    b. Thermal stress
    c. Defects during manufacturing process
    d. Mechanical stress due to fault in bearings
    e. Starting of the motor often

Figure 5. Types of stator winding faults

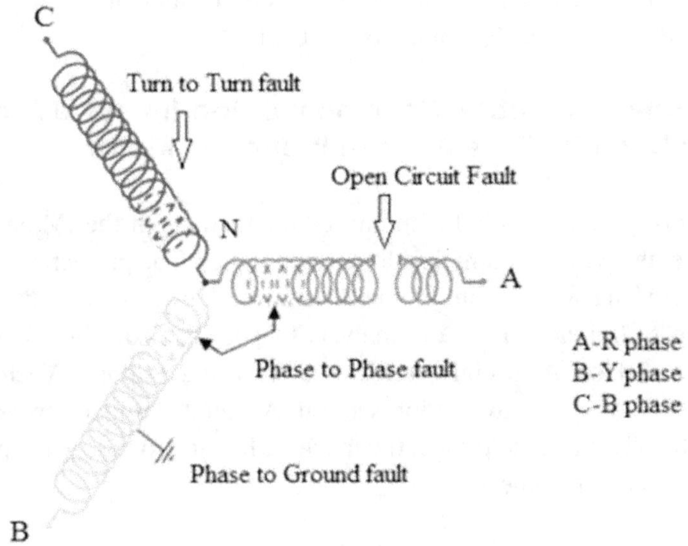

The presence of broken rotor leads to rotor asymmetry and changes in the magnetic field of the induction motor (Granda et al., 2017; Karmakar et al., 2016).

3. **Airgap Eccentricity:** Due to incorrect bearing positioning through assembly of IM, deflection of shaft, worn bearings and heavy load etc. this eccentricity fault may occur. The forms of airgap eccentricity are (Bhowmik, Pradhan & Prakash, 2013; Mehala & Dahia, 2007; Siddiqui et al., 2016):
    a. **Static Type:** Minimal radial airgap length has fixed in space for this type eccentricity. As well, an incorrect placing of stator and rotor at the time of commissioning phase can lead to the rise of static eccentricity. The minimal airgap leads to an unbalanced magnetic pull, which deflects the rotor and increases the amount of airgap eccentricity.
    b. **Dynamic Type:** There is no coincidence between center of rotor and center of rotation. The minimum airgap is not a fixed one in space but rotates with rotor. The cause of this phenomenon is bent shaft, wear of bearing and mechanical resonance at critical speed.
4. **Bearing Fault:** There are two sets of bearings placed at the ends of the rotor and they help the rotor to rotate freely without any friction. Each bearing consists of two races and balls between the races. If any one of the race or surface of the ball is getting damaged then it is termed as bearing fault. This fault may arise due to excessive load, corrosion, failure of lubricant, contamination,

fatigue failure and disarrangement of bearings. This fault tends to the increase of friction, which leads to the rise in temperature and vibration in the machine (Karmakar et al., 2016; Kompella et al., 2014).

Conventional Protection Scheme vs. Non-Invasive Protection Schemes Used for Three-Phase Induction Motor

The number of problems allied with unpredicted outage of the IM have addressed and implemented with different developed techniques. The protection of the IM has evolved from visual inspection, empirical analysis, usage of invasive techniques, non-invasive techniques based on analysis of spectrum of signals to the recent techniques based on Artificial Intelligence (AI) and Support Vector Machines (SVM) as mentioned in introduction section. All the techniques are attempting to offer a reliable diagnostic to protect the machine life and to ensure the profit of both manufacturers and customers.

1. Protection With Invasive Techniques

According to IEEE 1415 standard totally 41 invasive test can be used to test a new or repaired machine and to diagnose the faults. Some of the invasive tests are given in Table 1 for reference (IEEE, 2006).

The electromechanical and static relay based scheme of protection in industries are known as invasive methods. In addition, invasive schemes are required for costly sensors like temperature (RTD), vibration (velocity transducers) sensors for condition monitoring of the motor. However, installation of these sensors has reduced the possibility of accessing the machine for every application and has shown greater impact on protection of the machine. In Figure 6 some of the invasive techniques are depicted for fault diagnosis.

2. Protection With Non-Invasive Techniques

Noninvasive methods are the most preferable one because of their simple access on machine condition without disintegrating it for diagnosis. As well, these are suitable for online fault monitoring and diagnosis of the machine. More information that is useful have obtained from the stator current, vibration, magnetic field temperature and noise signals. The non-invasive techniques involved in mathematical analysis of these signals, to find out the failure of the induction motor (Drozdowski & Duda, 2014).

Table 1. IEEE given standard of invasive tests in induction motor

S.No	Test	Narration
1	AC high potential	Overvoltage test applied from conductor to ground.
2	Capacitance	AC test to get the line-to-ground insulation capacitance.
3	Core loss	Test to find out the shorted stator core lamination.
4	DC high potential	Overvoltage test applied line to ground test, to get leakage current and charging current.
5	Dissipation factor and power factor	AC test to measure the line-to-ground dissipation-capacitance
6	Grease analysis	Grease lubricated machine bearings, appearance, smell
7	Growler	Test for rotor core by producing magnetic field and monitors temperature, magnetic or current pattern of the rotor
8	Insulation resistance	Insulation resistance between the conductor and ground is measured.
9	Oil analysis	Analysis of oil for lubricant characteristics
10	Partial discharge	AC test to get line-to-ground partial discharge
11	Phase angle	Timed measurement of voltage and current angle in degrees
12	Phase balance	AC frequency test to measure stator line-to-line inductance or impedance balance
13	Single phase rotor test	Observes the AC current level during the machine is single phased at lower voltage level.
14	Polarization index	Ratio of ten minute IR to one minute IR.

Figure 6. Invasive techniques for fault diagnosis

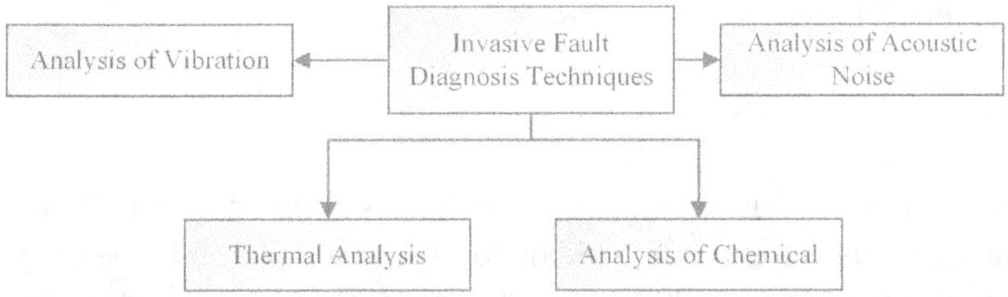

Generally the non-invasive techniques can be classified into three approaches namely Artificial Intelligence (AI) based approaches, signal processing based approaches and hybrid approaches (Chudasama & Shah, 2017).

1. **AI Based Approaches:** These approaches comprised of parameter estimation method with a mathematical model to understand the system dynamics. This AI approaches may be of the following types for fault detection.

   a. Fuzzy concepts
   b. Artificial Neural Network
   c. Support vector machine with back propagation method
   d. Neuro fuzzy system
2. **Signal Processing Based Approaches:** The fault detection schemes under these approaches are listed as follows
   a. Motor Current Signature Analysis
   b. Fourier Analysis (Fast Fourier Transform techniques and Short Time Fourier Transform analysis)
   c. Park's Vector Approach
   d. Wavelet Analysis
3. **Hybrid Approaches:** The Combination of above said methods may also be used to detect faults in better manner and they are as listed as follows
   a. Induction motor model and any signal processing method
   b. Induction motor model and any AI technique

Even though the approaches are novel and advantageous the following issues faced in these techniques are addressed as follows (Chudasama & Shah, 2012):

- A protection scheme with non-invasive technique should wrap all types of possible faults as well as it should offer the integration of operator investigative knowledge depend upon the condition.
- Detection of fault in inverter driven motors and application of artificial intelligence approaches to solve the issues such as frequent start, excessive start time, stalling etc.

3. Evolution of MCSA

The concept of Motor Current Signature Analysis is initiated in early 1970s. At first, this technique is used for inaccessible motor, which are placed in the risky areas of nuclear power plant (Korde, 2002). In 1982, Thomsan has initiated the research on MCSA based fault diagnosis. With the improvisation in digital signal processing the ability to produce more precise current spectra of the motor current to diagnose the abnormality. This technique consumes the spectral analysis of the supply current (stator current) of the IM to locate the failure during online condition of the motor without interrupting the production. For locating the common motor faults, MCSA has used as a predictive maintenance tool so that the catastrophic failures and production outages can be prevented at an early stage and the motor lifetime can be extended. The spectrum analyzer and MCSA instruments are available to produce current spectra which contains more useful information about condition of the IM are used in industries in a wider manner (Thomson & Culbert, 2017). These instruments have to be recognized because the initial source of information obtained from these tools are used to identify whether the problem is existing or not.

FUNDAMENTAL CONCEPTS OF MCSA

The main intension of MCSA technique is to obtain the motor current spectral differences from normal condition of the IM as well as abnormal condition of the IM. From this analysis the fault diagnosis is possible against the stator, rotor and bearing related problems. MCSA offers the following methods to analyses the spectrum of signal (Merizalde, Hernandez-Callejo & Duque-Perez, 2017; Nandi, Toliyat & Li, 2005).

- Parametric analysis method (time domain, frequency domain analysis)
- Non-parametric analysis method (FFT technique)
- Sub-space method (autocorrelation matrix)

For dynamic signal, bispectrum and trispectrum methods can be used to obtain the frequency spectrum.

Principle of Operation of MCSA

The basic principle of MCSA technique is the IM circuit can be viewed as a transducer. By placing a Hall Effect, current sensor on primary or secondary of the circuit the transients of the motor current can be monitored (Shukla, Jha & Qureshi, 2014). The current monitoring system has shown in Figure 7.

Generally, single stator current monitoring system is involved in MCSA ie the supply current of any single phase of the IM has monitored. Here the transducer is stator winding of IM, which captures the signal from the rotor. The current sensor has used to sense and record the current signal in time domain. Then this signal has passed to the spectrum analyzer or MCSA instrument for analyzation. In real case, the motor current consists of harmonics rather than ideal case that has the presence of pure sinewave. However, if any fault in motor leads to a change in magnetic field and imbalance in self and mutual inductance of the motor. These changes arose as sidebands around the supply frequency of the current harmonics as shown in Figure 8 (Mehla & Dahiya, 2007).

Figure 7. Block diagram of MCSA

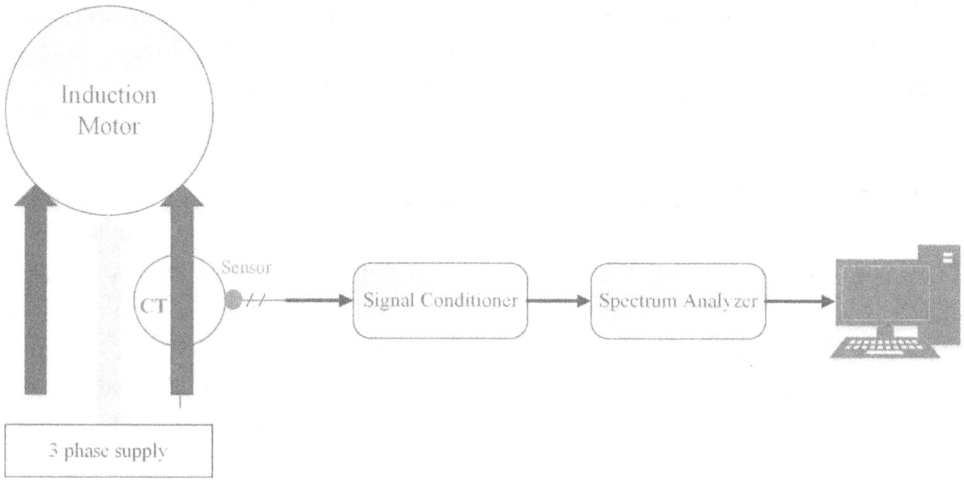

Figure 8. Current spectrum of IM (healthy-blue, faulty-red)

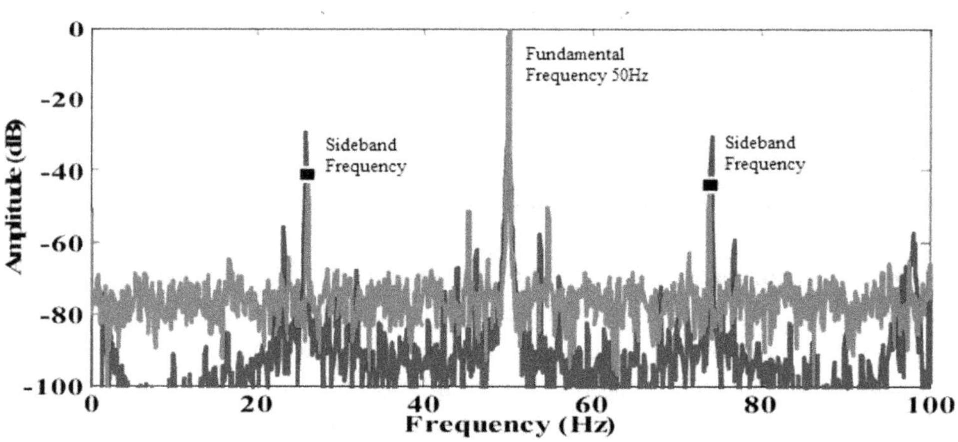

**For a more accurate representation see the electronic version.*

Online Condition Monitoring and Fault Detection Using MCSA

Condition monitoring involves in various activities to be performed by the operators. Some activities may be critical and they can be performed with the help of computers. These activities are very less in numbers and they can be easily located. This approach is as simple in finding out the common elements of the condition monitoring of machines. In large machines mostly mechanical sensors such as vibration sensors are used which are more expensive. In most of the circumferences, vibration monitoring methods are used to identify the presence of any glitches in the machine. However, many researchers suggested that, stator current monitoring system can imply the same indications without need for access to the machine (Chaturvedi, Iqbal & Singh, 2013). The focus have been made on non-invasive techniques which can detect faults with the help of major data without any direct access of the machine. A typical condition monitoring system receive the input from sensors and execute mathematical operations for further diagnosis process (Alwan, Farhan & S-Al-Sabbagh, 2017).

In MCSA, based online condition monitoring and fault detection there is no indeed of extra sensors. In addition, this method does not fond of the direct accessibility of the motor for fault detection. The fault detection procedures are given in Figure 9 (Merizalde, Hernandez-Callejo & Duque-Perez, 2017; Irfan et al., 2017). This technique is involved in online analysis of the current signal to detect the fault in three-phase induction motor while the motor is in operation and in service. The analysis of current signal leads to obtain the characteristic spectral signature from which, the abnormal operating conditions can be discriminated and the failures are identified. From the stator current spectral analysis, the rotor and stator faults can be identified.

As per the requirement, the MCSA instrumentation consists of the following for online monitoring (Thomson & Gilmore, 2003):

- A sensor (Current Transformer, Clamp probe) to sense the current signal.
- A resistive shunt across the current transformer output.
- A spectrum analyzer to present the current signature.

Figure 9. Fault detection procedures

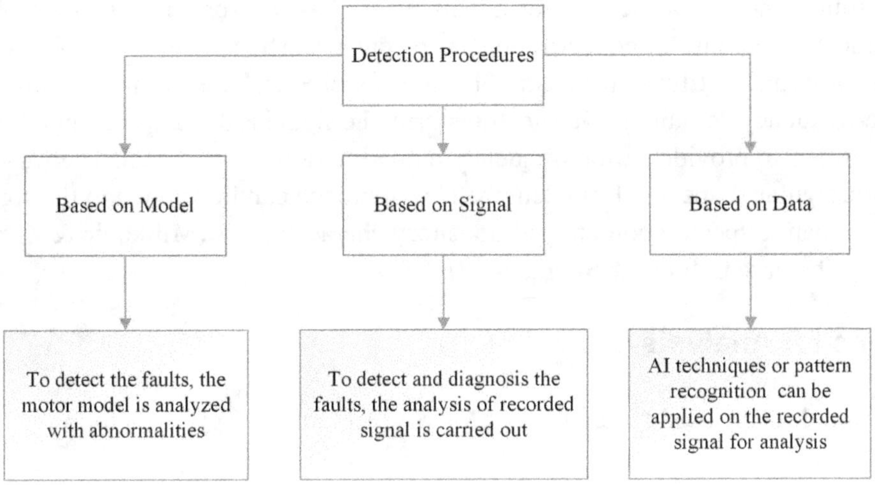

For the application of MCSA, the following key elements are required (Thomson & Gilmore, 2003):

- The basic understanding on design and working of three phase induction motor.
- MCSA instrument to record the current signature.
- Expert knowledge base within the consumer's organization.
- The drive operator who makes the final decision whether to remove the motor for repair or allows it to run and plan for the outage to diagnose the faults.

For the detection of faults from the current spectral analysis, several techniques involved with MCSA are given in figure 10. As the motor faults alter the harmonic content of the stator current, these techniques are used to extract the features of the current signal to detect the faults (Merizalde, Hernandez-Callejo & Duque-Perez, 2017; Narwade, Kulkarni & Partil, 2013; Miljkovic & Zagreb, 2015). Using Fast Fourier Transform (FFT) the frequency signature of the machine is identified which tends to better interpretation of motor current spectra. The motor current signature FFT based analysis is a consistent tool for the detection of abnormalities in induction motor. The spectrum obtained from this technique consists of only magnitude data of the frequency components and they can be analyzed in simpler way than the time domain analysis. This approach has higher accuracy than the traditional Discrete Fourier Transform (DFT) technique and has low computational cost. Later, the grid frequency and machine slip both are utilized in tuning of parameters involved in zoom Fast Fourier Transform (ZFFT) approach. This can be applicable for rotor faults too (Chaturvedi, Iqbal & Singh, 2013).

Some researchers used a novel harmonic analysis tool for analysis of current spectrum named as wavelet transform. This method has high sensitivity and reliability. In order to overcome the concern about the recent powerful mathematic tool, wavelet transform can be utilized in all sort of loading conditions. It performs the analysis in time-frequency domain. In wavelet transform, the signal is decomposed into family of wavelets to provide a time frequency representation of the signal. The wavelets have irregular shape and finite length. This approach can be applied to the signals with dynamic spectral content and transitory characteristics (Miljkovic & Zagreb, 2015; Chaturvedi, Iqbal & Singh, 2013).

Steps for Analysis

The simple steps used for analysis of MCSA technique are given as follows:

Figure 10. Techniques involved with MCSA for fault detection

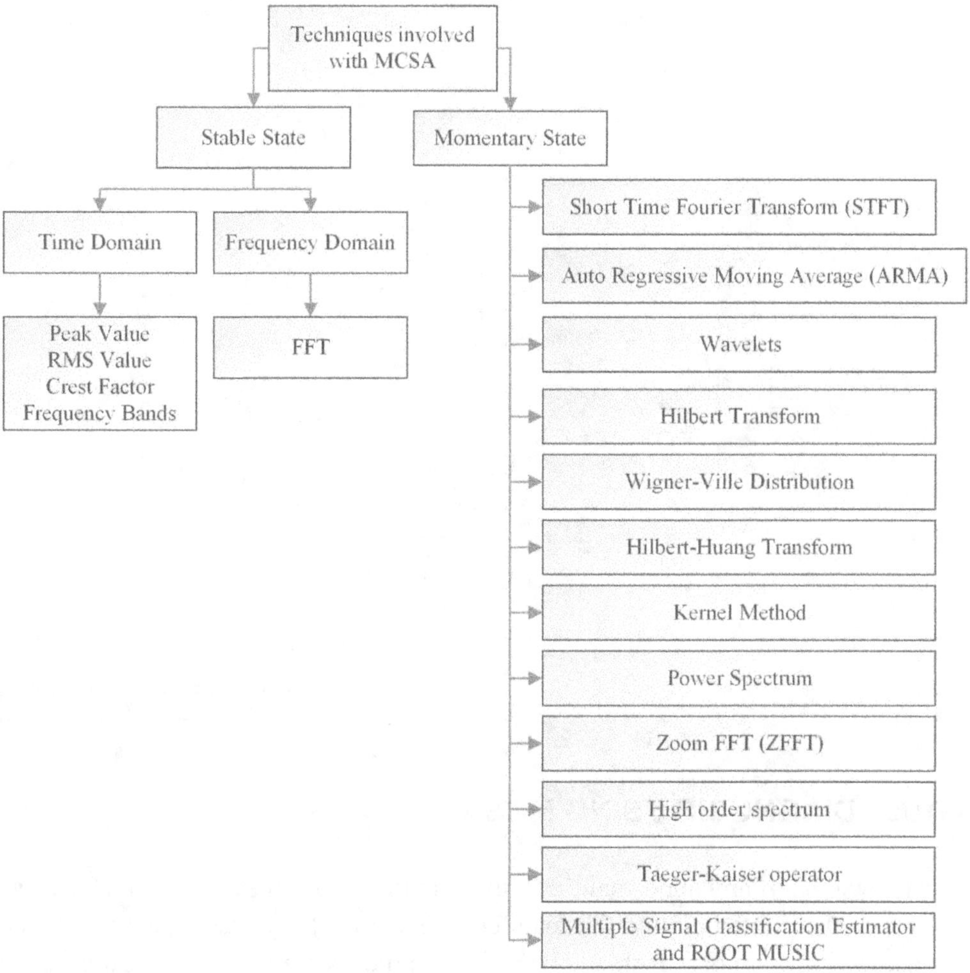

- Mapping of the system to be analyzed.
- Finding out the reason for improper working of machine through analysis.
- Searching for other data can be involved in analysis.
- Collection of data.
- Review and analysis of data.

Snapshot of the current has monitored for a specific time - period and the low frequency demodulated current has reviewed to monitor the rotor in order to locate the load related issues. Then high frequency demodulated current and voltage have reviewed to monitor electrical and mechanical failure.

The flow of the steps for analysis of MCSA technique is given in Figure 11.

Figure 11. Steps for analysis of MCSA technique

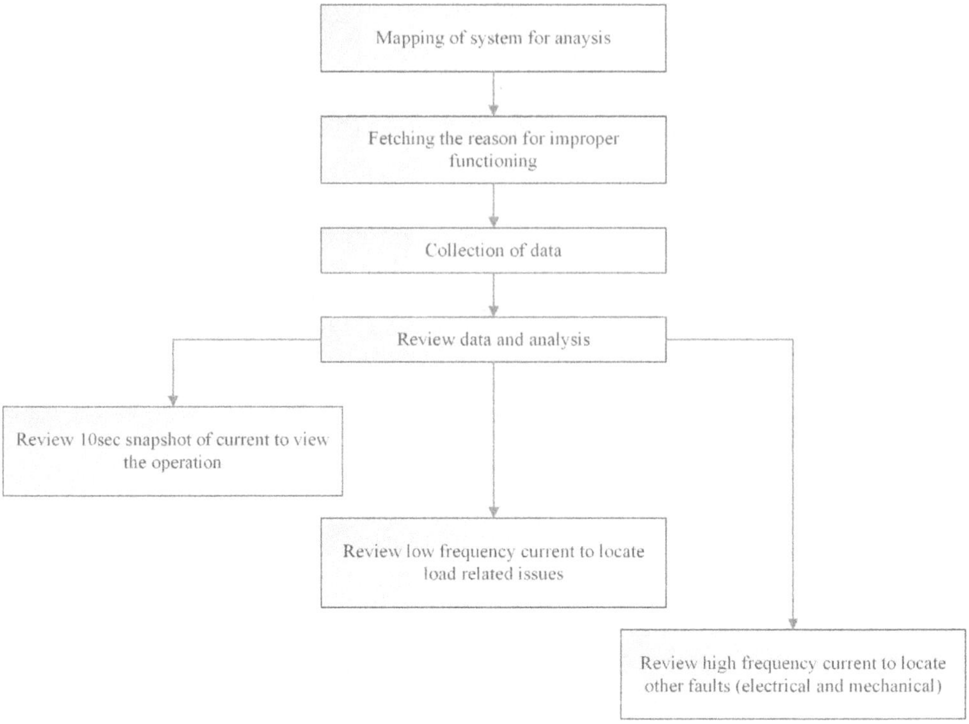

FAULT DIAGNOSIS USING MCSA

It is known that if any abnormalities present in the motor, they will be reflected as sidebands distributed around the main component of the current spectrum of the signal. During diagnosis, it should be noted that the spacing in the spectrum can be changed due to the factors like load variation.

MCSA technique can detect and diagnose the following types of faults (Bhowmik, Pradhan & Prakash, 2013; Miljkovic & Zagreb, 2015):

- Stator faults (abnormal connection of stator windings).
- Broken bar or cracked end rings of rotor.
- Static/Dynamic airgap eccentricities.
- Bearing and gearbox faults.
- Bent shaft which produce a rub between rotor and stator.
- Loading issues.

Detection of Stator Winding Faults

Most of the stator failure are due to stator winding short circuit issues. Normally, short circuit take place between the turns of one phase or turns of two phases or turns of all phases. In addition, short circuit occurs between stator core and winding conductors. These shorted turns lead to production of excess heat in stator winding as well as imbalance in current. MCSA technique has revealed that the resultant frequency components have induced in stator winding, due to the rotating flux (Messaoudi & Sbita, 2010; Mehala & Dahia, 2007; Prakasam & Ramesh, 2015; Miljkovic & Zagreb, 2105). The induced harmonic frequencies in the stator current is given by

$$f_{st} = \left[\frac{n}{p}\left(1-s\right) \pm k\right] f_g \quad (1)$$

Where: n = 1,2,3,.... and k = 1,3,5....

f_{st} : Frequency component related to shorted turn

f_g : Grid frequency

p: number of pole pairs

s: slip

Detection of Broken Bars

The reason for broken bars faults are pulsating mechanic load and direct online starting duty cycle. During normal working conditions of IM, the supply voltage with frequency f_s has fed to the stator winding to produce forward rotating magnetic field. If asymmetry has occurred in rotor then backward rotating magnetic field, arise at slip frequency with respect to forward rotating rotor. As a result, an emf has induced and current in the stator winding is at the following frequency component (Messaoudi & Sbita, 2010; Mehala & Dahia, 2007; Fenger, Lloyd & Thomson, 2003; Miljkovic & Zagreb, 2105):

$$F_{sb} = f_s(1 - 2s)Hz \quad (2)$$

Due to the broken bars, this is reported as lower twice slip frequency sideband. Henceforth, speed oscillations at twice the slip frequency ($2.s.f_s$) has produced because of the cyclic current variation. This speed oscillation has reduced the current magnitude of $f_s(1-2s)$ but the rotor oscillation has induced an upper sideband current component at $f_s(1+2s)$ and so on in stator winding. Finally, the induced harmonic components in the stator current because of the broken rotor bars has given as follows:

$$f_{bb} = f_s\left(1 \pm 2.k.s\right) \tag{3}$$

Where, k = 1, 2, 3… k ∈ N

f_{bb}: Frequency of broken rotor bars

f_s: supply frequency

s: slip

Granda et al (2017) have contributed a work on broken bar fault diagnosis using motor current signal for an IM with no load. In order to bring out the effectiveness of this MCSA the performance of the faulty IM at full load has also analyzed. In both cases, the fault identification has proved with the appearance of oscillations and fluctuations in the current spectrum.

Detection of Airgap Eccentricity

When the airgap between the stator and rotor is not uniform, this has reported as airgap eccentricity. This results in production of unique spectral pattern in the current spectrum. Here the magnetic flux in airgap has considered as product of magneto motive force and permeance. The alignment change in rotor bar tends to change in stator current because of change in stator magnetic field (Messaoudi & Sbita, 2010; Sonje & Monje, 2011; Miljkovic & Zagreb, 2105): [4, 5, 26].Hence, the appeared sideband frequency components determined by the following equation:

$$f_{ec} = f_g\left\{\left(R \pm n_d\right)\left[\frac{1-s}{p}\right] \pm n_{ws}\right\} \tag{4}$$

f_{ec} - Airgap eccentricity frequency components

f_g - Grid frequency

R - Number of rotor slots

n_d - ± 1

S - Slip

P - Pole pairs

n_{ws} - 1, 3, 5, 7….etc

Detection of Bearing Fault

When bearing fault occurs, vibrations also produced that affects the airgap eccentricity between rotor and stator. This results in induction of fault frequency in stator current signal. Bearing faults have classified into the following types (Kompella et al., 2017).

- Cyclic faults
    - Outer race defect
    - Inner race defect
    - Ball defect
    - Cage fault
- Non cyclic faults

Because of cyclic faults, vibration has produced between the bearing and raceway and these vibration frequencies have calculated with the following expressions. From these expressions, it is known that the estimation of cyclic fault frequencies have obtained from the vibration signal spectral analysis.

$$\text{Outer race: } f_0 = \frac{N_b}{2} f_r \left(1 - \frac{d}{D} \cos \varnothing \right) \quad (5)$$

$$\text{Inner race: } f_i = \frac{N_b}{2} f_r \left(1 + \frac{d}{D} \cos \varnothing \right) \quad (6)$$

Cage: $f_c = \frac{1}{2} f_r \left(1 - \frac{d}{D} \cos\varnothing\right)$ (7)

Ball: $f_b = \frac{d}{2D} f_r \left(1 - \left(\frac{d}{D}\right)^2 cos^2\varnothing\right)$ (8)

f_0 – Outer race frequency

f_i – Inner race frequency

f_c – Cage fault frequency

f_b – Ball defect frequency

N_b – Number of bearing balls

f_r – rotor speed

d – ball diameter

D – pitch diameter

The fault frequency of bearing failure in the stator current has given by the below expression (Mehala & Dahia, 2007; Fenger, Lloyd & Thomson, 2003; Kompella et al., 2017):

$f_{bearing} = \left|f_s \pm m.f_v\right|$ (9)

Where, m=1, 2, 3…harmonic number

f_v – fault frequency

f_s – supply frequency

Hence, the mechanical problems that leads to dynamic rotor disturbances cab be identified from the current spectrum changes.

Loading Effects

Due to load torque variation, frequency components have raised in the current spectrum of the stator current signal and this change in load torque coincide with the results of rotor position. Stator flux linkage is purely sinusoidal when the machine is under normal operating condition. However, if any fluctuation in load torque with rotational speed multiples (m f_r) lead to the production of stator current at frequency as mentioned below (Messaoudi & Sbita, 2010; Miljkovic & Zagreb, 2015):

$$f_l = f_g \pm m f_r = \left[1 \pm m\left[\frac{1-s}{p}\right]\right] f_g \tag{10}$$

Where, m = 1, 2, 3…, m$\in$N,

f_l – Frequency associated with load torque fluctuations

f_g – Grid frequency

f_r – Frequency of mechanical rotor speed

s – slip

p – Number of pole pairs

Unbalanced Supply Voltage

The unexpected opening of one of the three phases, tend to unbalance in stator supply voltage. The application of this asymmetrical supply voltage on three-phase IM lead towards the reduction of torque and increase in losses. Consequently, high rotor current has provoked and ended with heating of the motor. The unbalance in supply voltage have obtained from the following expression (Messaoudi & Sbita, 2010):

$$unbalance\ \% = 100 * Max\left[\frac{V_h - V_{av}}{V_{av}}, \frac{V_{av} - V_l}{V_{av}}\right] \tag{11}$$

Where,

V_h - highest voltage

V_l – lowest voltage

$$V_{av} = \frac{V_1 + V_{2+} V_3}{3} \tag{12}$$

The unbalance in supply voltage, prompt in stator current spectrum of the current signal at the below frequencies:

$$f_{un} = \left(1 + 2.k\right) f_g \tag{13}$$

Where, k=1,2,3…k∈N,

f_{un} – frequency of unbalanced supply voltage

f_g – grid frequency

Impact of Gearboxes

As mentioned above, the mechanical oscillations lead to the production of additional frequency components in the current spectrum of stator current signal. This may occur due to the low speed gearboxes, and this current spectrum is similar to that of the broken bar components. Hence, to make effective diagnosis towards rotor-winding problems in motor that connected to gearbox, the impact of gearbox frequency components in current spectrum has also taken for consideration (Messaoudi & Sbita, 2010; Fenger, Lloyd & Thomson, 2003). Mainly, the production of sideband frequency components due to the slow revolving shafts has given in equation (f_l). Here the mechanical rotor speed frequency of the shaft, rotating at N_r rpm has calculated as follows:

$$f_r = \frac{N_r}{60} \tag{14}$$

INTELLIGENT TECHNIQUES APPLICATION FOR INDUCTION MOTOR FAULT DIAGNOSIS

Overview of the Induction Motor Fault Diagnostic AI Techniques

Though advanced technologies have developed for failure diagnosis, it is important to improve expert system, which comprises the usage of incomplete data, probability functions and nonlinear systems. In this context, AI techniques have extended as a decision-making tool towards the MCSA results for condition monitoring and fault diagnosis of induction motors. The current spectrum obtained from MCSA results can be used for further diagnostic procedures with the usage of AI techniques. The following AI techniques are used for fault diagnosis (Merizalde, Hernandez-Callejo & Duque-Perez, 2017; Miljkovic & Zagreb, 2015; Agrawal, Yadav & Saini, 2015):

- Artificial Neural Network (ANN)
- Fuzzy Logic
- Neuro-Fuzzy Systems
- Genetic Algorithm
- Support Vector Machine (SVM)
- K-nearest neighbors (KNN)
- Data Mining
- Case-Based Reasoning (CBR)
- Expert Systems

AI based diagnostic approaches have minimum requirements for fault diagnosis and they are as follows:

- Having minimum number of rules, neurons and indicators
- Required prior minimum knowledge
- Usage of both AI and soft computing technique for fault diagnosis

The usage of AI techniques have made the fault monitoring system as more precise, flexible and accurate. The information provided by practicing operators may occupy a significant role in robust diagnosis of IM failure. The online monitoring process of fault diagnosis has given in the block diagram Figure 12.

Fuzzy Logic

Fuzzy systems are much more helpful in high complex systems whose attribute has not defined well and where an approximate solution have assured. Fuzzy is comprised of programs and mathematics involved in bringing up a model by learning and experience for representing high complex concepts. Fuzzy logic is implicative of human thinking process and the decisions have made based on information. Normally linguistic variables are condition of motor, fuzzy subsets describes stator current amplitude and rule base or data base supports fuzzy inference. The diagnosis of IM has carried out based on the rules of fuzzy inference mechanism (Miljkovic & Zagreb, 2015; Chaturvedi, Iqbal & Singh, 2013).

In this section, a case study have proposed based on hybrid monitoring technique inclusive of MCSA and fuzzy logic system has covered for the protection of three phase IM [30]. Here, MCSA has utilized for detection of fault and fuzzy logic system involves in diagnosis of IM. The block diagram of the proposed scheme has given in Figure 13.

In this technique, the common rotor faults in IM namely Rotor eccentricity, broken rotor bar and Bearing damage have considered for protection. The frequency spectrum for these faults have obtained from MCSA and signature analysis has done by FFT algorithm.

As discussed earlier, section 4 depicts the types of faults and their frequencies are given in expressions. From the frequency expressions, the frequency range of the faults such as rotor eccentricity (Expression 4), broken rotor bar (Expression 3) and bearing damage (Expression 9) has calculated. From the simulation the following figure 14 is obtained. By comparing the amplitudes of the stator current with the fault frequency range the occurrence of fault has ensured. As well, the deviation of

Figure 12. Block diagram for online monitoring process of fault detection

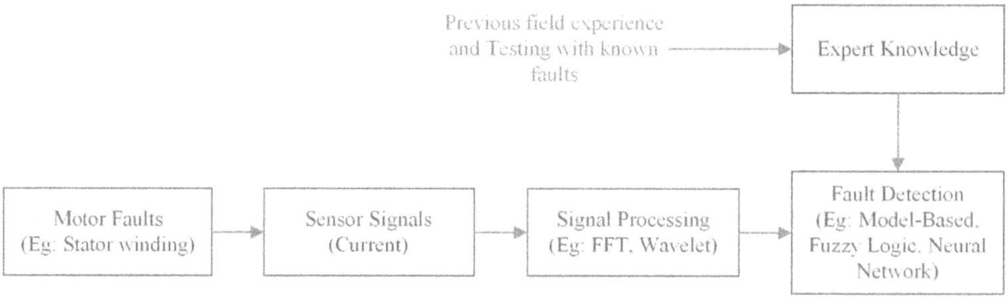

Figure 13. Block diagram of the proposed scheme

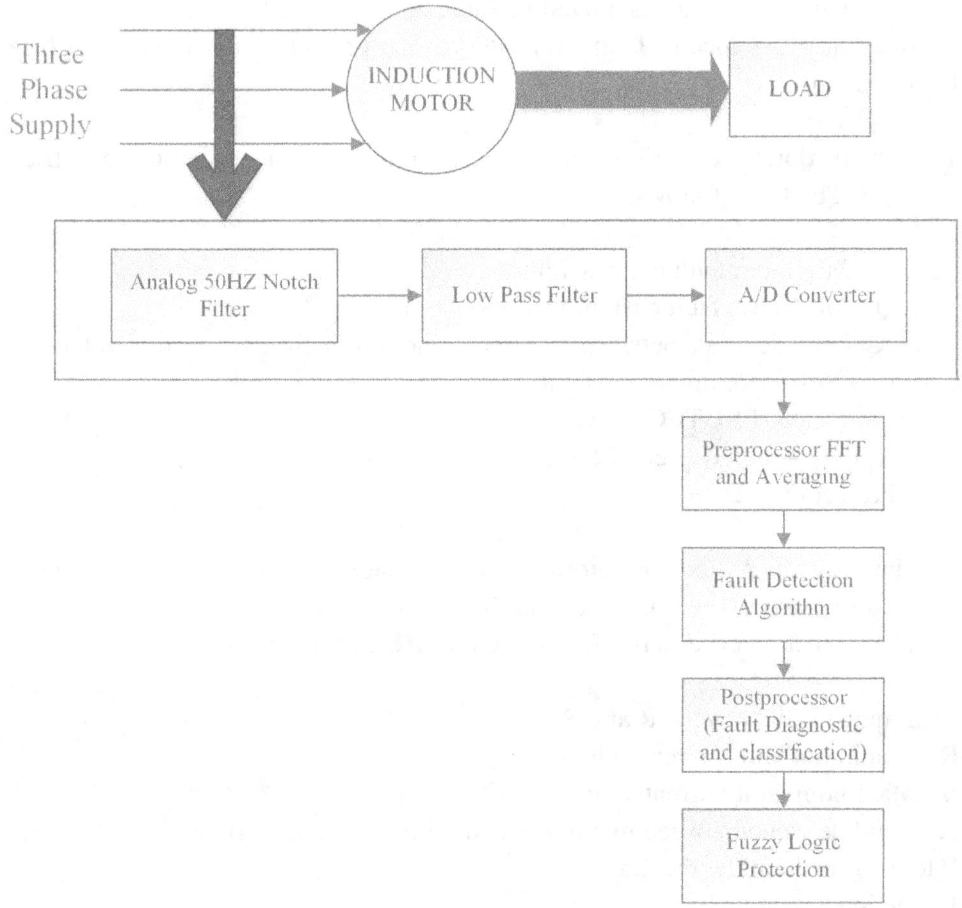

Figure 14. Simulation on fault identification of broken rotor bar and rotor eccentricity

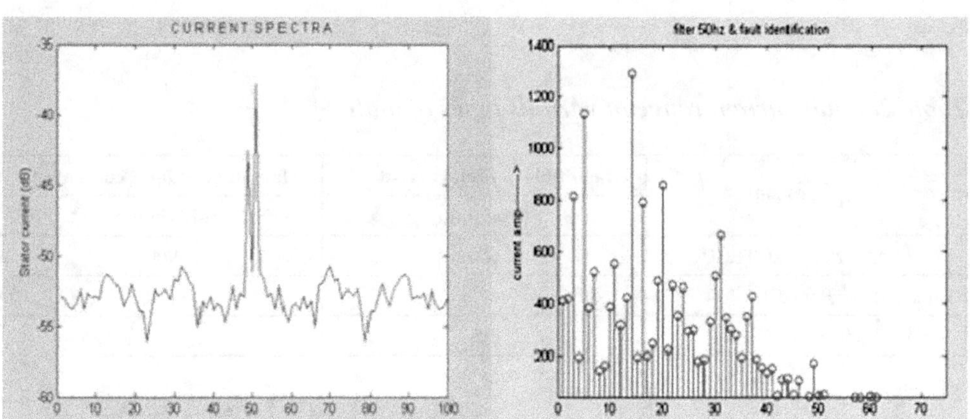

fault frequency from the predetermined level the types of fault also ensured. The obtained fault current has tabulated in Table 2.

After the assurance of fault from MCSA, the protection has given with fuzzy logic system.

1. **Protection Criteria:** The protection criteria has formulated for rotor eccentricity (ROTECC) as follows:

   P – Minimum fault current value
   Q - Maximum fault current value
   X_1 – middle value between maximum and minimum fault current value
   P to X_1 – medium degree fault
   Protection (PROTEC) – Alarm
   X_1 to Q – high degree of fault
   Protection – Trip

Similarly, the protection criteria for broken rotor bar (BB) and bearing damage (BD) formulated in the same way and they are given below:

The protection criteria has formulated for BB as follows:

Fault range is in between R and S
R – Minimum fault current value
S - Maximum fault current value
X_2 – middle value between maximum and minimum fault current value
P to X_2 – medium degree fault
Protection – Alarm
X_2 to Q – high degree of fault
Protection – Trip

Table 2. Fault current range of various types of fault

Type of Fault	Magnitude of fault current (A)	Magnitude of fault current (A)
	Minimum	Maximum
Rotor Eccentricity (ROTECC)	600	900
Broken Rotor Bar (BB)	40	160
Bearing Damage (BD)	65	120

The protection criteria has formulated for BD as follows:

Fault range between T and U
R – Minimum fault current value
S - Maximum fault current value
X_3 – middle value between maximum and minimum fault current value
P to X_3 – medium degree fault
Protection – Alarm
X_3 to Q – high degree of fault
Protection – Trip

2. **Fuzzy Variables:** Two fuzzy variables namely "medium" and "high" have chosen as input variables. Likewise, "alarm" and "trip" have chosen as output variables.
3. **Membership Function:** After the performance comparison made between different types of membership functions, trapezoidal and triangular membership function have chosen for input and output respectively.
4. **Fuzzy rules and Fuzzy Inference:** For the safe operation of relay, degree of fault have defined. Here, 18 rules have framed to find the degree of fault in IM.
    a. If (ROTECC is MEDIUM)then(PROTEC is ALARM)(1)
    b. If (ROTECC IS HIGH)then(PROTEC IS TRIP)(1)
    c. If (BB is MEDIUM)then(PROTEC is ALARM)(1)
    d. If (BB is HIGH)then(PROTEC is TRIP)(1)
    e. If (BD is MEDIUM)then(PROTEC is ALARM)(1)
    f. If (BD is HIGH)then(PROTEC is TRIP)(1)
    g. If (ROTECC is MEDIUM)and(BB is MEDIUM)then(PROTEC is ALARM)(1)
    h. If (ROTECC is MEDIUM)and(BB is HIGH)then(PROTEC is TRIP)(1)
    i. If (ROTECC is HIGH)and(BB is MEDIUM)then(PROTEC is TRIP)(1)
    j. If (ROTECC is HIGH)and(BB is HIGH)then(PROTEC is TRIP)(1)
    k. If (BB is MEDIUM)and(BD is MEDIUM)then(PROTEC is ALARM)(1)
    l. If (BB is MEDIUM)and(BD is HIGH)then(PROTEC is TRIP)(1)
    m. If (BB is HIGH)and(BD is HIGH)then(PROTEC is TRIP)(1)
    o. If (ROTECC is MEDIUM)and(BD is MEDIUM)then(PROTEC is ALARM)(1)
    p. If (ROTECC is MEDIUM)and(BD is HIGH)then(PROTEC is TRIP)(1)
    q. If (ROTECC is HIGH)and(BD is MEDIUM)then(PROTEC is TRIP)(1)
    r. If(ROTECC is HIGH)and(BD is HIGH)then(PROTEC is TRIP)(1)

Minimum inference has used in all the rules.

5. **Defuzzification:** For defuzzification, center of gravity method has used and range of defuzzified value is 0.5 to 1.

Based upon the minimum inference centroid value, the threshold value has found as 0.739 and this value is the protection deterministic criteria for Alarm and Trip. Hence, the protection has evolved as follows:

Alarm – 0.5 to 0.739

Trip – 0.739 to 1.0

As a whole, with the combination of fault current and the degree of fault value it has framed whether the fault diagnosis is an alarm signal or it is a trip signal from the relay. Based on the severity level of the fault current, the fuzzy protection system has given with proper decision on alarm and trip. The fault current level of each type of fault and the respective degree of fault with fuzzy based protection of IM has given in the Table 3.

Table 3. Protection Criteria based on degree of fault level calculation

ROTECC	BB	BD	Degree of fault level between 0 to 1	Protection	
				Alarm < 0.739	Trip > 0.739
690	61	-	0.637	Alarm	-
638	90	-	0.709	Alarm	-
779	74	-	0.753	-	Trip
855	113	-	0.756	-	Trip
631	-	74	0.637	Alarm	-
714	-	83	0.678	Alarm	-
783	-	77	0.754	-	Trip
869	58	113	0.756	-	Trip
-	68	106	0.637	Alarm	-
-	93	80	0.746	-	Trip
-	64	94	0.637	Alarm	-

Henceforth, from this case study, the MCSA and Fuzzy logic based protection scheme for IM rotor faults mainly rotor eccentricity, broken rotor bar and bearing damage have developed and simulation results have shown the effectiveness of the protection scheme.

ANN Based Protection for Induction Motors Using MCSA Technique

Artificial Neural Network is an efficient technique to diagnose the faulted machine based on the formulation of knowledge base system with prior inputs and outputs. ANN with learning is suitable for the execution of pattern classification. Proper training of the neural network, with data obtained from experimental results or from the simulation results on a faulted machine, provides a better diagnostic system (Chaturvedi, Iqbal & Singh, 2013). The training of datasets in ANN is mostly, based on supervised learning with normal and faulty conditions. If the fault diagnosis is inclusive of both MCSA and ANN means, the fault detection has done with MCSA and diagnosis based on ANN. Here, the detection of side band frequency components in stator current spectrum has carried out with MCSA. This data has given to ANN to perform supervised learning on given data to represent the faulty condition of the IM (Miljkovic & Zagreb, 2015).

T. Aroui et al (2007) depicted a supervised neural network for classification of severity of the fault. For this multilayer feed forward neural network has proposed for the diagnosis of problems associated with rotor. The data of the current spectra has measured for training the neural network. The experiment has conducted at different loading conditions and the arrived results prove that ANN technique is effective not only on detection of fault but also classification and diagnosis of fault through monitoring the current spectra of the stator current signal.

Neural networks comprises of simple elements inspired by biological nervous system, used to operate in parallel. A large network has framed by making connection among these elements. The training of neural network has carried out by adjusting the connection weights among the element connected in parallel (Greety & Jose, 2013). The ANN based fault diagnosis has the following steps in diagnosis of IM

- Collection of appropriate data set for training of neural network
- Selection of appropriate structure for neural network
- Neural network training
- Test pattern evaluation

Hayder O. Alwan et al (2017) proposed a combined protection approach of MCSA with ANN for inter turn fault diagnosis in induction motor (Alwan, Farhan

& S-Al-Sabbagh, 2017). In this approach MCSA has used for online monitoring and detection of inter turn fault in motor. To differentiate the amplitude of the frequencies in frequency spectrum the assumption has taken a turn shorts for 10 turns initially and 30 turns. The data from the frequency spectrum gas given to the neural network for training the ANN and finally the diagnosis obtained for induction motor protection.

A sample MCSA and neural network based online monitoring and diagnosis of induction motor has given as flowchart in Figure 15.

The basic need for the implementation of neural network based fault diagnosis is the readiness of data with respect to the fault. Henceforth, the input data for the neural network is the main pointer of fault. The selection of data from the infinite information is a complicated assignment. However, the selection of appropriate fault indicator is to find the factor, which provide a detailed information on the condition of the machine. By considering these factors, the mixed eccentricity inherent harmonic detection is efficient. The extension of this procedure is applicable to other faults for diagnosis (Guedidi et al., 2011).

Amongst many other online monitoring techniques MCSA is the most effective non-invasive technique as it requires only one input source of stator current signal. Then a hybrid method of fuzzy min max (FMM) neural network has proposed for the condition monitoring of induction motor (Seera et al., 2014).

Support Vector Machine Technique

In classification and pattern recognition problems a machine learning method called Support Vector Machine (SVM) is applied. It is based on statistical learning theory. SVM approach is applicable for MCSA also because of its high classification efficiency on problems. This SVM approach has the following steps (Miljkovic & Zagreb, 2015):

- Data acquisition
- Feature extraction
- Feature selection
- Fault classification

In feature extraction the parameters are calculated from the obtained current signal. Feature selection is performed with the help of Principal Component Analysis (PCA). With this SVM approach stator inter-turn faults and dynamic eccentricity can be identified (Miljkovic & Zagreb, 2015).

Figure 15. Flowchart for online monitoring and diagnosis of induction motor based on MCSA and neural network technique

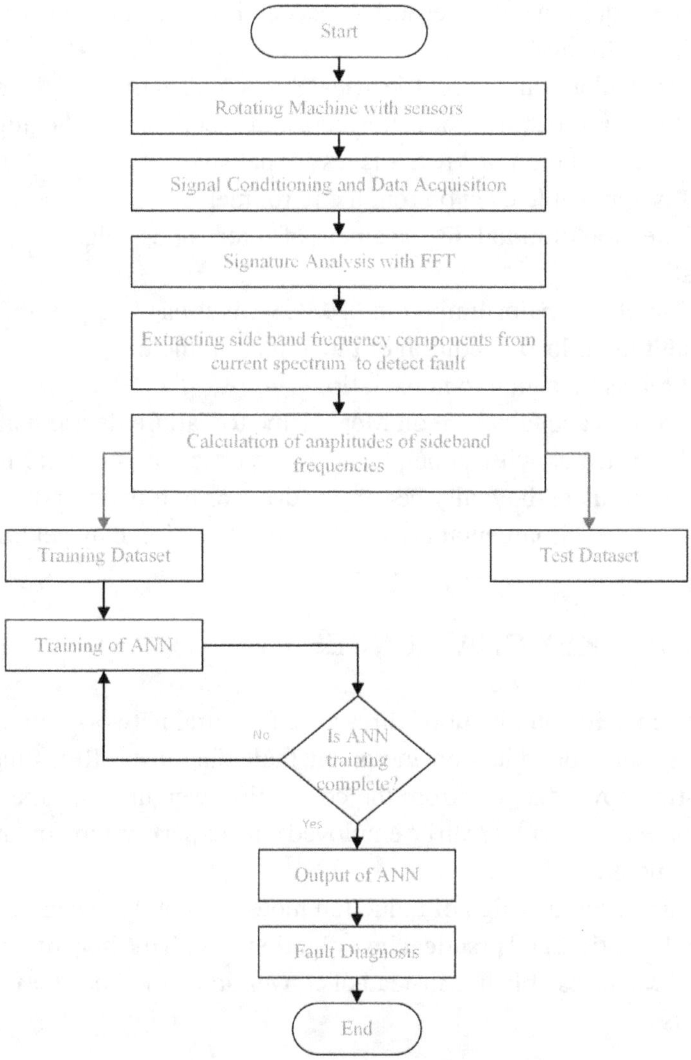

Advantages and Limitations of MCSA

So many merits and demerits have proved on usage MCSA for fault detection and some of them have listed below (Soomro et al., 2016; Corne et al., 2015):

1. Advantages:
   a. Fault detection is possible at early stage as it is a predictive maintenance tool.

   b. The processing of electrical signal that carries the current component is direct with the help of rotating flux components.
   c. The requirement of sensors is less and maintenance cost reduced with this technique
   d. It is a wider technique of Electrical Signature Analysis, and used in analysis of generators, power transformers and other electrical equipment.
   e. Implementation of MCSA is less expensive.
   f. It is applicable even on rough environments.
   g. Detection of radial displacement of rotor is possible.
2. Limitations:
   a. One of the major limitation of MCSA is it has less possibility to detect fault in no load condition of the induction motor.
   b. It reacts on signal to noise ratio.
   c. It has no acquaintance on Mean Time to Failure determination.
   d. For bearings with same physical properties, it is critical for MCSA to discriminate the faulty bearing at drive or non-drive end.
   e. Due to environmental issues, the fault detection may get affected.

FUTURE: THE KEY CHALLENGES

Expert knowledge in combination with artificial neural network can improvise the monitoring system for induction motor on fault diagnosis. Rules have obtained from heuristic knowledge and from the connection weight of a supervised neural network. This extracted rules, can be employed with expert system for fault diagnosis of induction motor.

For various size and design of induction motors, a database can be created with the usage of theoretical and practical investigations. With the help of this database, a new standard can be established instead of conventional methods for fault detection and diagnosis.

CONCLUSION

The induction motor failure leads to production downtime and involves in loss of maintenance and revenue. An overview of adequate faults in IM and the types of fault detection techniques and diagnosis procedures are discussed in this chapter. Amongst the techniques, MCSA technique has the potent to detect and diagnose faults through online monitoring of the induction motor. In addition, this scheme enables non-intrusive online analysis of frequency spectrum of the stator current signal for detection of faults in IM without interrupting its operation. As well, MCSA technique offers a highly sensitive and cost effective means for online monitoring of industrial run induction motors. For diagnosis purpose, this technique can be utilized in conjunction with other techniques. The advancements in theoretical concepts evolved in prediction of many faults, thus producing MCSA approach is a much more robust and less prone to error technology. The effectiveness of this technique depends upon the data collection and evaluation of the available system. MCSA is the most resourceful technique for online condition monitoring and fault diagnosis of induction motor. It ensures the safe working surrounding in industries with continuous process and resolves the shutdown problem. It has the capability to detect small changes in the motor parameters and involves in diagnosis of various failures occur in induction motor.

ACKNOWLEDGMENT

The authors are thankful to the authorities of Thiagarajar College of Engineering, Madurai – 625015, to do this research work. This research was supported by Ministry of Human Resource Development, Government of India through Technical Education Quality Improvement Programme (TEQIP) Scholarships under Ref: TCE/Dean-R&D/ TEQIP/ T&R Assistantship (2018-2019).

REFERENCES

Agrawal, D., Yadav, N., & Saini, S. (2015). Condition Monitoring of Slip-ring Induction Motor. *International Journal of Innovative Research in Advanced Engineering*, *2*(3), 78–84.

Alwan, H. O., & Farhan, N. M. (2017). Design of Data Aquistion interface circuit used in Detection Inter-turn Fault in Motor based on Motor Current Signature Analysis (MCSA) Technique. Int. *Journal of Engineering Research and Application*, *7*(6), 79–89. doi:10.9790/9622-0706017989

Aroui, T., Koubaa, Y., & Toumi, A. (2007). Application of feedforward neural network for induction machine rotor faults diagnostics using stator current. *J. Electr. Syst.*, *3*, 213–226.

Bayar, H., & Terz, U. K. (2016). Grid-connected induction generator interturn fault analysis using a PCA-ANN-based algorithm. *Turkish Journal of Electrical Engineering and Computer Sciences*, *24*, 1340–1350. doi:10.3906/elk-1312-145

Bhardwaj, K., & Agarwal, A. (2012). Fault Diagnosis of Three Phase Induction Motor Using Fuzzy Logic Controller and Sequence Analyzer. *MIT International Journal of Electrical and Instrumentation Engineering*, *2*(2), 112–118.

Bhowmik, P. S., Pradhan, S., & Prakash, M. (2013). Fault Diagnostic and Monitoring Methods of Induction Motor: A Review. *International Journal of Applied Control Electrical and Electronics Engineering*, *1*, 1–18.

Chaturvedi, D. K., Iqbal, S. M., & Singh, M. P. (2013). Health Monitoring Techniques of Induction Motor: An Overview. *Proceedings of Int. Conf. on Emerging Trends in Engineering and Technology*, 469-477.

Chudasama, K. J., & Shah, V. (2012). Induction Motor Noninvasive Fault Diagnostic techniques: A Review. *International Journal of Engineering Research and Technology*, *1*(5), 1–7.

Corne, B., Vervisch, B., Debruyne, C., Knockaert, J. & Desmet, J. (2015). Comparing MCSA with Vibration Analysis in order to detect Bearing Faults - A Case Study. *IEEE Explore*, 1366-1372.

Drozdowski, P., & Duda, A. (2014). A. Influence of magnetic saturation effects on the fault detection of induction motors. *Arch. Electr. Eng*, *63*, 489–506. doi:10.2478/aee-2014-0035

Fenger, M., Lloyd, B.A., & Thomson, W.T. (2003). Development of a tool to detect faults in induction motors via current signature analysis. *Cement Industry Technical Conference IEEE Explore*, 37-46.

Geethanjali, M., & Hemavathi, R. (2010). Fuzzy Logic Based Protection for Three Phase Induction Motor Rotor faults Using Motor Current Signature Analysis based DSP Algorithm. *Proceedings of the 34th National Systems Conference*, 85.

Granda, D., Aguilar, W.G., Arcos-Aviles, D. & Sotomayor, D. (2017). Broken Bar Diagnosis for Squirrel Cage Induction Motors Using Frequency Analysis Based on MCSA and Continuous Wavelet Transform. *MDPI Mathematical and Computational Applications, 22*, 30. doi:10.3390/mca22020030

Greety, J., & Jose, V. (2013). Induction Motor Fault Diagnosis Methods: A Comparative Study. *Proceedings of International Conference on Electrical Engineering*, 63-69.

Guedidi, S., Zouzou, S. E., Laala, W., Sahraoui, M., & Yahia, K. (2011). *Broken Bar Fault Diagnosis of Induction Motors Using MCSA and Neural Network*. IEEE Explore. doi:10.1109/DEMPED.2011.6063690

IEEE. (2006). *IEEE 1415TM-2006 IEEE Guide for Induction Machinery Maintenance Testing and Failure Analysis*. IEEE. Retrieved from: https://www.maintenance.org/fileSendAction/fcType/0/fcOid/399590942962795456/filePointer/399590942964800333/fodoid/399590942964800331/Pages_from_1415-2006-2.pdf

Irfan, M., Saad, N., Ibrahim, R., Asirvadam, V. S., Alwadie, A. S., & Sheikh, M. A. (2017). *An Assessment on the Non-Invasive Methods for Condition Monitoring of Induction Motors*. Fault Diagnosis and Detection Mustafa Demetgul, IntechOpen; doi:10.5772/67917

Karmakar, S., Chattopadhyay, S., Mitra, M., & Sengupta, S. (2016). *Induction Motor and Faults. Induction Motor Fault Diagnosis: Approach through Current Signature Analysis*. Springer. doi:10.1007/978-981-10-0624-1_2

Kompella, D. K. C., Gopala Rao, M. V., Rao, R. S., & Sreenivasu, R. N. (2014). Estimation of bearing faults in induction motor by MCSA using Daubechis wavelet analysis. *Proceedings of international conference on smart electric grid (ISEG14)*, 1–6.

Kompella, D. K. C., Gopala Rao, V. M., & Rayapudi, S. R. (2017). *Bearing fault detection in a 3 phase induction motor using stator current frequency spectral subtraction with various wavelet decomposition techniques. Ain Shams Engineering Journal*. doi:10.1016/j.asej.2017.06.002

Korde, A. (2002). On-Line Condition Monitoring of Motors Using Electrical Signature Analysis. *Recent Advances in Condition Based Plant Maintenance*, 17-18.

Kucuker, A., & Bayrak, M. (2015). Detection of stator winding fault in induction motor using instantaneous power signature analysis. *Turkish Journal of Electrical Engineering and Computer Sciences*, *23*, 1263–1271. doi:10.3906/elk-1304-72

Mehala, N., & Dahia, R. (2007). Motor Current Signature Analysis and its Applications in Induction Motor Fault Diagnosis. *International Journal of Systems Applications, Engineering and Development*, *2*(1), 29–35.

Mehla, N., & Dahiya, R. (2007). An Approach of Condition Monitoring of Induction Motor Using MCSA. International Journal of Systems Applications. *Engineering & Development*, *1*(1), 13–17.

Merizalde, Y., Hernández-Callejo, L., & Duque-Perez, O. (2017). State of the Art and Trends in the Monitoring, Detection and Diagnosis of Failures in Electric Induction Motors. *MDPI Energies*, *10*(7), 1056. doi:10.3390/en10071056

Messaoudi, M., & Sbita, L. (2010). Multiple faults Diagnosis in Induction Motor Using the MCSA Method. *International Journal of Signal and Image Processing*, *1*(3), 190–195.

Miljković, D. & Zagreb, H. (2015). *Brief Review of Motor Current Signature Analysis*. Academic Press.

Nandi, S., Toliyat, H., & Li, X. (2005). Condition monitoring and fault diagnosis of electrical motors—A review. *IEEE Transactions on Energy Conversion*, *20*(4), 719–729. doi:10.1109/TEC.2005.847955

Narwade, S., Kulkarni, P., & Partil, C. (2013). Fault detection of induction motor using current and vibration monitoring. *Int. J. Adv. Comput*, *3*, 272.

Prakasam, K., & Ramesh, S. (2016). Testing and Analysis of Induction Motor Electrical Faults Using Current Signature Analysis. *Scientific Research-Circuits and Systems*, *7*(09), 2651–2662. doi:10.4236/cs.2016.79229

Seera, M., Lim, C. P., Nahavandi, S., & Loo, C. K. (2014). Condition monitoring of induction motors: A review and an application of an ensemble of hybrid intelligent models. *Expert Systems with Applications*, *41*(10), 4891–4903. doi:10.1016/j.eswa.2014.02.028

Shukla, S., Jha, M., & Qureshi, M. F. (2014). Motor Current Signature Analysis for Fault Diagnosis and Condition Monitoring of Induction Motors using Interval Type-2 Fuzzy logic. International Journal of Innovative Science. *Engineering & Technology*, *1*(5), 84–95.

Siddiqui, K. M., Sahay, K., Giri, V. K., & Gothwal, N. (2016). Diagnosis of airgap eccentricity fault in the inverter driven induction motor drives by transformative techniques. *Perspectives in Science, Elsevier*, *8*, 127–131. doi:10.1016/j.pisc.2016.04.014

Sonje, D. M., & Monje, R. K. (2011). Rotor Cage Fault Detection in Induction Motors by Motor Current Signature Analysis. *International Conference in Computational Intelligence Proceedings*, 22-26.

Soomro, A. A., Kalwar, I. H., Kazi, K., Khoo, S. K., & Ansari, S. (2016). A Hybrid Monitoring Technique for Diagnosis of Mechanical Faults in Induction Motor. *Indian Journal of Science and Technology*, *9*(47), 1–9.

Thomson, W. T., & Culbert, I. (2017). Current Signature Analysis for Condition Monitoring of Cage Induction Motors. Wiley Publishers.

Thomson, W. T., & Gilmore, R. J. (2003). Motor Current Signature Analysis to detect faults in Induction Motor Drives-Fundamentals, Data Interpretation and Industrial Case Histories. *Proceedings of the 32nd turbomachinery Symposium*, 145-156.

Chapter 2
Fault Detection and Isolation in Power Systems Using Unknown Input Observer

Hassan Haes Alhelou
Tishreen University, Syria

ABSTRACT

Unknown input observers (UIO) find application in many cases for successful fault detection and isolation (FDI). In this chapter, a scenario where the unknown input observer is applied to load frequency control loops in interconnected power systems is analyzed. A UIO was chosen because load demand is not always constant and it can be considered to introduce an unknown disturbance to the system. Mathematical formulations on how to detect and isolate sensor faults are presented which are then implemented in MATLAB Simulink for simulations. Based on this historical survey on the application of UIO, a thesis on UIO application in FDI in distributed generation is done.

INTRODUCTION

Generally, FDI methods can be classified in two broad categories the model based methods and the process history/ data based methods. Figure 1 shows how these methods can then be further divided into different categories. The model-based methods can be broadly classified as qualitative or quantitative. The model is usually developed based on some fundamental understanding of the physics of the process or system. In quantitative models this knowledge is expressed as mathematical functional relationships between the inputs and outputs of the system while in qualitative model

DOI: 10.4018/978-1-5225-6989-3.ch002

Figure 1. The different types of methods

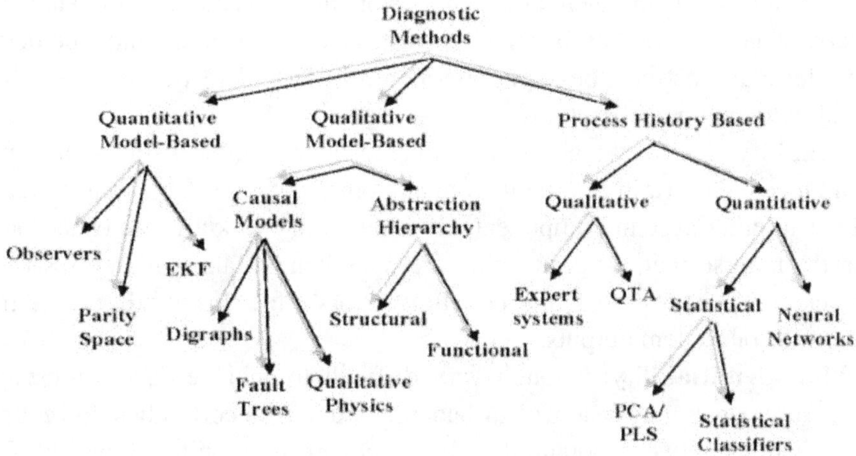

based methods these relationships are expressed in terms of qualitative functions centered on different units in a process.

In contrast to the model-based approaches in process history based methods only the availability of large amount of historical process data is assumed. This data is transformed and presented as a prior knowledge for fault diagnosis by a process called feature extraction. It should be noted however that within these methods there are overlaps.

The most commonly used FDI approaches include diagnostic observers, parity relations, Kalman filters and parameter estimation. Under observer methods there are unknown input observers (UIO), functional observers, Lueneberg observer, and proportional integral observer, sliding mode observers (SMO), state observers and many others. The motivation behind the development of model based methods is replacing direct redundancy with analytical redundancy. Most of the work on quantitative model-based approaches have been based on general input/output and state/space models. The main concern of observer-based FDI is the generation of a set of residuals which detect and uniquely identify different faults. These residuals should be robust that is they are not sensitive to unknown inputs as unstructured uncertainties like process and measurement noise and modelling uncertainties. Observers can be designed in such a way that it is sensitive to one fault and insensitive to the rest.

Generally, a good fault detection and isolation system (FDI) must at least have all or most the properties explained as below.

- **Early Detection and Diagnosis:** A fault diagnostic system must be able to detect and isolate incipient fault. The system must be able to detect faults in the early stages before they become a catastrophe. Small faults must be detected but false alarms must be kept under control. Early detection is a key issue in fault tolerant control systems.
- **Isolability:** This is the ability of a diagnostic system in distinguish, separate and locate the origin of a fault from any other potential fault sources that can exist in other system components. It is necessary to know the exact location of the fault so that corrective measures can then be taken. Uncertainties such as external disturbances and modelling errors affect Insolubility since it also depends on system outputs.
- **Fault Identifiability:** Though very difficult to achieve due to presence of measurement noise, system disturbances, modeling uncertainties, and coupling/ interactions between potential fault sources in the monitored system. Faulty identifiability is a key requirement in estimating the degree and nature of fault.
- **Robustness:** Robustness to measurements noise, system disturbances, and modeling uncertainties is one of the most important issues of a diagnostic system intended for practical implementations. In residual generating systems the residuals must be robust.
- **Multiple Fault Identifiability:** This refers to the ability of a diagnostic system to identify and correctly classify multiple faults that may coexist in a system. Multiple fault identifiability is not easy due to coupling that exists between states, the fault complexity and general system nonlinearity.
- **Explanation Facility:** A diagnostic system should be able to explain where a fault originated and how it propagated in the system. Adaptability: Since the operating conditions of the system change due to aging, disturbances and environmental changes because of this the system components experience performance degradation as the in-service period extends. Therefore a fault diagnosis should be able to successfully adapt to these changes in order to maintain its diagnostic performance. Reasonable storage and computational requirements: Memory and computational requirements are the two fundamental characteristics of any algorithm intended for online, real-time implementation. Diagnostic algorithms, especially the ones intended for embedded on-board fault diagnosis, are by no means an exception. Therefore, while designing a fault diagnosis system, it is necessary to keep in mind that the computational and memory requirements must always meet the specifications of the application, also including power consumption specifications. Moreover, depending on the application, a reasonable compromise between these two requirements should be made.

Advantages of Observer-Based Approaches

The isolation task can be fulfilled via a structured residual set designed by a dedicated or a generalized observer scheme or a directional residual vector designed via a fault detection filter. In most observer methods reaction to incipient faults is very fast. The methods have been successfully used for detecting and isolating faults in actuators and sensors. There is a possibility of detecting and isolating faults in system parameters, although the method is complicated to achieve. Once the input and output are well established from the system dynamics the design procedure is systematic and simple. The developed algorithm is simple and can be easily implemented in DSP controllers. Multiple faults can be easily handled if the measurement number is sufficient. A Kalman filter can be used to produce the fault-free residual with minimum variance and, consequently reducing false and missed alarms due to noise and other external disturbances.

Redundancy can be categorized into two types; direct and analytical. In direct redundancy, actual physical hardware redundancy is available. In terms of sensors double and triple redundancy can be used and this is costly. However, in normal operation, only one sensor is sufficient but usually two or three sensors (a voting system) are required to ensure reliable measurements in the case of fault. In voting system if say two out of three sensors indicate a faulty then a fault is considered to have occurred. In observer systems analytical redundancy replaces direct redundancy, instead of having multiple sensors that measure the same signal, an observer that provides an estimate of the signals is used.

A REVIEW OF OBSERVER-BASED FDI APPLICATIONS

In Power Systems and Control

An unknown input observer (UIO) is used to detect sensor faults in wind turbines in (Odgaard, et al. 2009). A drive train model is used in the observer design process. The main concept of UIO is to make the disturbance to be decoupled from the input and in this case since wind speed is not always constant it is considered the unknown input. This makes the system not dependent on the actual speed but an advanced simulation model is used to mimic the system and the sensor faults can be detected by comparing the two outputs. Five different sensor faults in the drive train of a three blade horizontal axis wind turbine are detected using the UIO (Odgaard, et al. 2009). A one body model for gear box and a two body model for the system are used for sensor fault simulations. The idea of UIO design is presented in detail by R J Patton (Chen, et al. 2012 & Alhelou, et al. 2016).

In (Hwas, et al. 2012), an observer-based sensor fault detection and isolation scheme is developed for wind turbine. The model is designed to be sensitive to be sensitive to system faults and insensitive to disturbances. Here one observer is used unlike other cases where a set of observers will be required.

Detection and isolation of power system faults is of key importance since a power plant is an extremely expensive industry. Most faults are unpredictable hence are modelled as unknown inputs decoupled from inputs and outputs. An observer based model is used in detecting the faults for a system consisting of a synchronous generator an exciter, a turbine and speed-governing system, and a network of lines and loads. The case where faults occur on the transmission network is considered. The advantage of this method that only one observer is designed to detect any type of fault and their location (Aldeen, et al. 2006 & Alhelou, et al. 2018).

An observer based method for moisture content estimation of coal in a coal mill and fault detection is presented. Key variables to monitor for model development are coal flow into and out of the mill as well as the moisture content in the coal. But in real systems these variables are almost nearly impossible to measure directly. A method is developed to estimate the unknown inputs based on the energy model of the coal mill. Two different residual generation and detection are used for fault detection and moisture estimation. An UIO is normally used to estimate states systems with unknown inputs or disturbance but in this paper it has been re-modified since the aim is to estimate e unknown inputs. Serious fault that can occur in coal mills include but not limited to blocking of the raw coal inlet pipe, faults in the primary air supply both the fan and the temperature controller, and failures in the sensors. This leads to a decrease in the coal output which can subsequently lead to a tip in generating units hence fault detection of coal mills is of importance (Odgaard, et al. 2008). The same author presented another paper for fault detection in coal mills (Odgaard, et al. 2006 & Odgaard, et al. 2009). In a similar case a Kalman filter issued to control the coal flow based on flue gas measurements (Andersen, et al. 2006). Methods involving the use of advanced PID and fuzzy PID have been worked on as well.

Here the problem of cyber-attack detection on power system networks is investigated. The network is divided into parts and an unknown input observer is used for each part to detect cyber-attack such as the denial of service (DOS) attack and the men in the middle (MitM) attack by using information from PMU networks (Varmaziari, et al. 2017 & Alhelou, et al. 2015).

A bilinear state observer stabilized by using feedback of rotor currents is used to detect stator current sensor faults. The observer is used for residual generation and to supply signals for fault measurement. Doubly fed induction generators are mainly used in wind farms for electricity generation. Open loop, closed loop models and the bi linear behavior of induction machine on the state space model is discussed (Rothenhagen, et al. 2006).

Fault detection and reconfiguration of the control loops of a Doubly-Fed Induction Generator is presented. Here the stator voltage is measured and observed to detect faults. When there is no fault the residual is used for field controls. A PLL is used to measure the stator voltage angle an UIO observer is used to supply an approximate of input voltage in-order to reconfigure the power control loop. It's concluded that further research is needed to limit the deviation between the actual power the observed power (Rothenhagen, et al. 2007 Alhelou, et al. 2015).

Caliskan applied a robust sensor and actuator fault detection and isolation (FDI) method based on unknown input observers (UIO) to the load frequency control loops of interconnected power systems. Changes in load demand are modelled as the unknown disturbances in the system making the UIOs robust to these disturbances. Simulations are done for power system composed of two areas which shows successful isolation of sensor faults. The selection of input variables determined the success of the method. A logic scheme where information is given to the controller is also designed. It was noted however that there are also other possible faults that can be observed in an AGC loop, such as unresponsiveness of some generating units to raise or lower signals. This was considered an area for further work (Caliskan, et al. 2008).

Following his paper on a two area network Caliskan proposed a robust actuator fault detection and isolation method to detect and isolate the governor faults in a multi-area power system in which the frequency is regulated by an area-wise decentralized load frequency control system using UIOs. Simulations to the power system which consisted of 5 generating units were successfully carried out. From the method the operator can be notified on the faulty actuator (Genc, et al. 2011).

In Electric Machine and Drives

This paper presented the detection and isolation of open-switch faults in induction motor (IM) drives using an observer based method. Stator currents and mechanical velocities are considered as outputs thus making residuals insensitive to load torque and other operating conditions. A three phase pulse with modulation inverter (PWM) is considered as the power actuator. A dedicated observer for residual generation is used for the detection and isolation of single and multiple faults in semiconductor devices. An important feature about this FDI structure is that it does not depend on the supply and switching frequencies in the inverter thus the observer convergence does not rely on the operating frequencies of the power actuator as well. For future work, it is proposed to evaluate the FDI structure under a variable-speed regime and to extend the FDI algorithm for multilevel inverter topologies due to their importance in high-power applications (Campos-Delgado, et al. 2011).

Here and adaptive e observer is used for sensor fault detection and isolation for induction-motor drives. Closed-loop induction motor drives with voltage-source inverters use a speed or position, a dc-link voltage, and two or three phase-current sensors. In this scheme the estimated phase

currents and rotor resistance are sent to a decision-making unit which identifies the faulty sensor type and its phase based on a deterministic rule base. Here only one current observer with rotor-resistance estimation is used for isolation of all sensor faults unlike other case where a bank of observers will be required. The observer is used to estimate phase currents and rotor resistance and faulty sensor are singled out based on the observer outputs and measured signals. The method is simulated to verify its effectiveness (Najafabadi, et al. 2011).

In (Yan, et al. 2008), a general class of nonlinear systems with uncertain parameters is presented. Here an adaptive law is designed to identify unknown parameters under the assumption that the time derivative of some of the outputs is measurable after which a sliding-mode observer is proposed to estimate the system state variables. Using the sliding mode model the faulty can be assessed to any degree of accuracy. A simulated example is then given.

In (De Angelo, et al. 2009), a vector of specific residual generation strategy using a state observer is presented. The residual is generated from a decomposition of the current estimation error for fast detection of incipient faults. Stator voltages and currents are considered as outputs and the strategy is insensitive to power supply and load changes. Inter-turn faults can be detected and in this scheme speed can be estimated from the error without need of a sensor to measure speed. An online diagnosis, observer-based and parameter-estimation techniques have also been presented in (Kallesoe, et al. 2007). The strategy in (De Angelo, et al. 2011) is a modified for the strategy in (De Angelo, et al. 2009) which now uses a state observer to quickly detect and isolate incipient faults.

State observer models have also been used to detect inter-turn short-circuit fault in induction motors. Input voltage and output current measurements are used by the observer to estimate the stator current which is compared by the measured current to generate the residuals. The current estimation error is then decomposed into two terms one that is used to generate the vectorial residual and the other that is used to correct the observer parameters according to the operating conditions. Due to the adaptive nature of the observer, it has been reported to work even in the case of supply-voltage unbalances and speed changes. The main challenge with these model-based approaches is that if any factor was not modeled properly it might lead to huge errors in the detection of fault especially in the case of permanent magnet motors (PM) motors, which possess inherent nonlinearities (De Angelo, et al. 2009). Figure 2 shows the diagrammatic representation of induction motor inter-turn detection technique

Figure 2. the diagrammatic representation of induction motor inter-turn detection technique (De Angelo, et al. 2009)

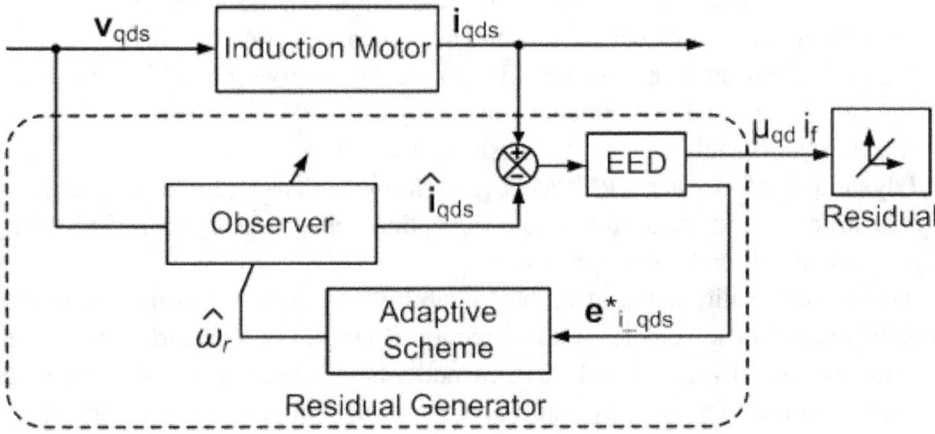

In Power Electronics

In (Espinoza-Trejo, et al. 2012), a PI observer-based FDI scheme was proposed to detect and isolate actuator faults in VSI for IM. The scheme showed robustness to model uncertainties. Dedicated non-linear PI observers were suggested for the identification of the DC component of the fault profiles. In the experimental validation, a robust estimation was achieved despite noise measurement and the presence of oscillatory components in the fault profiles. Stator currents and mechanical position or velocity were required as inputs to the model while the load torque needs to be measured in order to achieve concurrent fault isolation in the dq-frame. Meanwhile, in order to isolate the faults in the original abc-frame, a directional residual evaluation was suggested by relying on the normalization of the estimated DC terms. FDI strategy under parameter variations induced by a constant operation of the IM was validated. One important property of the FDI scheme is that it can isolate simultaneous faults in power semiconductors, hence reconfigurable strategies can be suggested in the actuator to improve the post-fault performance.

In this paper a fault detection method for modular multilevel converters which is can locate a faulty semiconductor switching device in the circuit based on SMO and the switching model of a half bridge is proposed. The method uses the converter arm currents and the cell capacitor voltages as the inputs and hence no additional sensors are required. In this paper it is shown that not only is the faulty cell, but also the faulty switching device can be located. This method has good immunity to conditions such as transient operation, degradation of capacitance over time, and measurement inaccuracies (Shao, et al. 2013 & Freire, et al. 2014).

In this paper (Qiao, et al. 2013), a novel sliding-mode observer (SMO) is used for sensor-less control of permanent-magnet synchronous motor (PMSM). Based on the back electromotive force (EMF) a model an observer model is built. Based on the simulations and experiments done using a 11-kW low-speed PMSM it was shown that the model can estimate position and the speed to achieve better static and dynamic performances. PMSM have high power density, high efficiency, and simple structure and are used in many fields thus sensor-less control of PMSM is key to reduce cost and size of drives.

Due to the requirements of higher reliability in power electronics converters applicable in electric vehicles, fault detection (FD) plays an important role. Here (Salehifar, et al. 2014 & Fini, et al. 2016), a model-based observer is used in parameter estimation of the open transistor fault diagnosis for a voltage-source inverter (VSI) feeding a five-phase permanent magnet motor drive. A sliding mode observer is designed to estimate the phase currents in an ideal model and compares them to real currents. A cross correlation factor is then obtained which is used to set the FD limit in VSI. The scheme is simple and is proved to be fast and can detect multiple open switch or open phase faults in contrast to conventional methods. The FD index is independent from the load parameters and it is robust against fast transients. The designed algorithm can be implemented in a DSP controller. Simulations results show that the method is very efficient in fault detection and isolation.

Area Considered

A system composed of two interconnected generating areas is considered. Load frequency control method is applied to maintain the system frequency and a bank of UIOs are used to detected and isolate faults that might occur. Faults in sensor and actuators can cause undesired frequency deviation which can easily result in the loss of generation. The two area network considered is describes by the equations below:

$$\dot{f} = \frac{f_0}{2H_i}(\Delta P_{gi} - \Delta P_{di} - \Delta P_{tie(i)} - D_i \Delta f_i) \tag{1}$$

$$\Delta \dot{P}_{gi} = -\frac{1}{Tt_i}\Delta P_{gi} + \frac{1}{Tt_i}\Delta X_{gvi} \tag{2}$$

$$\Delta \dot{X}_{gvi} = -\frac{1}{Tgv_i}\Delta X_{gvi} - \frac{1}{Tgv_i R_i}\Delta f_i + \frac{1}{Tgv_i}\Delta P_{ci} \tag{3}$$

$$\Delta P_{tie(i)} = \sum_{j=1}^{n} T_{ij}\left(\int \Delta f_i dt - \int \Delta f_j dt\right) \tag{5}$$

$$ACE_i = \Delta P_{tie(i)} + \beta_i \Delta f_i \tag{6}$$

$$\Delta P_{ci} = -K_{Ii}\int ACE_i dt \tag{7}$$

The system described by the set of equations above is a nineth order sytem which has a state space representation of the form:

$$\dot{x} = Ax + Bu + Ed$$

$$y = Cx \tag{8}$$

where

$$x = [\int \Delta P_{tie(i)} dt \;\; x_1 \;\; x_2]^T$$

is a vector representing the system states,
$d = [\Delta P_{d1} \;\; \Delta P_{d2}]^T$ is the unknown disturbance vector and

$$u = [\Delta P_{c1} \;\; \Delta P_{c2}]^T = -Kx$$

is the controller or input vector. $A\epsilon\,\mathbb{R}^{nxn}$; $B\epsilon\,\mathbb{R}^{nxr}$; $C\epsilon\,\mathbb{R}^{pxn}$ and $E\epsilon\,\mathbb{R}^{nxr}$ are matrices which are computed from system and given as below in box 1:

Unknown Input Observer Design for Sensor Fault Detection

In order to design an UIO for sensor fault detection and isolation our two area network is first modelled into state space equations. Assuming all conditions hold

Box 1.

$$A=\begin{bmatrix} 0 & T_{12} & 0 & 0 & 0 & -T_{12} & 0 & 0 & 0 \\ 0 & 0 & 1 & 0 & 0 & 0 & 0 & 0 & 0 \\ 0 & -\frac{f_0 T_{12}}{2H_1} & -\frac{f_0 D_1}{2H_1} & \frac{f_0}{2H_1} & 0 & \frac{f_0 T_{12}}{2H_1} & 0 & 0 & 0 \\ 0 & 0 & 0 & -\frac{1}{T_{t_1}} & \frac{1}{T_{t_1}} & 0 & 0 & 0 & 0 \\ 0 & 0 & -\frac{1}{T_{gv_1} R_1} & 0 & -\frac{1}{T_{gv_1}} & 0 & 0 & 0 & 0 \\ 0 & 0 & 0 & 0 & 0 & 0 & 1 & 0 & 0 \\ 0 & \frac{f_0 T_{12}}{2H_2} & 0 & 0 & 0 & -\frac{f_0 T_{12}}{2H_2} & -\frac{f_0 D_2}{2H_2} & \frac{f_0}{2H_2} & 0 \\ 0 & 0 & 0 & 0 & 0 & 0 & 0 & -\frac{1}{T_{t_2}} & \frac{1}{T_{t_2}} \\ 0 & 0 & 0 & 0 & 0 & 0 & -\frac{1}{T_{gv_2} R_2} & 0 & -\frac{1}{T_{gv_2}} \end{bmatrix}$$

$$B=\begin{bmatrix} 0 & 0 & 0 & 0 & \frac{1}{T_{gv1}} & 0 & 0 & 0 & 0 \\ 0 & 0 & 0 & 0 & 0 & 0 & 0 & 0 & \frac{1}{T_{gv2}} \end{bmatrix}^T, \quad E=\begin{bmatrix} 0 & 0 & -\frac{f_0}{2H_1} & 0 & 0 & 0 & 0 & 0 & 0 \\ 0 & 0 & 0 & 0 & 0 & 0 & -\frac{f_0}{2H_2} & 0 & 0 \end{bmatrix}^T$$

$$K=\begin{bmatrix} K_{I_1} & K_{I_1}\beta_1 & 0 & 0 & 0 & 0 & 0 & 0 & 0 \\ -K_{I_2} & 0 & 0 & 0 & 0 & K_{I_2}\beta_2 & 0 & 0 & 0 \end{bmatrix}$$

true as will be shown in this paper a function representing the sensor faults is then introduced to the state space equations. The system is also assumed to be free from actuator faults. Mathematically the system can be described by equations (8) describe a nominal system which is fault free but in the presence of an unknown disturbance *(d)*. To the system described by (8) a sensor fault is now introduced and the state space equations becomes:

$$\dot{x} = Ax + Bu + Ed$$

$$y = Cx + f_s \tag{9}$$

Where $f_s \epsilon R^m$ describe the sensor fault which is considered as an additional bias to the system output *y*. In some situations, the sensor faults can be considered as multiplicative. Since (y) is the vector output which comprises of a number of outputs, in the presence of sensor faults a series of vectors can then be described.

$$y^j = C^j x + f^j{}_s$$

$$y_j = C_j x + f_{sj} \tag{10}$$

Where C_j is the jth column of the matrix C and C^j is the new C matrix obtained by deleting the jth row from the original C matrix. The idea of deleting one row is to make the observer sensitive only to one sensor fault and insensitive to the rest. If y is an m-column vector it follows that j takes the range of all integer values from 1up to m. For fault detection and isolation, m-based UIO residual generator is then constructed from the principles of observer design. Since a column is deleted for every new C-matrix formed, it means that every residual generator is driven by all inputs and all outputs minus one which correspond to the deleted row. The structure of a generalized UIO is described with following equations:

$$\dot{z} = Fz + TBu + Ky$$

$$\hat{x} = z + Hy$$

$$r = \left(I - CH\right)y + Cz \tag{11}$$

m-based UIO for (j=1, 2…m) can then be designed as:

$$\dot{z}^j = F^j z + T^j Bu + K^j y^j$$

$$r^j = \left(I - C^j H^j\right)y^j + C^j z^j \tag{12}$$

For the observer design the following conditions must hold true:

$$\left(H^j C^j - I\right) = 0$$

$$T^j = I - H^j C^j$$

$$F^j = A - H^j C^j A - K^j{}_1 C$$

$$K^j{}_2 = F^j H^j$$

$$K^j = K^j{}_1 + K^j{}_2 \tag{13}$$

The equation ($F^j = A - H^j C^j A - K^j{}_1 C$) is only solvable if:

$$Rank\left(C^{j}E\right) = Rank\left(E\right)$$

1. $\left(C^{j}\ A_{1}\right)$ is a detectable pair

Where $A_{1} = A - E[(C^{j}E)^{T}C^{j}E]^{-1}(C^{j}E)^{T}C^{j}A$ [2]

Observability and Detectability

In order to see what is going on inside the system under observation (a system described by equation (8)), the system must be observable. If system (8) is expressed in (*n*) discrete form it can be said to be observable if in *n* time steps when the initial condition, x_{0}, can be recovered from a sequence of observations, $y_{0},\ldots,y_{n-1}$, and inputs, $u_{0},\ldots,u_{n-1}$. The observability problem is to find x_{0} from a sequence of observations $y_{0},\ldots,y_{k-1}$, and inputs $u_{0},\ldots,u_{k-1}$. Based on fundamentals of discrete time systems we can conclude that for continuous systems the knowledge of (t_{0}) is sufficient to determine (*t*) at any time instant, this is because a continuous system is general a discrete system with more iterations. That is to say:

$$x\left(t\right) = e^{A\left(t-t_{0}\right)}x\left(t_{0}\right) \tag{14}$$

For simplicity purposes the continuous system in (8), is assumed to be input free as described below in (15). Where $x\left(t\right)\epsilon\,\mathbb{R}^{n}$; $A\epsilon\,\mathbb{R}^{nxn}$; $y\epsilon\,\mathbb{R}^{p}$; $C\epsilon\,\mathbb{R}^{pxn}$

$$\dot{x}\left(t\right) = Ax\left(t\right) \text{ where } x\left(t_{0}\right) = x_{0} = unknown$$

$$y = Cx\left(t\right) \tag{15}$$

Since a time shift in the discrete-time corresponds to a derivative in the continuous time. An analogous technique in the continuous-time domain is obtained by taking derivatives of the continuous-time measurements this can be expressed as below:

$$y\left(t_{0}\right) = Cx(t_{0})$$

$$\dot{y}\left(t_{0}\right) = C\dot{x}\left(t_{0}\right) = CAx\left(t_{0}\right)$$

$$\ddot{y}\left(t_0\right) = C\ddot{x}\left(t_0\right) = CA^2x\left(t_0\right)$$

$$\vdots$$

$$y^{n-1} = Cx^{n-1}\left(t_0\right) = CA^{n-1}x\left(t_0\right) \tag{16}$$

In matrix form the above system of equations can be expressed as below

$$\begin{bmatrix} y\left(t_0\right) \\ \dot{y}\left(t_0\right) \\ \ddot{y}\left(t_0\right) \\ \vdots \\ y^{n-1}\left(t_0\right) \end{bmatrix} = \begin{bmatrix} C \\ CA \\ CA^2 \\ \vdots \\ CA^{n-1} \end{bmatrix} x\left(t_0\right) = Ox(t_0) = Y\left(t_0\right) \tag{17}$$

O is the observability matrix and the initial condition $x(t_0)$; can be obtained if and only if the observability matrix has full rank, that is $\left\{rank\left(O\right) = n\right\}$. Many texts have expressed the above set of linear equations (16), in a generalized single linear equation. In other words if given any system with known inputs or states and known observations which can be expressed as differential equations, with the initial condition, $x(t_0)$, then $\left(x\left(t\right)\right)$ can be calculated if and only if

$$rank\left(\begin{bmatrix} C \\ CA \\ CA^2 \\ \vdots \\ CA^{n-1} \end{bmatrix}\right) = n \tag{18}$$

For systems with higher-order derivatives the rank of the observability matrix is not increased since by the Cayley–Hamilton theorem for $k \geq n$ we have [4];

$$A^k = \sum_{i=0}^{n-1} \propto_i A^i \tag{19}$$

This means that for any additional equations they would be linearly dependent on the previously defined n-equations. In many texts $rank(O) = n$ is widely known as observability of (C, A).

Simulations

In order to detect the sensor faults for the 9th order system given in this paper the designed matlab Simulink model has the form given in Figure 3.

Figure 3. Robust sensor fault detection and isolation scheme

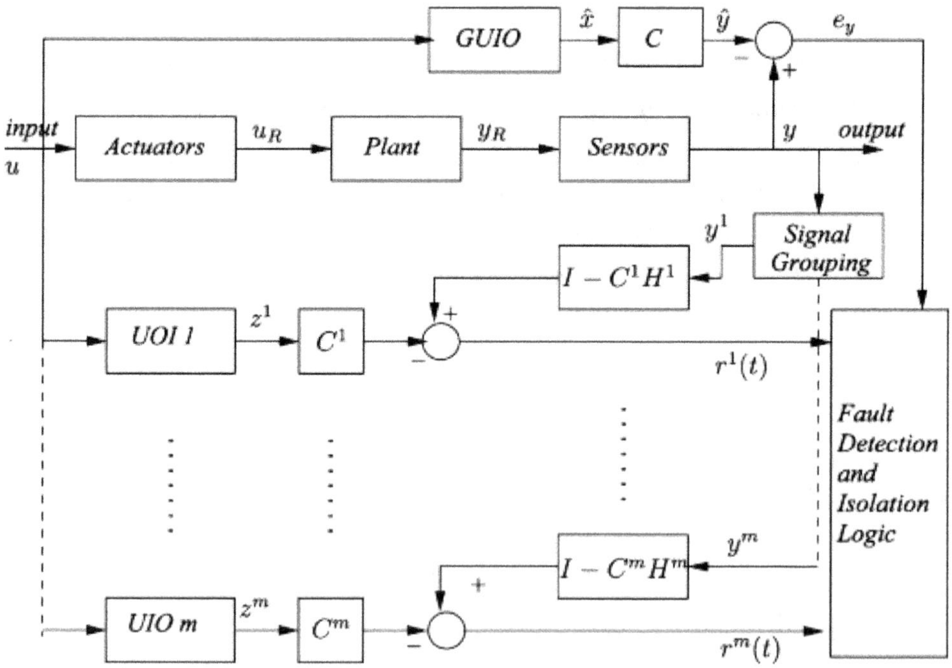

Simulation Case

For complete success in isolation and detection of sensor faults the vector y is chosen as $y = [\Delta f_1 \ \Delta f_2 \ \Delta P_{tie} \ \Delta f_1 + \Delta f_2]^T$ and from equation (8) C is calculated and used for the simulations. A fault is introduced in sensor 1 at t=5s, the fault is assumed to be a deviation of 0.017. The results obtained are given below. Figure 1 shows no sensitivity to fault since it was made insensitive to fault by removing the row corresponding to its input hence it shows that the fault occurred in sensor one. Squared residuals in Figures 4, 5 and 6 shows the effect of the faults in sensor one hence its detection and isolation.

Figure 4. Squared residual 1

Figure 5. Squared residual 2

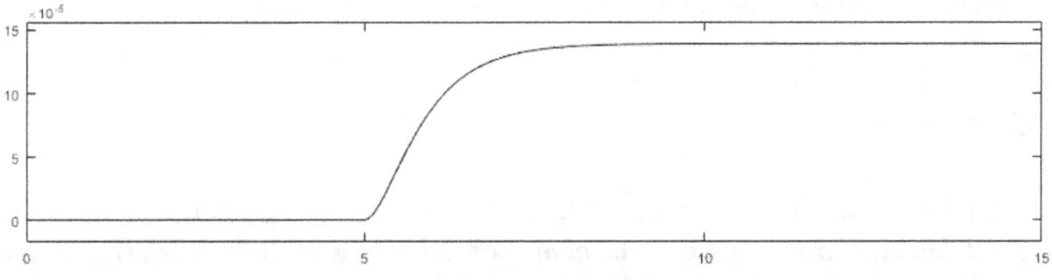

Figure 6. Squared residual 3

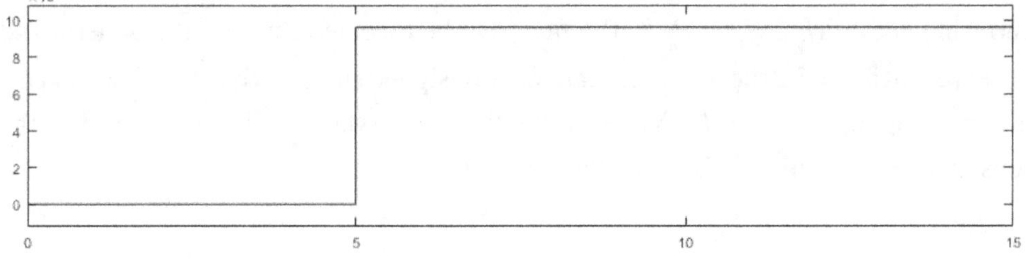

Figure 7. Squared residual 4

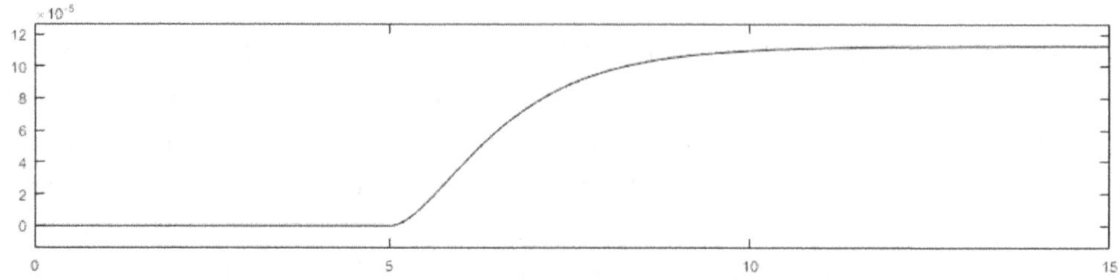

FUTURE RESEARCH DIRECTIONS

In this chapter, a comprehensive study on fault detection and isolation methods have been presented. Different types of fault diagnosis methods that can be used for power system including wind power plants have also been discussed. Likewise, some simulation scenarios and case studies have been presented. This chapter provided a comprehensive comparison between fault diagnosis methods for future power systems.

As a future research direction of this chapter, different types of evolutionary optimization methods can be used for fault diagnosis methods to achieve the optimal fault diagnosis method. Also, a comparison between the well-known diagnosis methods such as observers, kalman filters, and ANN can be studied. Furthermore, the researchers are directed to investigate the feasibility of using GPS system to get the accurate fault location point in the wide area power systems.

CONCLUSION

The UIO based method for fault detection and isolation is a well-developed method in fault detection and isolation. The choice of the output vector (y) determines the detectability and Isolability of the faults. In this chapter, the case where successful detection and isolation of faults happens is given. However, from other simulations done it is noted that in some cases the full set of observers cannot be designed and also in some cases the faults can neither be detected nor isolated. For example, choosing $y = [\Delta f_1 \ \Delta f_2 \ \Delta P_{tie}]^T$ the fault can be detected through the generalized observer but for *j=1,2* the observers cannot be designed since $rank(C^j * E) \neq rank(E)$. by choosing $y = [\Delta f_1 \ \Delta f_2 \ \Delta \dot{P}_{tie}]^T$ a bank of observers can be designed but the sensor – fault can neither be detected nor isolated.

REFERENCES

Aldeen, M., & Crusca, F. (2006). Observer-based fault detection and identification scheme for power systems. *IEE Proceedings. Generation, Transmission and Distribution*, *153*(1), 71–79. doi:10.1049/ip-gtd:20045101

Alhelou, H. H., & Golshan, M. E. H. (2016, May). Hierarchical plug-in EV control based on primary frequency response in interconnected smart grid. In Electrical Engineering (ICEE), 2016 24th Iranian Conference on (pp. 561-566). IEEE. doi:10.1109/IranianCEE.2016.7585585

Alhelou, H. H., Golshan, M. H., & Askari-Marnani, J. (2018). Robust sensor fault detection and isolation scheme for interconnected smart power systems in presence of RER and EVs using unknown input observer. *International Journal of Electrical Power & Energy Systems*, *99*, 682–694. doi:10.1016/j.ijepes.2018.02.013

Alhelou, H. S. H., Golshan, M. E. H., & Fini, M. H. (2015, December). Multi agent electric vehicle control based primary frequency support for future smart micro-grid. In *Smart Grid Conference (SGC)*, 2015 (pp. 22-27). IEEE.

Andersen, P., Bendtsen, J. D., Pedersen, T. S., Mortensen, J. H., & Nielsen, R. J. (2006). Observer-based coal mill control using oxygen measurements. IFAC Proceedings Volumes, 39(7), 171-176.

Bachir, S., Tnani, S., Trigeassou, J. C., & Champenois, G. (2006). Diagnosis by parameter estimation of stator and rotor faults occurring in induction machines. *IEEE Transactions on Industrial Electronics*, *53*(3), 963–973. doi:10.1109/TIE.2006.874258

Caliskan, F., & Genc, I. (2008). A robust fault detection and isolation method in load frequency control loops. *IEEE Transactions on Power Systems*, *23*(4), 1756–1767. doi:10.1109/TPWRS.2008.2004831

Campos-Delgado, D. U., & Espinoza-Trejo, D. R. (2011). An observer-based diagnosis scheme for single and simultaneous open-switch faults in induction motor drives. *IEEE Transactions on Industrial Electronics*, *58*(2), 671–679. doi:10.1109/TIE.2010.2047829

Chen, J., & Patton, R. J. (2012). *Robust model-based fault diagnosis for dynamic systems* (Vol. 3). Springer Science & Business Media.

De Angelo, C., Bossio, G., Giaccone, S., Garcia, G. O., Solsona, J., & Valla, M. I. (2006, November). Model based stator fault detection in induction motors. In *IEEE Industrial Electronics, IECON 2006-32nd Annual Conference on* (pp. 1095-1100). IEEE. 10.1109/IECON.2006.347895

De Angelo, C. H., Bossio, G. R., Giaccone, S. J., Valla, M. I., Solsona, J. A., & Garcia, G. O. (2009). Online model-based stator-fault detection and identification in induction motors. *IEEE Transactions on Industrial Electronics*, *56*(11), 4671–4680. doi:10.1109/TIE.2009.2012468

Espinoza-Trejo, D. R., Campos-Delgado, D. U., Barcenas, E., & Martınez-Lopez, F. J. (2012). Robust fault diagnosis scheme for open-circuit faults in voltage source inverters feeding induction motors by using non-linear proportional-integral observers. *IET Power Electronics*, *5*(7), 1204–1216. doi:10.1049/iet-pel.2011.0309

Fini, M. H., Yousefi, G. R., & Alhelou, H. H. (2016). Comparative study on the performance of many-objective and single-objective optimisation algorithms in tuning load frequency controllers of multi-area power systems. *IET Generation, Transmission & Distribution*, *10*(12), 2915–2923. doi:10.1049/iet-gtd.2015.1334

Freire, N. M., Estima, J. O., & Cardoso, A. J. (2014). A voltage-based approach without extra hardware for open-circuit fault diagnosis in closed-loop PWM AC regenerative drives. *IEEE Transactions on Industrial Electronics*, *61*(9), 4960–4970. doi:10.1109/TIE.2013.2279383

Genc, I., & Caliskan, F. (2011). Robust Actuator Fault Detection and Isolation in a Multi-Area Interconnected Power System. *Proceedings of the World Congress on Engineering*, 2.

Hwas, A., & Katebi, R. (2012, September). Model-based fault detection and isolation for wind turbine. In *Control (CONTROL), 2012 UKACC International Conference on* (pp. 876-881). IEEE. 10.1109/CONTROL.2012.6334747

Kallesoe, C. S., Izadi-Zamanabadi, R., Vadstrup, P., & Rasmussen, H. (2007). Observer-based estimation of stator-winding faults in delta-connected induction motors: A linear matrix inequality approach. *IEEE Transactions on Industry Applications*, *43*(4), 1022–1031. doi:10.1109/TIA.2007.900494

Makdisie, C., Haidar, B., & Alhelou, H. H. (2018). An Optimal Photovoltaic Conversion System for Future Smart Grids. In *Handbook of Research on Power and Energy System Optimization* (pp. 601–657). IGI Global. doi:10.4018/978-1-5225-3935-3.ch018

Najafabadi, T. A., Salmasi, F. R., & Jabehdar-Maralani, P. (2011). Detection and isolation of speed-, DC-link voltage-, and current-sensor faults based on an adaptive observer in induction-motor drives. *IEEE Transactions on Industrial Electronics, 58*(5), 1662–1672. doi:10.1109/TIE.2010.2055775

Odgaard, P. F., Lin, B., & Jørgensen, S. B. (2007). Observer-based and regression model-based detection of emerging faults in coal mills. Fault Detection, Supervision and Safety of Technical Processes 2006, 687-692. doi:10.1016/B978-008044485-7/50116-0

Odgaard, P. F., & Mataji, B. (2006). Fault detection in coal mills used in power plants. *IFAC Proceedings Volumes, 39*(7), 177-182.

Odgaard, P. F., & Mataji, B. (2008). Observer-based fault detection and moisture estimating in coal mills. *Control Engineering Practice, 16*(8), 909–921. doi:10.1016/j.conengprac.2007.10.008

Odgaard, P. F., Stoustrup, J., Nielsen, R., & Damgaard, C. (2009, March). Observer based detection of sensor faults in wind turbines. *Proceedings of European wind energy conference*, 4421-4430.

Qiao, Z., Shi, T., Wang, Y., Yan, Y., Xia, C., & He, X. (2013). New sliding-mode observer for position sensorless control of permanent-magnet synchronous motor. *IEEE Transactions on Industrial Electronics, 60*(2), 710–719. doi:10.1109/TIE.2012.2206359

Rothenhagen, K., & Fuchs, F. W. (2006, November). Current sensor fault detection by bilinear observer for a doubly fed induction generator. In *IEEE Industrial Electronics, IECON 2006-32nd Annual Conference on* (pp. 1369-1374). IEEE. 10.1109/IECON.2006.347390

Rothenhagen, K., Thomsen, S., & Fuchs, F. W. (2007, September). Voltage sensor fault detection and reconfiguration for a doubly fed induction generator. In *Diagnostics for Electric Machines, Power Electronics and Drives, 2007. SDEMPED 2007. IEEE International Symposium on* (pp. 377-382). IEEE. 10.1109/DEMPED.2007.4393123

Salehifar, M., Arashloo, R. S., Moreno-Eguilaz, M., Sala, V., & Romeral, L. (2014). Observer-based open transistor fault diagnosis and fault-tolerant control of five-phase permanent magnet motor drive for application in electric vehicles. *IET Power Electronics, 8*(1), 76–87. doi:10.1049/iet-pel.2013.0949

Shao, S., Wheeler, P. W., Clare, J. C., & Watson, A. J. (2013). Fault detection for modular multilevel converters based on sliding mode observer. *IEEE Transactions on Power Electronics, 28*(11), 4867–4872. doi:10.1109/TPEL.2013.2242093

Varmaziari, H., & Dehghani, M. (2017, May). Cyber-attack detection system of large-scale power systems using decentralized unknown input observer. In *Electrical Engineering (ICEE), 2017 Iranian Conference on* (pp. 621-626). IEEE. 10.1109/IranianCEE.2017.7985114

Yan, X. G., & Edwards, C. (2008). Adaptive sliding-mode-observer-based fault reconstruction for nonlinear systems with parametric uncertainties. *IEEE Transactions on Industrial Electronics*, *55*(11), 4029–4036. doi:10.1109/TIE.2008.2003367

Chapter 3
Diagnosis of the Broken Rotor Bars Faults by Root-MUSIC Method

Ahmed Hamida Boudinar
University of Sciences and Technology of Oran "Mohamed Boudiaf", Algeria

Azeddine Bendiabdellah
University of Sciences and Technology of Oran "Mohamed Boudiaf", Algeria

Noureddine Benouzza
University of Sciences and Technology of Oran "Mohamed Boudiaf", Algeria

ABSTRACT

This chapter describes a new diagnosis approach, the Root-MUSIC (RM) method, for identification of the progressive cracking in the rotor of induction motors. This method used initially in the domain of RADAR for localizing mobile targets is being applied to the domain of induction motors diagnosis. This approach has several advantages compared to the conventional power spectral density estimation (PSD) by periodogram technique. Indeed, the main advantage of this approach is its very good frequency resolution for a very short acquisition time, something impossible to achieve with the conventional method. However, in order to reduce the computation time which is the main drawback of the RM method, this method will be applied to only a specified frequency band: one that carries information about the sought fault.

DOI: 10.4018/978-1-5225-6989-3.ch003

INTRODUCTION

The induction motor is the most common electric machine in the industry. Its main advantage is the absence of sliding electrical contacts, which leads to a simple and robust structure easy to build with low cost. However, various faults can appear on the induction motor making the fault detection procedure necessary to prevent the interruption of the industrial process (Toliyat et al., 2013).

In order to diagnose these faults, several techniques have been developed, depending on the type of physical quantity measured. The most widespread is that based on the vibration analysis. Moreover, the analysis of the stator current is a very promising approach, since its main advantages lie in the instrumentation used, the ease of its implementation as well as the richness of information provided on the existence or not of the fault and its severity. To reveal this richness of information, several methods of signal processing have been developed.

Among these methods, the spectral analysis is the most approach used. In general, spectral analysis involves estimating the power spectrum of a signal from a limited number of samples. If the signal to be analyzed is deterministic, the main tool is the Fourier Transform (FT). But, if the signal to be processed is random as is the case for the stator current, the FT of a single measurement is often insufficient. For this reason, one must calculate or rather estimate the Power Spectral Density (PSD). The PSD thus makes it possible to recognize the nature of a signal, to possibly identify its origin and to detect the variations of its characteristics. To estimate this PSD, several techniques and algorithms have been developed, which can be classified in two main approaches:

- The non-parametric approach.
- The parametric approach.

The methods belonging to the non-parametric approach, are relatively simple to implement, they provide excellent results provided that the duration of the observed signal is not too limited. In the opposite case, parametric methods are generally more efficient. Their implementation is, however, much more complex and their success depends on the adequacy of the chosen method to the type of signal to be analyzed.

Amongst the existing analysis methods, the PSD by Periodogram of the stator current, is considered as a very popular technique and is widely used in industry. Several studies (Boudinar et al., 2016; El Bouchikhi et al., 2012) have demonstrated the reasons for its durability in the field of diagnosis and its preference to other recent methods. The main reasons are:

- Independent of the signal physical nature to be analyzed (vibration or electrical), provided that this signal is stationary.
- Simple to program.
- Fast calculation time due to the use of Fast Fourier Transform (FFT) algorithm.
- Easy to implement on FPGA cards for a real-time use.

However, this method presents two major drawbacks:

- Frequency resolution conditioned by the acquisition time: Indeed, in order to have an efficient diagnosis, one must increase the acquisition time to be able to distinguish two close harmonics from each other, which is not always easy.
- Identification for low harmonics quasi impossible: Indeed, the presence of side-lobes related to the type of the window function used in the PSD estimation by Periodogram (Boudinar et al., 2015) risks masking the frequency signatures of incipient faults. Unfortunately, even an adequate choice of this window does not improve the diagnosis for certain operating modes of the motor, such as the rotor faults diagnosis at very low load. Indeed, in this case, the incipient fault signature is embedded in that of the fundamental.

To eliminate this effect from the fundamental, some studies use the combined information found in the current and voltage, in order to calculate the instantaneous power (Ibrahim et al., 2008; Dzwonkowski & Swędrowski, 2012). However, this solution is not always accurate and, in addition, it is expensive, because it requires the use of sensors of currents and voltages.

In recent years, several high-resolution methods have been applied for the identification of faults in induction motors. These methods are based on a priori knowledge of the signal model. Indeed, instead of making the non-realistic hypothesis of considering the signal as zero outside the study interval, as is the case with classical spectral analysis methods, these methods assume that the signal is modeled inside and outside the study interval by the same mathematical model. In addition, these methods have higher frequency resolution than conventional methods under certain condition on the Signal to Noise Ratio (SNR) and the length of the signal, hence the name High Resolution.

For example, the methods, as MUSIC (Multiple SIgnal Classification) (Boudinar et al., 2016 ; Kia et al., 2007) and ESPRIT (Estimation of Signal Parameters via Rotationnal Invariance) (Kim et al., 2013; Xu et al.,2013), use the decomposition of the covariance matrix of the signal analyzed in two separate subspaces: signal subspace and noise subspace. They are more robust to noise but require an important computation time, due to the complexity of their algorithms. On other hand, the

PRONY method (Sahraoui et al., 2015), which is not based on the decomposition of space, is faster compared to MUSIC, but more sensitive to measurement noise.

The objective of this chapter is to present two approaches of spectral analysis allowing the estimation of the PSD:

- The first so-called classic is based on the technique of the Periodogram, widely used in industry.
- The second modern belonging to the family of high resolution methods such as the Root-MUSIC method which is a variant of the MUSIC method.

To illustrate the merits and effectiveness of the Root-MUSIC method compared to the classic Periodogram method, a series of experimental tests are carried out at the end of this chapter to diagnose rotor faults, for different modes of operation of a squirrel cage induction motor.

SPECTRAL ANALYSIS OF THE STATOR CURRENT BY PERIODOGRAM

Before tackling the classic method of the Periodogram used widely in the diagnosis of induction motors on industrial sites, it seems important to give some basic notions about spectral analysis.

Fourier Transform (FT)

The Fourier Transform (FT) is the basic mathematical tool for the spectral analysis of stationary type physical quantities. In the case of a stator current $i_s(t)$, this FT is expressed by (Hayes, 1991):

$$I_s(f) = TF\left\{i_s(t)\right\} = \int_{-\infty}^{+\infty} i_s(t)e^{-j2\pi ft}dt \tag{1}$$

Evidently, it is impossible to analyze this signal over an infinite period. It is therefore necessary to choose a weighting window for the analysis (rectangular window, Hanning ...) to correct the effects of a finite temporal windowing. Moreover, in addition to the shape, the size of the window has an influence on the maximum possible resolution.

Numerical Fourier Transform (NFT)

The Numerical *FT* (*NFT*) of a signal sampled at the period T_{sT}, is defined by taking: $t = n.T_{sT}$. In these conditions, equation (1) becomes (Hayes, 1991):

$$Is(f) = \sum_{n=-\infty}^{+\infty} i_s(n.T_{sT}).e^{-j2\pi.n.T_{sT}f} \tag{2}$$

Where T_{sT} is the sampling time

The major property of the *NFT* is its periodicity. Indeed, the spectrum of the stator current is periodic of period f_{sf} in the frequency domain because of the temporal sampling operation.

Discrete Fourier Transform (DFT)

DFT is not directly used in most practical applications because the transition from Fourier analysis of continuous signals to that of discrete signals involves three operations:

- The time signal sampling.
- The limitation of the duration of analysis of this signal.
- The discretization of the frequency for the digital spectral analysis.

These three seemingly trivial operations have consequences that are important to evaluate the extent. Indeed, the use of a digital computer to determine the *FT* of a signal only allows obtaining a limited number of values of the frequency *f* and this due to the limitation of its computing power. It is therefore necessary to replace the continuous variable *f* with a discrete variable. This defines the Discrete Fourier Transform (*DFT*).

As a result, and in the case of digital processing of the stator current, the number of samples of $i_s(n)$ cannot be infinitely large. We are therefore forced to take into account only part of the original signal. Mathematically, this truncation operation amounts to multiplying the signal is $i_s(t)$ by a window most often rectangular $w(t)$ of width T. The signal $i_s(t)$ will be recorded for a finite duration T by sampling N values of the signal $i_s(t)$.We obtain therefore for $T = N.T_{sT}$,the sequence of the following stator current: $\left[i_s(0), i_s(1.T_{sT}),. \quad ., i_s((N-1)T_{sT})\right]$. In this case, the *DFT* of the signal is written as:

$$Is(f) = \sum_{n=0}^{N-1} i_s(n.T_{sT}).e^{-j2\pi.n.\frac{f}{f_{sf}}} \tag{3}$$

Where f_{sf} is the sampling frequency and N the sampling number.

Frequency Discretization

The discretization of the frequency variable consists of replacing f by $k\Delta f$ where k is a relative integer. The frequencies $f_n = k\Delta f$ are then called harmonic frequencies of the *DFT*. As the *FT* of a digital signal is periodic of period f_{sf}, it suffices to be confined to a single period and to divide it by the N samples. (Hayes, 1991; Marple, 1987) prove that:

$$\Delta f = \frac{f_{sf}}{N} \tag{4}$$

Where Δf is called frequency resolution

Note that to improve the frequency resolution; it is advisable to increase the acquisition time in other words the number of samples. Unfortunately, this can affect the computation time of the analysis. Under these conditions, replacing f by its discretized form and for N sufficiently large, the *DFT* of the sequence $\left[i_s(0), i_s(1.T_{sT}),\ldots, i_s((N-1)T_{sT})\right]$ can be rewritten as follows (Hayes, 1991):

$$Is\left(k\Delta f\right) = \sum_{n=0}^{N-1} i_s(n.T_{sT}).e^{-j2\pi.n.\frac{k}{N}} \text{ For } k = 0, \ldots, N\text{-}1 \tag{5}$$

Fast Fourier Transform (FFT)

Fast Fourier Transform (*FFT*) is a fast calculation algorithm for the *DTF* developed in 1965 by J. W. Cooley and J. W. Tuckey. The basic algorithm of this transform uses a number of N_{TF} points equal to a power of 2, which makes it possible to obtain a saving in computation time, with respect to a calculation with the *DTF*, of:

$$Gain_{Temps} = \frac{N_{TF}}{\log_2(N_{TF})} \tag{6}$$

Do not be fooled by the meaning of the *FFT*: the *FFT* algorithm is not a new transformation. This is nothing more than a quick way to get the same results as those provided by the *DTF*. This fast Fourier transform is widely used when it is essential to obtain an "in line" frequency analysis in certain processes through a sliding observation window (Hayes, 1991; Marple, 1987).

Effect of the Frequency Spectrum Plot on a Semi-Logarithmic Scale

Most often, the spectrum of the signal is drawn not on a linear scale but on a semi-logarithmic scale (the frequency axis). The logarithmic spectrum shows information that is not detectable on the linear spectrum. The primary effect of the logarithm is to enhance the amplitude of weak harmonics (Marple, 1987). Thus, when one moves away from the fundamental, the information related to the harmonic of small amplitude which is undetectable on the linear spectrum, is raised on the logarithmic spectrum and gains importance. The second effect of the logarithm is the broadening of the frequency band. The transformation of the scale of amplitudes from the linear domain to the logarithmic domain makes it possible to change the appearance of the envelope of the spectrum of the different lines of the signal and thus this spectrum becomes wider (Hayes, 1991).

Estimation of the PSD by the Periodogram Technique

Most of the signals that one encounters in practice are not deterministic and one cannot know exactly their evolution over time. We can only make conjunctures on

Figure 1. Example of a stator current frequency spectrum plot

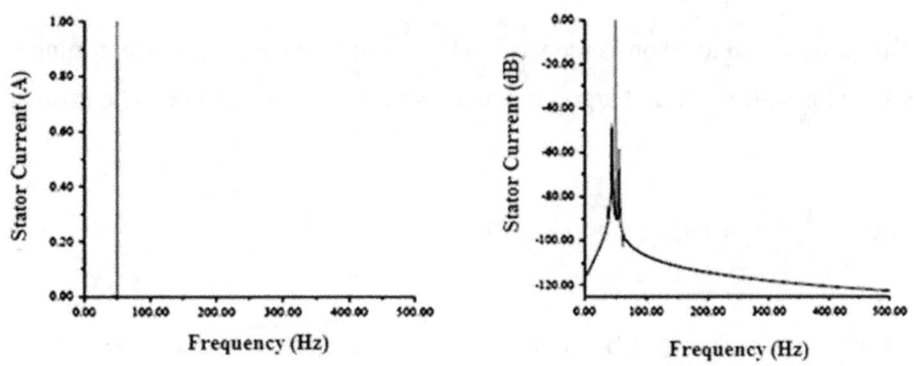

this evolution. For this, in practice, we choose certain duration of observation, trying to take it as we have all possible variations of the signal. For this reason, and in the case of the stator current, several periods are taken in steady state.

These signals are then described as a random sequence which consists of a set of possible realizations or observations and which has been observed to be realized. Theoretically, these processes are characterized by a probability density function and generally they are defined in the second order, i.e. by their correlation functions. Under these conditions, the notion of spectral density of a deterministic signal is then replaced by that of the average of the spectral densities obtained over several observation intervals. We then obtain a *Periodogram*. Also note that the method of the *Periodogram* is based on the Wiener-Kintchine theorem. This theorem is stated as follows (Marple, 1987).

Wiener-Kintchine Theorem: *The Power Spectral Density of a random process whose mean value and autocorrelation function are invariant in time (stationary in the broad sense) is equal to the Fourier transform of its statistical autocorrelation function.*

In other words:

$$PSD = TF\left\{R_i(m)\right\} = \sum_{m=-\infty}^{+\infty} R_i(m)e^{-j.2.\pi.m.f} \tag{7}$$

Knowing that the correlation function or more exactly the autocorrelation function is defined as follows:

$$R_i(m) = E\left\{i_S^*(n)i_S(n+m)\right\} \tag{8}$$

Since the signal to be processed $\left\{i_S\left(n\right)\right\}_{n=0}^{N-1}$ is limited in time and assuming that this signal is stationary and ergodic, the autocorrelation function can be estimated by:

$$\hat{R}_i\left(m\right) = \frac{1}{N}\sum_{n=-\infty}^{+\infty} i_S^*(n).i_S(n+m) \tag{9}$$

Of course, for $\hat{R}_i(m)$ to be equal to $R_i(m)$, we have to make N tend towards infinity.

Thus, the estimation of the *PSD* by *Periodogram* is defined as the direct *FT* of equation (9). We get the following relation (Hayes, 1991; Marple, 1987):

$$PSD_{per}\left\{i_s\left(n\right)\right\} = \frac{\left|\sum_{n=0}^{N-1} i_s\left(n\right) e^{-j2\pi nk\Delta f}\right|^2}{N} \tag{10}$$

However, this estimator is biased and has non-zero variance due to signal truncation. For this reason it is said that the *Periodogram* is not a consistent estimator of the *PSD* (Hayes, 1991). In order to reduce the variance of this estimator, one can use an averaged *Periodogram*. This consists in separating the signal into *K* slices (of length *N/K*), calculating the *Periodogram* on each slice and doing the average. Because of the *K* averages, the variance is almost divided by *K*. With this approach, the resolution decreases because the slices are shorter.

SPECTRAL ANALYSIS OF THE STATOR CURRENT BY ROOT-MUSIC

In general, High Resolution Spectral Analysis Methods (*HRSAM*), rely on the model of a sum of sinusoids or complex exponentials plus white noise. Moreover, these methods exploit the properties of the proper structure of the correlation matrix of the observation vector (stator current in our case). In this set of methods two families can be distinguished:

- General *HRSAM*
- Specific *HRSAM*

General *HRSAM* use simple algorithms and the resolution are improved over direct Fourier methods. However, since these methods use all the proper space of the correlation matrix, this results in a bias on the estimation of the desired frequencies. These methods are therefore very sensitive to noise. One can cite as general *HRSAM*: the *PRONY* method. As for the specific *HRSAM*, they exploit the fact that for the data model, the proper space of the correlation matrix can be decomposed into two orthogonal subspaces: signal subspace (desired harmonics) and noise subspace. These specific methods are more robust with respect to noise and allow having a very good resolution. One can cite as specific *HRSAM*: the method of *MUSIC* and the method *ESPRIT*. Nevertheless, one of the major drawbacks of these methods is the computation time since they use more complex algorithms (Marple, 1987).

In this section, the authors propose to study the specific *HRSAM*: *Root-MUSIC* which is only a variant of the *MUSIC* method. However, before the presentation and study of this method, a data model must be established in order to apply this method.

Data Model

It is well known that when the signal to be processed can be put into the form of a sum of weighted complex exponentials, the most efficient methods for extracting useful information are the so-called *HRSAM* methods (Marple, 1987). The *Root-MUSIC* method, which is part of this family, seeks to separate the initial space into a signal subspace, containing only useful information, and its orthogonal complement, classically called noise subspace. This decomposition makes the processing more robust and improves the separation capacity of the signal components. Consequently, the *Root-MUSIC* method allows a very good estimation of the frequencies in the case where the signals to be treated are brought back to the following model (Boudinar et al., 2016; Hayes, 1991; Marple, 1987):

$$i_s(n) = \sum_{i=1}^{L} I_i . e^{j(2\pi \frac{f_i}{f_{sf}} n + \varphi_i)} + w(n);\ n=0, 1,\ldots, N\text{-}1 \tag{11}$$

This method assumes that the discrete-time signal $i_s(n)$ can be represented by L complex sinusoids in a noise *w(n),and* where f_{sf} is the sampling frequency, N is the sampling number. I_i, f_i, ϕ_i represent the amplitude, frequency, and initial phase angle of the ith harmonic respectively, *w(n)* is a white noise. (11) represents well the stator current model in the discrete time-domain. In matrix form, equation (11) is written (Cupertino et al., 2003; Hayes, 1991):

$$\underline{I}_s = S\underline{A} + \underline{W} \tag{12}$$

Where S is the Vandermonde matrix: $S = [s_1 \ldots s_i \ldots s_L]$

$$\text{With } s_i = \begin{bmatrix} 1 & e^{\frac{j.2\pi.f_i}{f_{sf}}} & e^{\frac{j.4\pi.f_i}{f_{sf}}} & \ldots & e^{\frac{j.2\pi.(N-1).f_i}{f_{sf}}} \end{bmatrix}^T \tag{13}$$

NB: The matrix S contains the information inherent to the frequency position of the spectral lines on the unit circle, while the vector A preserves the energy information.

And A is the powers matrix of harmonics defined as:

$$A = diag\left[I_1^2 \quad I_2^2 \quad \dots \quad I_L^2\right] \tag{14}$$

H represents the Hermitian operator. σ_w^2, and I are the noise variance of the white noise and the identity matrix of size (N x N) respectively.

Principle of Root-MUSIC Method

Knowing that the noise $w(n)$ is a parasitic random process; its presence requires the introduction of statistical hypotheses capable of managing interactions between:

- On the one hand the elements A, which are a function of the random variable ϕ_i and
- On the other hand, between W and S.

These assumptions are defined as follows:

- The frequency positions of the harmonics do not vary during the measurement, which supposes that the matrix S has deterministic coefficients
- The amplitudes A_i of the different harmonic are assumed to be random and not correlated with each other
- Noise is assumed to be independent of harmonics. In the present study, we will limit ourselves to the case of decorrelated centered noises and of the same variance $\sigma_w^{\ 2}$.

Taking into account all these assumptions, the first step for the implementation of this method, as well as all the *HRSAM* based on orthogonal subspace decomposition, consists in the proper analysis of the autocorrelation matrix R_i of the observation vector which in our case is the stator current $i_s(n)$. This matrix is therefore estimated as follows:

$$R_i = E\left[\underline{I}_s.\underline{I}_s^H\right] \tag{15}$$

The autocorrelation matrix R_i of the noisy signal $i_s(n)$ can be considered as being the sum of two autocorrelation matrices that of the signal and that of the noise.

Taking into account equation (12), the estimated autocorrelation matrix R_i of measurement signal $i_s(n)$ represented by (11) can be written as follow:

$$R_i = R_s + R_w = S.A.S^H + \sigma_w^2.I \tag{16}$$

Where I is the identity matrix of size (N x N)

The estimated autocorrelation matrix R_i is the sum of two autocorrelation matrices; these are the signal matrix R_s and the noise matrix R_w (Boudinar, et al., 2016; Marple, 1987; Sahraoui et al., 2015).

The Eigen decomposition of the autocorrelation matrix of the signal Ri is given by the following relation:

$$R_i = \sum_{k=1}^{N} \lambda_k.\underline{u}_k.\underline{u}_k^H = \underbrace{\sum_{k=1}^{L} \left(\lambda_k + \sigma_w^2\right).u_k.u_k^H}_{R_S} + \underbrace{\sigma_w^2 \sum_{k=L+1}^{N} u_k.u_k^H}_{R_B} \tag{17}$$

By its construction and the hypotheses seen previously, the R_s matrix is Hermitian, defined as non-negative (Eigen values are real positive or null) and its Eigen vectors are orthogonal (Boudinar, et al., 2016; Kia et al., 2005; Marple, 1987). Consequently, the Eigen vectors of R_i are the same as those of R_s, and the Eigen values are shifted by σ_w^2. Under these conditions, R_i can be rewritten as follows:

$$R_i = \underbrace{U_s.D_s.U_s^H}_{R_s} + \underbrace{U_w.D_w.U_w^H}_{R_w} \tag{18}$$

Where the matrices U_s and U_w are composed by the Eigen vectors u_k related to Eigen values arranged in descending order λ_k. The matrices D_s and D_w are diagonal matrices made by Eigen values λ_k.

$$\begin{aligned} U_s &= [u_1 \;\; \ldots \;\; u_L] \;\; ; \;\; D_s = diag[\lambda_1 \;\; \ldots \;\; \lambda_L] \\ U_w &= [u_{L+1} \;\; \ldots \;\; u_N] \;\; ; \;\; D_w = \sigma_w^2.I_{N-L} \end{aligned} \tag{19}$$

Concept of signal space and noise space: Given equation (17) and (18), the authors can affirm that the Eigen value associated with the autocorrelation matrix R_i can be divided into two orthogonal subspaces (Boudinar et al., 2016):

- **Signal Subspace E_s:** Composed of the Eigen-values corresponding to the L largest Eigen values.
- **Noise Subspace E_w:** Composed of the Eigen-values corresponding to the N-L remaining Eigen-values.

Theoretically, all Eigen values corresponding to the noise subspace are equal to σ_w^2 (Marple, 1987).

$$\underbrace{\lambda_1 \geq \lambda_2 .... \geq \lambda_L}_{\substack{\text{These values genarates} \\ \text{E}_\text{s}}} \geq \underbrace{\lambda_{L+1} = ... = \lambda_N = \sigma_w^2}_{\substack{\text{These values genarates} \\ \text{E}_\text{w}}} \tag{20}$$

For this reason, D_w is written in the form given by (19). By comparing (11), (18) and (19), one can write:

$$R_i.U_w = S.A.S^H.U_w + \sigma_w^2.U_w \tag{21}$$

This implies that:

$$S^H.U_w = 0 \tag{22}$$

The problem comes down therefore to compute the solutions of (22). These solutions or roots appear accordingly on (or near) the unit circle. Unfortunately in reality, this is not too simple since the localization is not easy in the noise signal presence, because the roots corresponding to the sought frequencies are no more situated on the unit circle but mixed with the other solutions. To keep away the parasites roots from the unit circle and hence decrease the error probability, *Root-MUSIC* method considers that the whole set of the Eigen vectors are associated to the noise space E_w in order to obtain a better robustness with respect to the noise. (Boudinar et al., 2016; Hayes, 1991; Xu et al., 2013) show that it is necessary to solve the following equation:

$$s_i^H U_w.U_w^H.s_i = 0 \tag{23}$$

Frequencies Estimation

The phases of the L closest roots to the unit circle will correspond to the sought frequencies. These frequencies will be computed directly from the following relation (Boudinar et al., 2016):

$$f_i = \frac{f_{sf}}{2.\pi}.\arg(e^{j.2.\pi.\frac{f_i}{f_{sf}}}) \text{ with } i=1,...,L \tag{24}$$

Amplitudes Estimation

Knowing the frequencies of the sought harmonics, the amplitudes of these components can be determined by using the following equation:

$$R_s = S.A.S^H = \sum_{k=1}^{L} (\lambda_k + \sigma_w^2) u_k . u_k^H \tag{25}$$

Noted that it is easier to inverse R_s rather than to inverse S. In these conditions, the amplitudes are estimated by (Boudinar et al., 2015):

$$A^{-1} = \frac{1}{S^H \; R_s^{-1} \; S} \tag{26}$$

Where:

$$\begin{cases} R_s^{-1} = \sum_{k=1}^{L} \frac{1}{\lambda_k + \sigma_w^2} u_k . u_k^H \\ \sigma_w^2 = \frac{1}{N - L} \sum_{k=L+1}^{N} \lambda_k \end{cases} \tag{27}$$

The problem of the harmonics identification is thus solved, knowing the frequencies and the amplitudes of the various components. However, for *Root-MUSIC* to be effective, the matrix R_i must be non-singular (Marple, 1987). This condition is realized if averaging is done on a sufficiently large number of measurements. However, this condition is not generally verified in practice because most often the estimation of the desired frequencies must be done from a single data vector "a single measurement". For this reason, the most used averaging to ensure the complete rank of the matrix makes use of spatial smoothing (Hayes, 1991; Marple, 1987). Under these conditions, the autocorrelation matrix R_i is estimated by the following relation:

$$\hat{R}_I = \frac{1}{N - M + 1} D.D^H \tag{28}$$

With

$$D = \begin{bmatrix} i_s(0) & . \; . \; . & i_s(N-M) \\ . & . \; . \; . & . \\ . & . \; . \; . & . \\ . & . \; . \; . & . \\ i_s(M-1) & . \; . \; . & i_s(N-1) \end{bmatrix} \tag{29}$$

This matrix is constructed using a sliding window of length *M*, which is moved along the data sequence. (*N-M+1*) vectors are thus obtained, which can be considered as the new observation vectors. However, it should be noted that the effective band of these vectors is reduced compared to the band of the initial vector, because only *M* samples are taken into account for each of them. The columns of the matrix *D* thus constitute the observation space, of dimension *M*. In this case, the dimension of the vectors of the signal subspace and the noise subspace will also be limited to *M*. It is obvious that this procedure of spatial smoothing will result in more computation time; this disadvantage is nevertheless subordinate to the performances of the results one desires to obtain (Hayes, 1991).

Choice of parameter M: There is no precise rule concerning the choice of the dimension of the observations space. Some authors empirically advise a value of *M* between:

$$\frac{1}{2}N \leq M \leq \frac{2}{3}N \tag{30}$$

Improvements of the Root-MUSIC Method

The major drawbacks of *Root-MUSIC* method are (Boudinar et al., 2015; Hayes, 1991):

- The number of the sought harmonics estimation (*L*).
- The computation time which is very important and that increases with the samples number and the model order *L* (*L* represents the sought faults).

To solve these drawbacks, the authors propose the following solutions:

First Solution

The *RM* performances can be completely deteriorated by choosing a wrong value of the model order *L*. In fact, if L is too small, then the harmonics of low amplitudes can be lost. By cons, if *L* is too large, then the spectrum may contain spurious harmonics. To determine this parameter, various estimation criterions have been developed (Boudinar et al., 2015; Hayes, 1991; Marple, 1987).

- Chi-deux test.
- Akaike Information Criterion (AIC).
- Minimum Description Length (MDL).
- Efficient Detection Criterion (EDC).

Second Solution

Several studies have demonstrated that the signature of each type of faults are localized on a well precise frequency band of the stator current spectrum; the suggested idea consists therefore in processing the data on a given frequency band and not on all the spectrum of the stator current. This solution will hence enable to reduce the length of the spectrum and consequently will reduce the computation time (Boudinar et al., 2015; Kia et al., 2005).

Hence, *Root-MUSIC* method will be applied solely to a frequency band defined by a low cut-off frequency f_l and a high cut-off frequency f_h. These cutoff frequencies will be selected on the spectrum width $[0, f_{sf}/2]$, depending on the type of the studied fault. With this solution, the processing will be done on $2.N.f_p \,/\, f_{sf}$ samples where $f_p = f_h - f_l$ and not on the *N* samples, these reducing the computation time.

To confirm the positive contribution of this solution with respect to the computation time, several tests have been carried out. Thus, on the basis of several new tests, the computation time is equal to 147.53 s for the fundamental identification only (*L=1*), with the original *Root-MUSIC* method, see Table 1. Note that, these results are obtained by considering a signal of 10000 samples and using a PC equipped with a dual-core processor of 2 GHZ and 3 GB of RAM.

Table 1. Speed computation comparison

Method	Data length	Fundamental Identification (Frequency / Magnitude)	Computation time (s)
Original Root-MUSIC	10000	50.003 Hz / 16.44 dB	147.53
Improved Root-MUSIC	136	50.004 Hz/ 16.15 dB	0.4

Unfortunately if the size of the signal increases a hundred thousand samples, then the computation time quickly becomes prohibitive. This problem occurs because of the roots calculation (see equation (23)) and the matrix inversion (see equation (26)). This problem becomes even more complex, if the number of the sought harmonics (L) is higher. By cons, our solution, which is to seek the desired harmonics only within a defined band, avoids this problem. In fact, Table 1 shows that the improved Root-MUSIC allows to reduce the data length (136 samples) and the computation time (0.4 s) by considering an analysis band [*40 Hz 60 Hz*] for the identification of the fundamental ($L = 1$).

EXPERIMENTAL VALIDATION OF ROOT-MUSIC PERFORMANCE

To verify the performances of the *Root-MUSIC* method on real cases, the authors built a test bench to measure the stator currents in the presence of rotor faults in order to analyze them. But before seeing these tests and the results obtained, it is important to know how the presence of a rotor fault in the stator current occurs.

Frequencies Signatures of a Rotor Fault

Several theoretical studies (Boudinar et al., 2015; El Bouchikhi et al., 2014; Toliyat et al., 2013), have shown that the stator current has a rich spectrum which not only allows to indicate the possible faults existence, but also the nature and the severity of these last ones in relation to the harmonics positions and their amplitudes. Actually each type of fault is characterized by well determined frequencies. Several authors (Kliman et al., 1988; Toliyat et al., 2013) have shown that the frequency signature of the broken rotor bar faults is characterized by the appearance of sideband harmonics around the fundamental (*50Hz*). This harmonic sequence is given by the following equation:

$$f_b = (1 \pm 2ks) f_s \text{ with } k=1,2,3,... \quad (31)$$

Where f_b represents the harmonics sequence associated to the broken rotor bars faults, f_s is the line supply frequency and s is the motor slip. The lower sideband is defined by all the harmonics characterizing the fault that appear on the left of the fundamental, and the higher sideband by those harmonics which are on the right of the fundamental (see Figure.2). It is to note that the analysis of the sidebands amplitudes enables an effective diagnosis of the fault severity.

Figure 2. Sideband frequencies around the fundamental line frequency (ideal results)

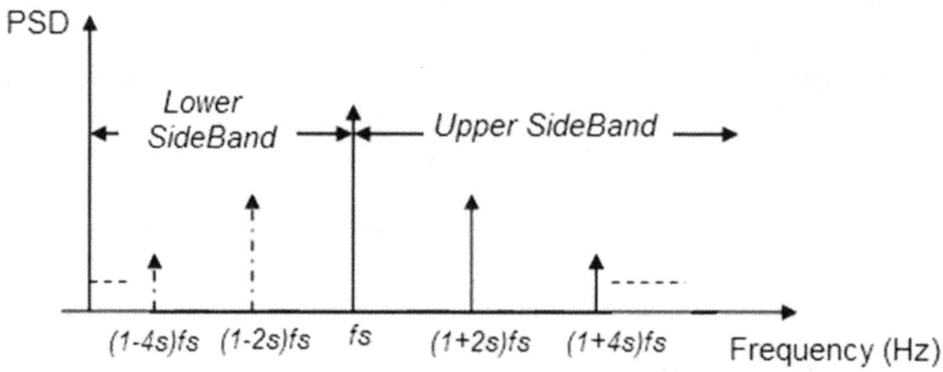

Test Rig and Acquisition Data Parameters

The main experimental tests that are going to be presented and discussed in this chapter were carried out by the Diagnosis Group of the Electric Drives Development Laboratory at the University of Sciences and Technology of Oran (USTO-MB) Algeria. The employed motor in these tests is a three phase squirrel cage induction motor mechanically coupled to a DC generator used as a load. The motor parameters are given in the appendix. The measuring system includes three current Hall Effect sensors, an anti-aliasing filter with a *400 Hz* adjustable cut-off frequency chosen for our tests and an acquisition card. The whole set is connected to a computer for viewing the processed sensed signal as shown in Figure 3. In addition, a tachometer

Figure 3. Photo of the experimental set-up

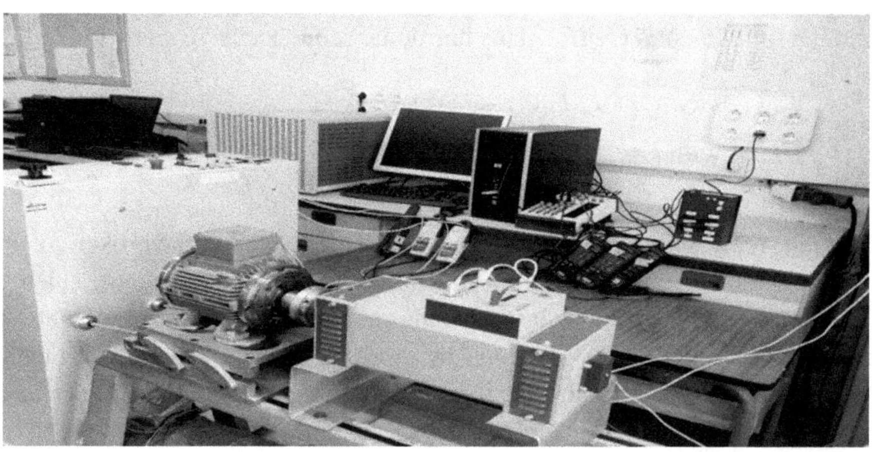

is used for measuring the actual shaft speed of the motor. All the acquisitions were carried out in the steady state mode. The acquisition period of the data is 20 seconds with a sampling frequency of *1.5 KHz* that gives a frequency resolution of *0.05 Hz*.

The treated rotor bars fault are created artificially by drilling a *1mm* hole to simulate an incipient fault and hence to see the same situations as those real faults. Figure 4 shows two rotors with faults in the bars.

The various modes of operation investigated in order to validate the diagnostic procedure are:

- Motor operation of the motor with a healthy rotor.
- Motor operation with a faulty rotor (one broken bar).
- Motor operation with a faulty rotor (two broken adjacent bars).

For a more reliable analysis and due to the randomness of the measured signals, several acquisitions were made for each operation. Moreover and since the authors are focusing in this work on the diagnostic of broken rotor bars faults, the study will consequently be done around the fundamental harmonic (*50 Hz*). For experimental results validation, the stator current spectrum will be analyzed solely on the frequency band [*40 Hz, 60 Hz*]. The choice of this frequency range is dictated by the fact that the slip of induction motors can never exceed *6%* in reality even when overloaded.

EXPERIMENTAL RESULTS

Motor Operation With a Healthy Rotor

In this first test, the authors will analyze the stator current in the case where the rotor is healthy. This analysis will be considered as the benchmark for all future tests. Theoretically, for this operating mode, the stator current spectrum should

Figure 4. Photo of a rotor with broken bars faults

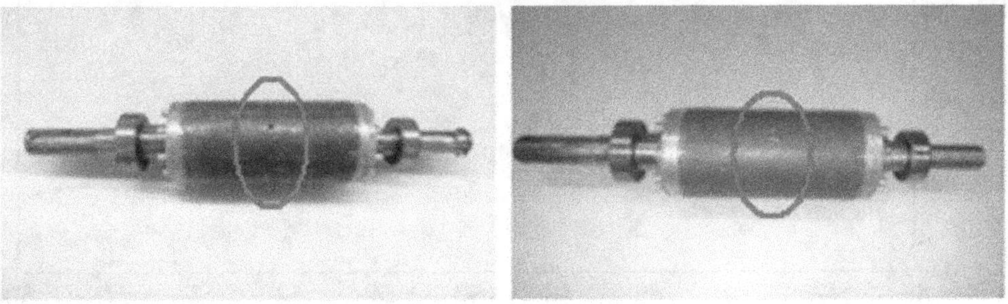

be composed solely of the fundamental component (*50 Hz*) on the selected zone [*40 Hz, 60 Hz*]. The obtained results carried by the classical *PSD* (estimation by *Periodogram* algorithm) in the case of the no-load tests (see Figure.5) show that none of the harmonics appear around the fundamental. However, the authors notice the appearance of weak variations around this frequency but the identification of these events remains difficult, see impossible. In fact, at no-load, the motor slip is too weak (*0.2%*) such that the *PSD* classical method cannot enable the identification of the sought harmonics.

For the experimental tests at nominal load, an estimated 20 Nm torque is considered, Figure 6, shows the results obtained by the classical *PSD*. Noted that besides the fundamental; only one harmonic appears at the frequency of (*54.67 Hz)* but no other one is visible or detectable on the low sideband of the fundamental. In these conditions, there is no meaningful information that can be extracted on the state of the squirrel cage rotor.

The following figures represent graphically the obtained experimental results by the *Root-MUSIC* method for the case of both no-load and nominal load charges.

Figure 5. Stator current analysis by the classical PSD (No-load tests with Healthy rotor)

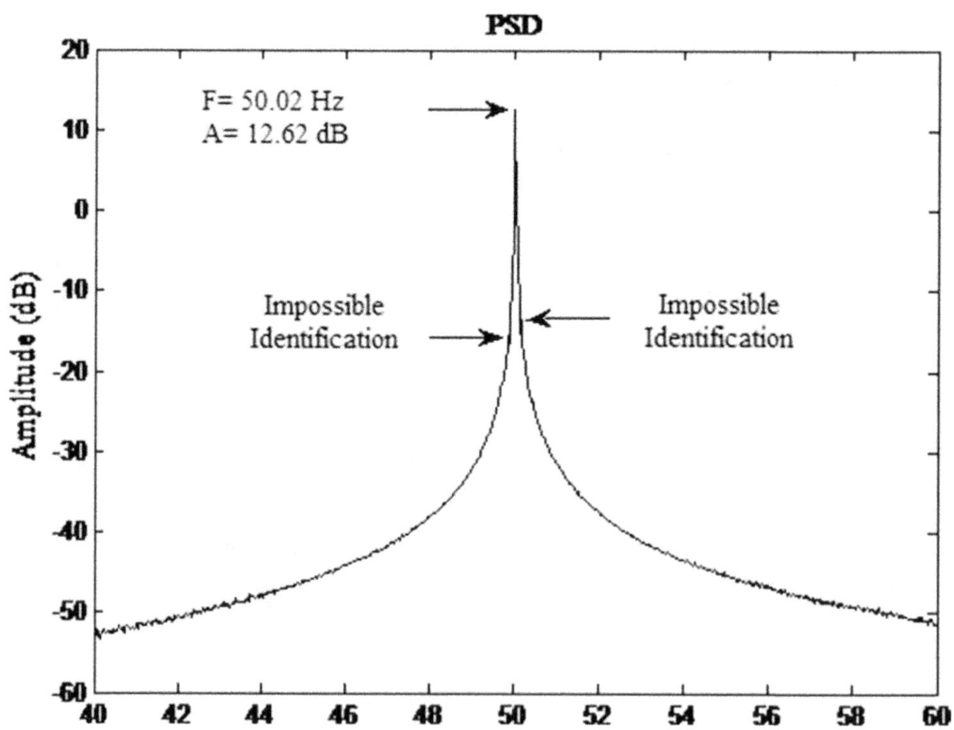

Figure 6. Stator current analysis by the classical PSD (Nominal load tests with Healthy rotor)

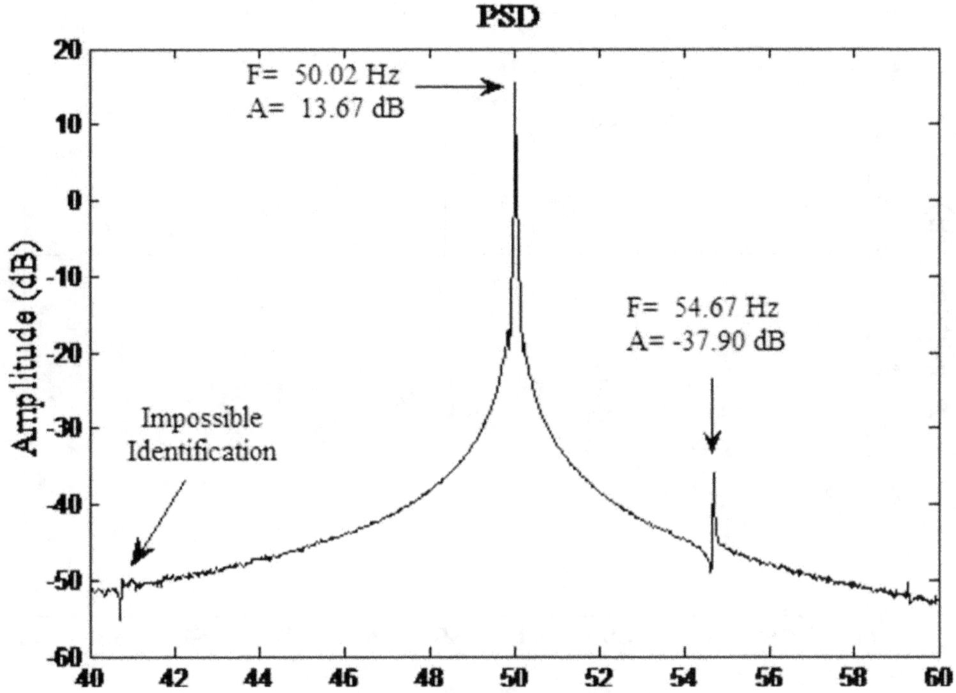

For the no-load tests case, the *Root-MUSIC* method whose resolution power is well superior to that of the classical *PSD* shows the presence of two frequencies (*49.88 Hz and 50.16 Hz*) besides the fundamental one (see Figure 7). The authors suppose that these frequencies which could not be detectable with the *PSD* method represent the signature of an unbalanced rotor circuit for a very low slip. This is probably due to the stage of cage casting where some air bubbles subsist, that is the case of every real motor even as a new healthy one.

Table 2 confirms that the harmonics estimated by the *Root-MUSIC* method represent the signature of a broken rotor bar comparatively to theoretical frequencies obtained in according to (31) and with a measured mechanical speed of *1497 rpm* which corresponds to a slip of *0.2%*.

However, for the nominal load tests case (see Figure 8), Table 3 shows that the estimated harmonics by the *Root-MUSIC* method are very close to those obtained theoretically and which represent the signature of a rotor broken bars faults. These results are obtained for a measured mechanical speed of 1431 rpm which corresponds to a slip of *4.6%*.

Figure 7. Stator current analysis by Root-MUSIC (No-load tests with Healthy rotor)

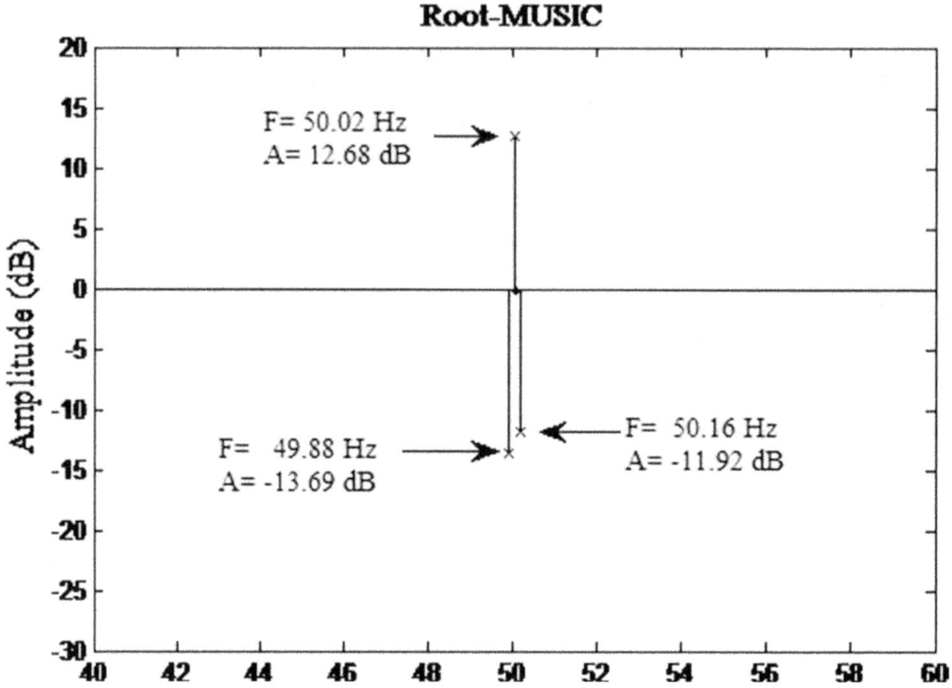

Furthermore, one notices the appearance of two harmonics very close to the fundamental (*49.70 Hz and 50.15 Hz*). This is due to an overestimation of the number of harmonics of the used *AIC* algorithm (these are spurious harmonics). This is the main drawback of the high resolution techniques. For that, in the domain of *RADAR* for localizing mobile targets, these overestimated harmonics are called 'false alarms'.

In the light of these first obtained results for the case of a healthy rotor for both no-load and nominal load, this first identification step is very important and the carried set of tests are very necessary, because all the next set of tests and measures are going to be compared to that first identification step known as 'the reference step'.

Motor Operation With a One Broken Rotor Bar Fault

Theoretically, at nominal load, the mechanical speed is equal to *1440 rpm* and the slip is around *4%*. With such a slip and in the presence of a one broken rotor bar, its signature should appear (according to (31)) at the following frequencies: (*42, 46, 54* and *58 Hz*) over the band [*40Hz, 60Hz*]. These four harmonics are localized around the fundamental. Table 4 gives the estimated harmonics by using *Root-MUSIC*.

Figure 8. Stator current analysis by Root-MUSIC (Nominal load tests with Healthy rotor)

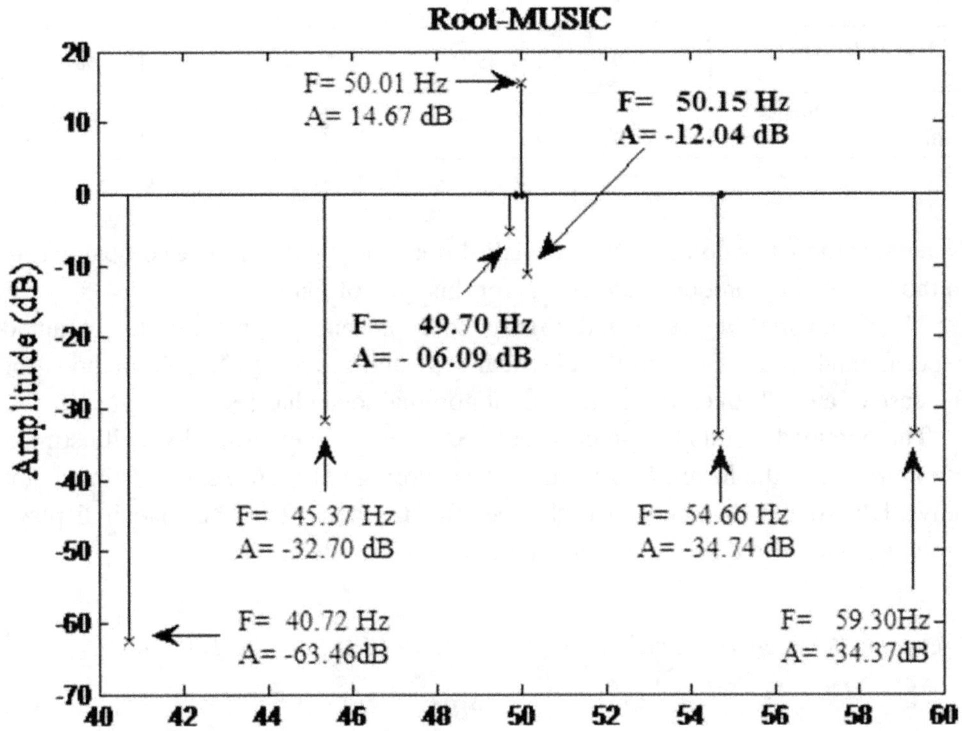

Table 2. Harmonics estimation by Root-MUSIC '' No-load tests with healthy rotor ''

Theoretical frequencies (Hz)	49.8	50	50.2
Estimated frequencies (Hz)	49.88	50.02	50.16
Estimated amplitudes (dB)	-13.69	12.68	-11.92

Table 3. Harmonics estimation by Root-MUSIC ''Nominal load tests with healthy rotor''

Theoretical frequencies (Hz)	40.80	45.40	50			54.60	59.2
Estimated frequencies (Hz)	40.72	45.37	49.70	50.01	50.15	54.66	59.30
Estimated amplitudes (dB)	-63.46	-32.70	-6.09	14.67	-12.04	-34.74	-34.37

Table 4. Harmonics estimation by Root-MUSIC ''Case of a one broken bar''

Theoretical frequencies (Hz)	42	46	50	54	58
Estimated frequencies (Hz)	41.91	45.94	49.97	54.00	58.05
Estimated amplitudes (dB)	-40.34	-26.25	14.99	-26.95	-43.87

Noted that these harmonics reflect well the broken bar fault presence comparatively to the obtained theoretical frequencies for this type of fault.

The following Figure 9 and Figure 10 represent graphically the obtained experimental results by both the classical *PSD* and the *Root-MUSIC* methods for the case of a one broken rotor bar fault at nominal load charges.

The obtained results by the classical *PSD* are surely exploitable, as illustrated in Figure 9. But the *Root-MUSIC* method, according to Figure 10, can estimate not only all the sought harmonics over the specified frequency band but also, it displays these harmonics in a very clear manner to be easily detectable.

Figure 9. Stator current analysis by the classical PSD (One broken bar)

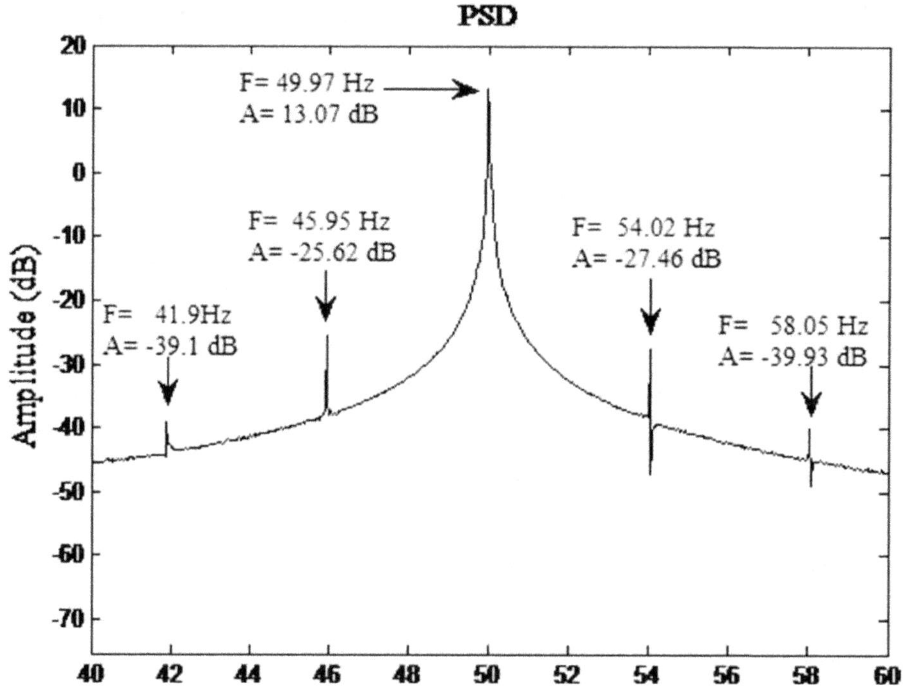

Figure 10. Stator current analysis by Root-MUSIC (One broken bar)

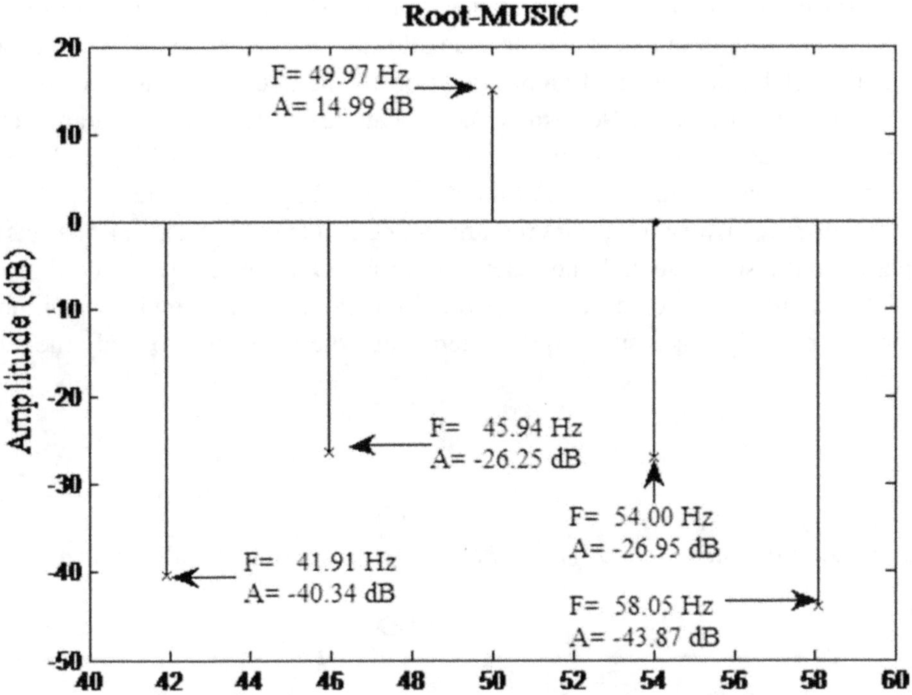

Motor Operation With a Two Broken Rotor Bars Fault

Theoretically, at nominal load, the slip is around *4%*. In these conditions, the frequency signature of two broken bars fault should; according to (31); appears at the following frequencies: (*42, 46, 54* and *58 Hz*) within the defined band [*40Hz, 60 Hz*]. Table 5 gives the estimated harmonics by using the *Root-MUSIC* method. Noted that these harmonics reflect well the rotor bars faults presence comparatively to the obtained theoretical frequencies.

This slight difference between the obtained and theoretical frequencies is due to the frequency resolution and the measurement error of the shaft speed frequency.

Table 5. Harmonics estimation by Root-MUSIC " Case of two broken bars"

Theoretical frequencies (Hz)	42	46	50		54	58
Estimated frequencies (Hz)	42.40	46.19	49.81	49.97	53.74	57.53
Estimated amplitudes (dB)	-40.14	-23.33	-20.48	15.37	-22.46	-40.72

Figure 11 shows that the obtained results by the classical *PSD* for the case of a two broken bars fault are surely usable. But by using *Root-MUSIC*, see Figure 12, one can easily note that this technique allows a very best identification (good estimation of frequencies and amplitudes) of all the sought harmonics over the specified frequency band. Nevertheless, one can also notice the appearance of a false alarm (that is a spurious harmonic) at the frequency of *(50.12 Hz)* and which is surely due to an overestimation of the number of sought harmonics.

Furthermore, a very important remark can be deduced from the various sets of experimental tests presented and discussed so far. One can easily notice that the amplitudes of the harmonics as well as the fundamental increase with the number of broken bars. So this result is quite interesting when one tries to study the fault severity.

Figure 11. Stator current analysis by the classical PSD (two broken adjacent bars)

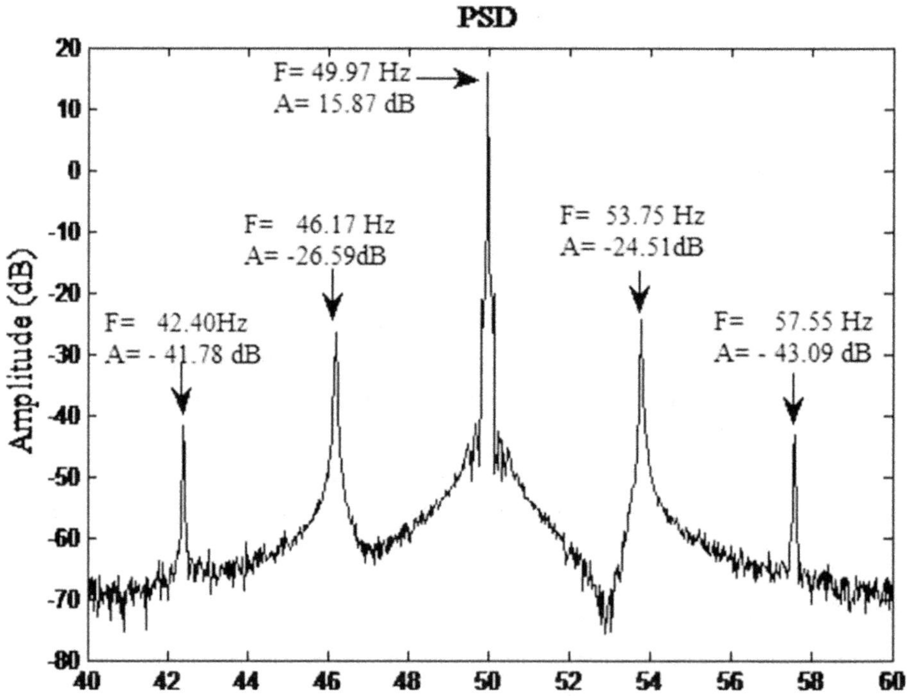

Figure 12. Stator current analysis by Root-MUSIC (Two broken adjacent bars)

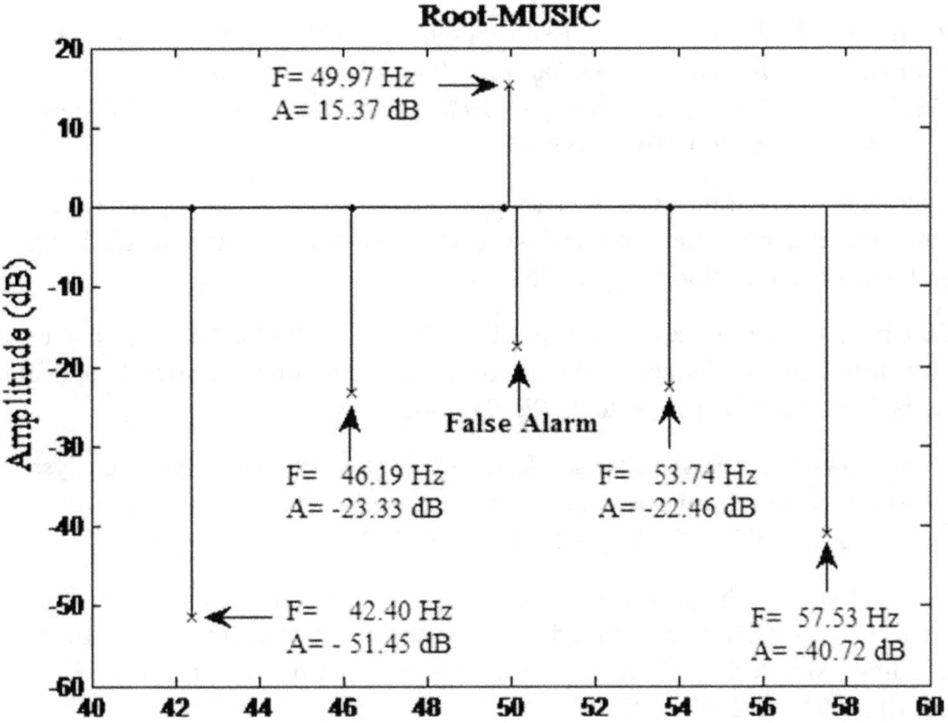

CONCLUSION

The *Root-MUSIC* method is a powerful tool for the harmonics identification, but its major drawback is the computation time. Effectively, the computation time becomes very rapidly important when the length of the signal or the number of the sought harmonics increase. The proposed approach is being introduced in this paper so as to enable overcoming this disadvantage and to make the *Root-MUSIC* method faster and more accurate in extracting the sought frequencies over a given specified bandwidth. Moreover, the obtained results based on experimental data, clearly indicate that the *Root-MUSIC* method has better discrimination capability and is more robust compared to the classical *PSD* method. These results also prove that the *Root-MUSIC* method is very effective in the case of incipient rotor fault. These results also show that the harmonics as well as the fundamental amplitudes increase with the increase number of broken rotor bars. This observation may lead to a very interesting and useful result in order to determine the fault characterization and hence its severity.

REFERENCES

Boudinar, A. H., Benouzza, N., & Bendiabdellah, A. (2015). *Induction motor cracked rotor bars fault analysis using an improved Root-MUSIC method. 3rd International Conference on Control, Engineering and Information Technology (CEIT)*, Tlemcen, Algeria. 10.1109/CEIT.2015.7233035

Boudinar, A. H., Benouzza, N., Bendiabdellah, A., & Khodja, M. E. A. (2016). Induction motor bearing fault analysis using Root-MUSIC method. IEEE Trans. on Industry Applications, 52(5), 3851-3860.

Cupertino, F., Martorana, G., Salvatore, L., & Stasi, S. (2003, September). Diagnostic start-up test to detect induction motor broken bars via Short-Time MUSIC algorithm applied to current Space-Vector. *EPE*, *03*, 2–4.

Dzwonkowski, A., & Swędrowski, L. (2012, September). Uncertainty analysis of measuring system for instantaneous power research. *Metrology and Measurement Systems*, *19*(3), 573–582. doi:10.2478/v10178-012-0050-7

El Bouchikhi, E., Choqueuse, V., & Benbouzid, M. E. H. (2012, December). Current Frequency Spectral Subtraction and Its Contribution to Induction Machines' Bearings Condition Monitoring. *IEEE Transactions on Energy Conversion*, *28*(1), 135–144. doi:10.1109/TEC.2012.2227746

El Bouchikhi, E., Choqueuse, V., & Benbouzid, M. E. H. (2014). Induction machine faults detection using stator current parametric spectral estimation. Mechanical Systems and Signal Processing, 52-53, 447-464.

Hayes, M. H. (1991). *Statistical digital signal processing and modelling*. John Wiley & Sons.

Ibrahim, A., Badaoui, M. E., Guillet, F., & Bonnardot, F. (2008, December). A New Bearing Fault Detection Method in Induction Machines Based on Instantaneous Power Factor'. *IEEE Transactions on Industrial Electronics*, *55*(12), 4252–4259. doi:10.1109/TIE.2008.2003211

Kia, S. H., Henao, H. G., & Capolino, A. (2005). Zoom-MUSIC frequency estimation method for three phase induction machine fault detection. *The 32nd Annual Conference of IEEE,* 2603-2608.

Kia, S. H., Henao, H., & Capolino, G. A. (2007, August). A high-resolution frequency estimation method for three-phase induction machine fault detection. *IEEE Transactions on Industrial Electronics*, *54*(4), 2305–2314. doi:10.1109/TIE.2007.899826

Kim, Y. H., Youn, Y.-W., Hwang, D.-H., Sun, J.-H., & Kang, D.-S. (2013, September). Resolution Parameter Estimation Method to Identify Broken Rotor Bar Faults in Induction Motors. *IEEE Transactions on Industrial Electronics, 60*(9), 4103–4117. doi:10.1109/TIE.2012.2227912

Kliman, G. B. (1988). Non-invasive detection of broken rotor bars in operating induction motors. *IEEE Trans. on Energy Conversion, 3*(4), 873-879.

Marple, S. L. (1987). *Digital Spectral Analysis with Applications.* Prentice-Hall Signal Processing Series.

Sahraoui, M., Cardoso, A. J. M., & Ghoggal, A. (2015). The use of a modified prony method to track the broken rotor bar characteristic frequencies and amplitudes in three-phase induction motors. *IEEE Transactions on Industry Applications, 51*(3), 2136–2147. doi:10.1109/TIA.2014.2375384

Toliyat, H. A., Nandi, S., Choi, S., & Meshgin-Kelm, H. (2013). Electric machines: Modeling, condition monitoring and fault diagnosis. New York: Taylor & Francis Group.

Xu, B., Sun, L., Xu, L., & Xu, G. (2013, March). Improvement of the hilbert method via esprit detecting rotor fault in induction motors at low sli. *IEEE Trans. Energ. Conv., 28*(1), 225–233. doi:10.1109/TEC.2012.2236557

APPENDIX

Induction Motor Parameters

Rated Power 3 KW
Supply frequency 50 Hz
Rated voltage 380 V
Rated current 7A
Rotor speed 1440 rev/min
Number of rotor bars 28
Number of stator slots 36
Power factor 0.83
Number of pair of poles 2

Chapter 4
Advanced Fault Diagnosis Monitoring Scheme in Asynchronous Motor Using Soft Computing Method

K. Vinoth Kumar
Karunya Institute of Technology and Sciences, India

Ramya K. C.
Sri Krishna College of Engineering and Technology, India

Muhammad Irfan
Najran University, Saudi Arabia

ABSTRACT

This chapter deals with the implementation of a PC-based monitoring and fault identification scheme for a three-phase induction motor using artificial neural networks (ANNs). To accomplish the task, a hardware system is designed and built to acquire three phase voltages and currents from a 3.3KW squirrel-cage, three-phase induction motor. A software program is written to read the voltages and currents, which are first used to train a feed-forward neural network structure. The trained network is placed in a Lab VIEW-based program formula node that monitors the voltages and currents online and displays the fault conditions and turns the motor. The complete system is successfully tested in real time by creating different faults on the motor.

DOI: 10.4018/978-1-5225-6989-3.ch004

INTRODUCTION

A three phase induction motor consists of stator and a rotor. The stator is fabricated from laminated sheet steel stampings having slots on the inner periphery. The rotor may squirrel cage or wound rotor. A squirrel cage or (simply) cage rotor has a number of conducting bars housed in slots on the outer surface of rotor core presented in Acosta.et.al., 2004. These bars are short circuited at both ends by conducting end rings. The three-phase squirrel-cage induction motor is frequently used, owing to its ruggedness, robust construction, and low cost, it can however show defects. The flow of phase currents in the stator winding produces a rotating magnetic field. This magnetic field has a constant amplitude and rotate at synchronous speed assume that the rotor is initially stand still. The rotating stator field induces an EMF in the rotor conductors deals in Benbouzid (2000).

Since the rotor circuit is closed (either because of the end rings in cage rotor or because the slip rings are short circuited in a wound rotor), a current flows in the rotor circuits. This rotor creates a rotor field. The interaction of stator on rotor produces a torque which causes the rotation of rotor in the direction of stator rotating field. The three-phase induction motor is a workhorse of the manufacturing industry. They can come either from stator, rotor or a mechanical problem. Knowing that industrial constraints are even stronger, reliability and a safe operating system have to be considered. As a matter of fact, the stresses of induction motor generally bring about a breakdown with a range of few minutes to several days in industrial processes deals in Blodt.et.al., 2006.

It is well known that many defects come from a rotor problem. A broken rotor bar is an example. The brazen connection between the end ring and the rotor bars, induced by several consecutive start up, is another one. Moreover, some defects can come from an eccentricity of the rotor which can be static or even dynamic. These ones are incipient, and they, slowly but surely, induce some perturbations like an increase of vibrations of the motor process, the appearance of torque ripples, and/or an increase of the temperature of the induction motor. These types of defects can lead to dismantling. A predictive maintenance cunningly allows one to avoid expensive repair due to damages.

Consequently, monitoring and fault severity have been focused on for some decades. The most popular approach is the monitoring of the induction motor, owing to the spectral analysis of the stator line current absorbed by the induction motor, this paper is devoted to the monitoring and fault detection of an induction motor drive using two current sensors. The diagnostic approach we propose in this paper is based on fuzzy fault detection, owing to human knowledge of the process. The main goal of this approach is to inform an operator of symptoms by means of indicator lights presented in Çalis.et.al., 2007. It could be the presence (or not) of an incipient rotor defect or the presence of one rotor broken bar. This motor may encounter several fault conditions, which can damage the motor. These conditions include overload, unbalanced supply voltage, locked rotor, single phasing, under-voltage, and overvoltage. A 3.3 kW squirrel-cage, three-phase induction motor is used for this purpose presented in Casimir.et.al., 2006.

ANN STRUCTURE AND FAILURE MODES

Stator Winding Failure

NNS is a simulator for artificial neural networks (ANN). This simulator is designed with a new graphical user interface. In this technique, Feed Forward Neural Network (FFNN) structure is used with back propagation algorithm is shown in figure 1. This ANN involves with one input layer and one output layer. In this technique, five neurons are used in the hidden layer. The Activation function used in hidden layer and output layer is Log Sigmoid function.

The ANN structure is trained through momentum coefficient (α) and learning coefficient (η) with 0.1 to 0.9 values and the best results obtained for momentum coefficient (α) and learning coefficient (η) is found to be 0.2. With the momentum coefficient (α) and learning coefficient (η) of 0.2, the error graph and log file is obtained. In this proposed work, Neural Network Simulator is used to diagnosis the condition of stator and rotor of a three phase squirrel cage induction motor. For training first open the NNS program, then choose create / layer from the tool menu presented in Didier.et.al.,2007.

Here choose the number of neurons in the input, hidden and output layer and also select activation function, then choose feed forward connection to create a network. After that load a pattern set. Then open a control panel in tool menu. Here, one can choose the learning function, set its parameters, number of learning cycles and finally perform network initialization and learning. During training error graph displays the error curve and similarly that error can be written on log file.

In this technique, NNS (ANN) structure is trained with the derived positive and negative sequence component for normal and known faults condition of the machine. Total of 10 input patterns corresponding to different faults and normal condition as described in Table 1, are used to train ANN to predict the stator condition. Figure 2 (a). Illustrate the ANN structure trained with the data sets obtained from R phase stator winding short circuited by a low resistance between turns. After training, weights and bias values are stored in log file. Figure 2 (b) illustrates the Error graph and log file of the trained ANN to identify the stator condition. For stator inter turn fault, total of 10 patterns considered.

Figure 1. Measurement of Sequence Component

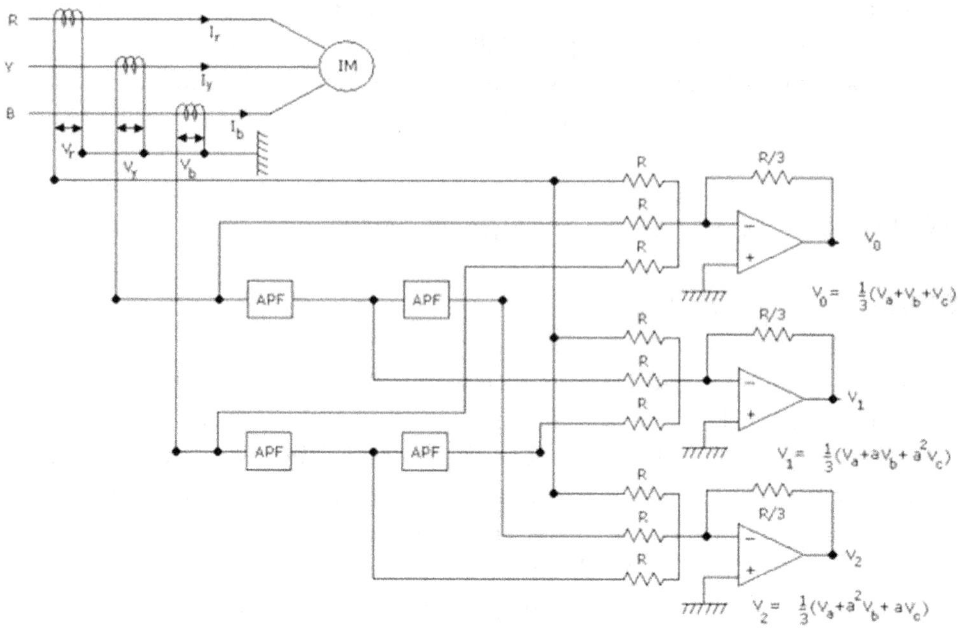

Figure 2. (a) Trained ANN Structure for R-Phase Stator Winding Short Identification (b) Error Graph and Log File

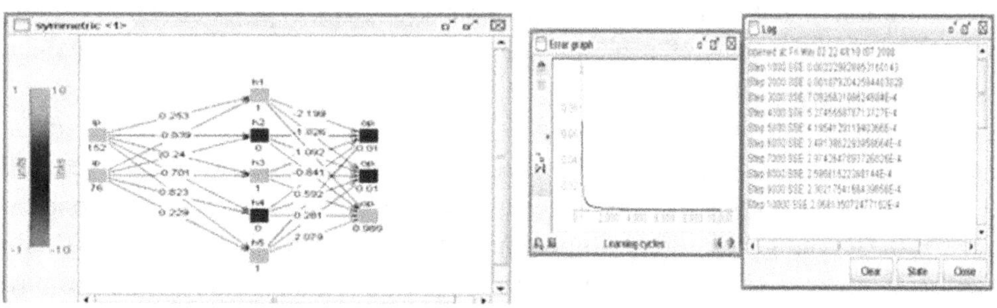

(a) (b)

Rotor Failure

Motor Current Signature Analysis (MCSA) is well known tool to identify the different fault in the electrical system as well as in the machines. The types of faults using MCSA are normally analyzed through the amplitude difference between the line frequency amplitude in dB and amplitude of the first pole pass side band below the

Table 1. Input and desired output for a stator inter turn fault

Input		Desired output	Conditions
Positive sequence magnitude in mA	Negative sequence magnitude in mA		
438	84	0 0 0	Normal Case
152	76	0 0 1	R phase short
160	120	0 1 0	Y phase short

line frequency in dB. To diagnose the rotor condition Motor current signal spectrum is captured in terms of amplitude in dB versus frequency spectrum in Hz and the amplitude difference between line frequency amplitude in dB and the amplitude of the first pole pass sideband below line frequency in dB is measured by using Tektronix TDS 2014 power analyzer as in Figure 3. In this technique, Artificial neural network structure is trained with measured amplitude difference between the line frequency amplitude in dB and amplitude of the first pole pass side band below the line frequency in dB values for good rotor, rotor bar with one crack and rotor bar with two crack condition.

Figure 3. MCSA Arrangement in Induction Motor

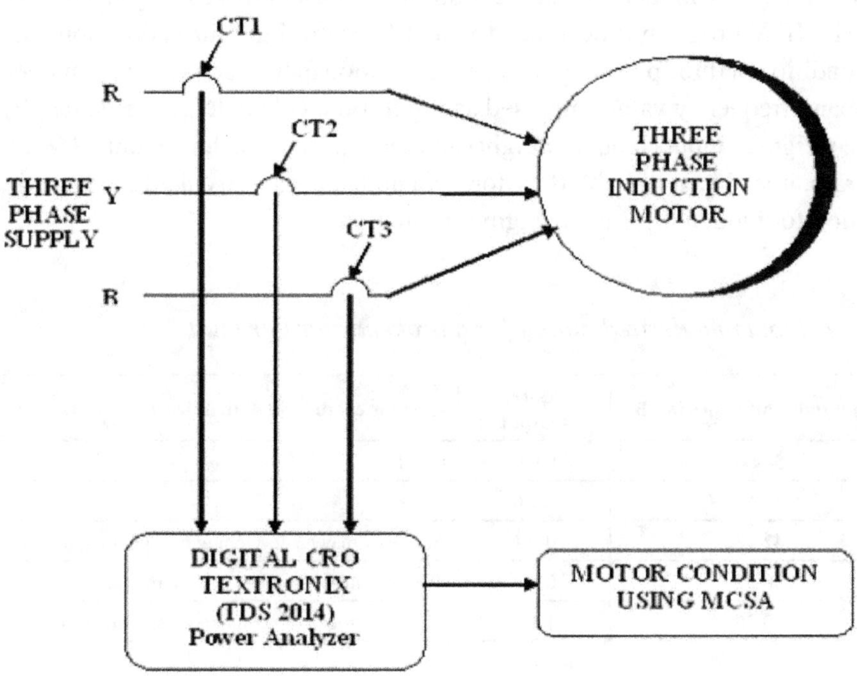

Figure 4. (a) Trained ANN Structure for Single cracked (broken) rotor bar using NNS (b) Error Graph and Log File

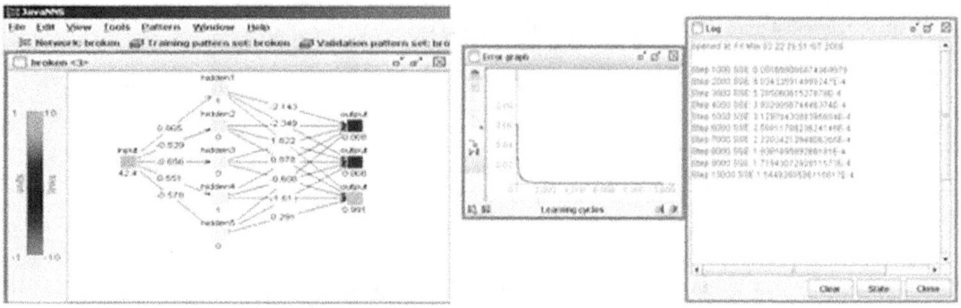

(a) (b)

A total of 10 input patterns corresponding to good rotor and different rotor faults condition as described in table 2. Are used to train the ANN to predict the rotor condition. Figure 4 (a). Illustrate the ANN structure trained with the data sets obtai4 (b). Illustrates the error graph and log file of a trained ANN to identify the rotor condition. Trained network weights and bias values from NNS and the necessary inputs for identify the particular faults are placed in LABVIEW Formula Node. Then a LabVIEW program is designed to monitors the machine condition. In figure.5 illustrates how the induction motor condition can be identified using LabVIEW. [11]

LabVIEW Program is designed to identify Stator inter turn condition and rotor bar condition of three phase squirrel cage induction motor. Sequence component and side band frequency values are used as input. First calculate the net value, by sum of the weighted input (input × weight) and the bias (b) which is obtained from NNS program .it is placed in LabVIEW formula node in order to calculate the activation function for hidden layer using sigmoid function.

Table 2. Input and desired output for a broken rotor bar fault

Amplitude Difference in dB	Desired Output	Rotor condition (With at least 75% of rated load)
54-60	0 0 0	Good
48-54	0 1 0	Moderate
42-48	0 0 1	Bar crack may be developing or high resistance joints
36-42	0 1 1	Two bars may be broken or high resistance joints likely
30-36	1 1 1	Multiple broken or open bars or end ring probable

Output from hidden layer is used as input to output layer. Once again calculate the activation function for output layer in similar method. The output goes to one if fault condition exits, otherwise it is zero. LED display the status of a machine whether it is fault or not. [12]

RESULTS AND DISCUSSIONS

Finally the online monitoring results for stator Inter turn fault and broken rotor bar fault is compared with the ANN desired output shown in Table 3 and Table 4. This proposed work reported that the method of identifying the stator turn and rotor bar condition in three phase induction motor and the results are shown in table 3 and table 4 respectively. It is observed that artificial neural network approach using NNS is very efficient in monitoring the faults.

Figure 5. Overall Block Diagram of ANN based Induction motor fault detection

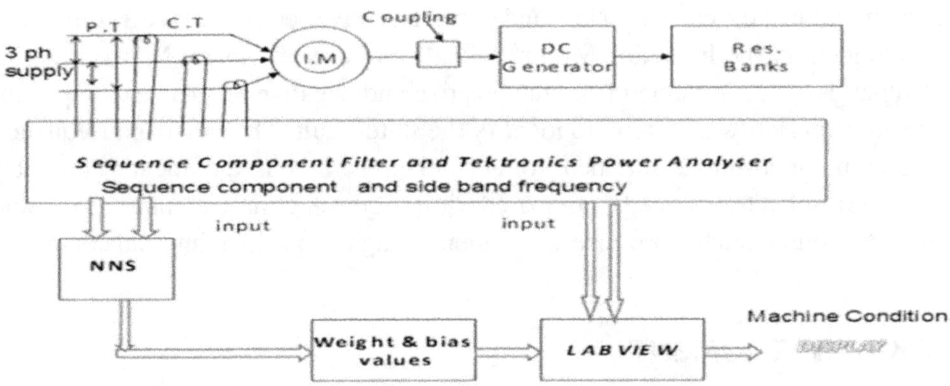

Table 3. Comparison of ANN and LabVIEW of stator inter turn fault identification

Name of the Fault	Desired output			ANN output			LabVIEW output		
Healthy case	0	0	0	0.007	0.007	0.007	0.055	0.041	0.053
R phase short	0	0	1	0.008	0.008	0.991	0.061	0.069	0.964
Y phase short	0	1	0	0.001	0.989	0.001	0.056	0.965	0.076

Table 4. Comparison of ANN and LabVIEW of Broken rotor bar fault identification

Fault	Desired output			ANN output			LabVIEW output		
Good rotor	0	0	0	0.007	0.007	0.007	0.007	0.089	0.056
Moderate rotor	0	1	0	0.007	0.992	0.007	0.039	0.945	0.041
Single cracked	0	0	1	0.008	0.008	0.991	0.045	0.044	0.921
Double cracked	0	1	1	0.007	0.992	0.992	0.080	0.981	0.966
Multi cracked	1	1	1	0.992	0.989	0.991	0.943	0.949	0.945

CONCLUSION

This work concluded the implementation and testing of a real time based fault identification scheme for a three-phase induction motor using the ANN. To accomplish this, a hardware system was designed and built to acquire three-phase voltages and currents from an induction motor. A 1/3 HP squirrel-cage three-phase induction motor was used to create faults and collect data for training and testing the ANN based system. The LabVIEW program was written that read the voltages and voltage representation of the currents. NNS and LabVIEW have been used as soft computing tools to identify the Induction Motor faults. Feed Forward Neural Network where the Input data's are obtained from the positive and negative Sequence Component derived from Hardware circuit to identify the stator fault. The side band frequency of input motor current is obtained from Power analyser is used to identify the rotor fault. The result thus obtained is compared with the conventional technique results and have been found much more accurate in identifying the machine internal condition.

ACKNOWLEDGMENT

Authors may acknowledge to Karunya University, India that supported to any part of study.

REFERENCES

Acosta, G. G., Verucchi, C. J., & Gelso, E. R. (2004). A current monitoring system for diagnosing electrical failures in induction motors. *Mechanical Systems and Signal Processing*, *20*(1), 953–965.

Benbouzid, M. E. H. (2000). A Review of induction motors signature analysis as a medium for faults detection. *IEEE Transactions on Industrial Electronics*, *47*(5), 984–993. doi:10.1109/41.873206

Blodt, M., Chabert, M., Regnier, J., & Faucher, J. (2006). Mechanical Load Fault Detection in Induction Motors by Stator Current Time-Frequency Analysis. *IEEE Transactions on Industry Applications*, *42*(6), 1454–1463. doi:10.1109/TIA.2006.882631

Çalis, H., & Çakir, A. (2007). Rotor bar fault diagnosis in three phase induction motors by monitoring fluctuations of motor current zero crossing instants. *Electric Power Systems Research*, *77*(5), 385–392. doi:10.1016/j.epsr.2006.03.017

Casimir, R., Boutleux, E., Clerc, G., & Yahoui, A. (2006). The use of features selection and nearest neighbours rule for faults diagnostic in induction motors. *Engineering Applications of Artificial Intelligence*, *19*(2), 169–177. doi:10.1016/j.engappai.2005.07.004

Didier, G., Ternisien, E., Caspary, O., & Razik, H. (2007). A new approach to detect broken rotor bars in induction machines by current spectrum analysis. *Mechanical Systems and Signal Processing*, *21*(2), 1127–1142. doi:10.1016/j.ymssp.2006.03.002

Chapter 5
A Mobile Augmented Reality System for Enhancing Electrical Machine Supervision

Jesús Gimeno
University of Valencia, Spain

Sergio Casas
University of Valencia, Spain

Cristina Portalés
University of Valencia, Spain

ABSTRACT

Electrical machines are used almost everywhere, and our daily life depends on them. For this reason, it is important to articulate mechanisms to control, supervise, and perform proper maintenance of these machines, especially those used in critical industrial process. The SCADA protocol is one of the technologies that eases the operation and supervision of electrical machines. However, the absence of a spatial connection between the SCADA signals and the machines being supervised suggests the use of augmented reality (AR) to fill this void. This chapter describes SIMARA: A Mobile AR application based on a dual computer-vision system (QR-codes and fiducial markers). SIMARA provides a robust client application for the integration of AR and SCADA signals by means of virtual panels shown on top of real SCADA machines. An authoring tool is also provided in order for users to customize the application to their particular needs, allowing to create, by means of web services, customized virtual panels, and links between SCADA signals and the virtual information shown in the AR application.

DOI: 10.4018/978-1-5225-6989-3.ch005

INTRODUCTION

Electrical machine supervision is a crucial task, since our daily life depends heavily on machines. This dependency is likely to grow in the future. For this reason, it is increasingly important to articulate new mechanisms to control, supervise and perform proper maintenance of these machines. These mechanisms need to be fast, safe, reliable, non-invasive (Sheikh, Nor, Ibrahim, & Bakhsh, 2017), easy-to-use or even unsupervised, when possible (Sheikh, Nor, Ibrahim, & Irfan, 2017), especially for those machines used in critical industrial processes. In this regards, Augmented Reality (AR) may be one of the technologies that enhance electrical machine supervision.

The possible applications of AR are very diverse and the potential of this technology is enormous, since AR is now a rather mature field. Not surprisingly, AR is a technology that has been applied to many different scientific, social and industrial environments. A substantial amount of research can be found dealing with the application of AR for different purposes in different areas. However, not many of these applications are used beyond an initial research or prototype phase and become successfully applied in real environments. Electrical machine supervision, and industrial application in general, is a good example of this, since there is a lack of AR tools in this area that are successfully and productively used in industrial applications outside a laboratory environment.

AR is a computer-based technology by which virtual information can be added to the user's perception (typically a view) of the real world. Most of the AR applications deal with visual information by merging virtual images, computationally generated in real time, with an interactive view of the real word. The image of the real world comes usually from a camera and it is visualized on a screen with the virtual information added. Nevertheless, special displays can also be used to depict only the virtual information while the user sees the real world through the display (see-through displays) or reflected on it (Augmented Reality Mirrors (Giner Martínez & Portalés Ricart, 2005; Portalés, Gimeno, Casas, Olanda, & Giner, 2016)). Sometimes, even the real-world itself can be used to project synthetic images on top of it (Spatial Augmented Reality).

As a result of the rapid development of mobile technologies, AR systems based on mobile devices (Mobile Augmented Reality) represent the most common setup since the cameras of regular smartphones or tablets can be used to create AR scenes. This makes the AR paradigm almost ubiquitous, and it is one of the reasons why AR has become so popular in recent years. However, this popularity does not cause that all the AR applications be successfully used in real environments, especially in industrial environments, where conditions are particularly harsh (electrical noise, uncontrolled illumination, etc.), so robust solutions are hard to build.

Unlike what some people may think, the virtual information in AR applications is not just placed on top of the real world's image. Both worlds must be properly synchronized and aligned for the system to be considered as an AR system. For this reason, AR systems must recognize the real world, usually by means of markers placed on real objects, so that the AR application is able to perform an essential 3D registration process in order to place the virtual objects in the right location with respect to the real ones, depending on the view of the real world the user is experiencing. The AR system should be able to reposition the virtual objects whenever this view of the real word changes.

However, not all the applications that show virtual information based on a view of the real world really need this registration step (necessary for AR) and simpler applications based on simpler markers that do not provide 3D registration capabilities may be of interest for some purposes. In fact, markers can also be used to trigger specific actions or communicate some kind of information. In this regard, this chapter presents an application of Mobile AR that uses both fiducial markers (to perform the AR registration process) and QR-codes markers (to provide a much simpler but effective way of communicating information between the real and the virtual world). Mobile AR provides many potential benefits since it can be used in a wide variety of scenarios: learning and training systems (Dunleavy & Dede, 2014), medical treatment and rehabilitation (Tang, Kwoh, Teo, Sing, & Ling, 1998), construction (Gimeno, Morillo, Casas, & Fernández, 2011), entertainment (Díaz, Boj, & Portalés, 2016), education (Wu, Lee, Chang, & Liang, 2013), culture heritage (Galatis, Gavalas, Kasapakis, Pantziou, & Zaroliagis, 2016), business, etc. One application area of AR that has also been widely explored is their use in industrial processes, but there is a lack of successful cases that are effectively used beyond the prototype phase. Many industrial corporations use SCADA (Supervisory Control And Data Acquisition) systems to control and supervise most, if not all, of their electrical machines (Boyer, 2009).

In this chapter we will be presenting and describing SIMARA, a novel Mobile Augmented Reality system designed to work with SCADA in an integrated manner. SIMARA is built and designed to provide additional value to the SCADA system and save substantial amounts of money to industrial corporations through a better maintenance, supervision and repair of their electrical machines, thus solving many of the aforementioned inconveniences and substantially advancing regarding to the current available technologies.

THE SCADA SYSTEM

The acronym SCADA represents a software protocol that allows operators and practitioners to control and monitor industrial processes remotely. The SCADA architecture was developed as a universal means of accessing, supervising and controlling machines and devices of different types. One key attribute of the SCADA protocol is that it allows operators to supervise the operation of a variety of different proprietary devices (from different manufacturers). For this reason, it has become a *de facto* standard for the interoperability and control of industrial processes. SCADA is widely used in various industries, such as sugar plants (Acebes, Merino, Alves, & Prada, 2009), cement (Harchandani & Bindu, 2014), oil and gas industries (Parvez, Ali, Ahmed, & Farhan, 2015), textiles (Teraiya, Solanki, & Raval, 2016), power plants (Yuvaraj, Ranjith Kumar, & Krishnan, 2016), biotechnology centers (Bárzaga Martell, Paneque, Robert, & Valdés slope, 2016), etc. This software protocol is designed with the objective of providing an easy-to-use real-time feedback interface between sensors, actuators and human supervisors, so that different industrial devices can be monitored and controlled automatically in a centralized and systematic manner. For this reason, SCADA-compatible machines are often designed without gauges, buttons, levers, displays or indicators, since all the relevant information for machine operation and supervision is sent through the SCADA network to and from the SCADA server. Therefore, supervisors do not usually perform machine supervision *in situ*, since the important information is typically depicted in a centralized control room.

A SCADA system collects all the information that an industrial process generates. This information is typically recorded and stored in a central server. This way, data can be managed, logged, manipulated and be used for control purposes. In addition, the server can send this information to a web site or to SCADA-compatible mobile devices, so that a real-time *in situ* supervision of any machine is also possible. These latter devices are expensive and there is a lack of spatial connection with the supervised machine if a simple web-based inspection system is used. The benefits of the utilization of the SCADA system for supervising machine operation are multiple: (i) it allows a real-time feedback of the status of the machines; (ii) it eases the automatization of the control process, since all the control data is centralized and available for the control algorithms; (iii) the industrial processes are under a strict and global supervision that allows a complete control of the production system; (iv) early-warning and fault-detection tasks are easier to implement; (v) automatic self-diagnostics are easier to design and the whole industrial process is safer. All these features boost productivity and increase the efficiency of the industrial processes.

However, regarding the *in situ* supervision of machines and devices in real industrial environments, the SCADA system still presents several limitations that hinder or slow down these supervision processes. These limitations are mainly due to a lack of direct relationship between the element of the real environment (sensor, actuator, device, wiring, etc.) and its representation (or indicator) on the mobile device that displays the SCADA data, when it exists (SCADA is typically used in a control room and not with mobile devices). Thus, for instance, the operator must first identify the area where there has been an alarm, and then identify the sensor or actuator that has caused such alarm, to finally identify the damaged or defective item. This has a negative impact due to the economic loss that this fact causes, both by increasing the time that operators dedicate to the process and by reducing productivity because of the possible malfunctions of the system. Therefore, it is necessary to find a way to provide supervising information within the SCADA system that is able to overcome this problem, providing a meaningful spatial connection between the SCADA devices and their representations. As we will be showing in this chapter, we argue that AR is the right solution for this problem.

RELATED WORK

This section is dedicated to review the literature on the use of AR in industrial applications, with special attention to its application in machine supervision and maintenance in general. First, the AR technology is properly described, reviewing key properties, features, benefits and definitions about Augmented Reality. Then, the review will focus on industrial processes in which AR has been proposed. Finally, the use of AR for machine supervision in combination with the SCADA system is explored.

Augmented Reality

AR is not a new technology. The term "Augmented Reality" was first coined by Caudell and Mizell (1992) in order to describe a special kind of display used by electricians in the repair and maintenance of aircraft. This display was able to show virtual information on top of real images. Since this first conception of AR was established, the technology has evolved rapidly and now it is no longer linked to the particular hardware display through which it is applied. Therefore, the definition of AR is now much more precise and broad at the same time. When Caudell and Mizell coined the term, the technology was closely linked to a special kind of device (Head Mounted Displays, HMD). This specific kind of display was the only visualization hardware in which the technology had been applied. Now, the definition of AR includes additional elements other than just blending virtual and real information. Azuma (1997) defined Augmented Reality as the technology that simultaneously combines real and virtual objects which are registered in 3D and are interactive in real-time. This means that the virtual objects are not just placed on top of the real objects, as in some of the early works working with HMDs. They have to be registered so that the integration of virtual and real objects is spatially consistent from the user's perspective.

Given the relatively long history of AR, it is now a widespread technology which is applied in multiple knowledge areas and scientific fields, such as entertainment (Giner Martínez & Portalés Ricart, 2005; Vera, Gimeno, Coma, & Fernández, 2011), cultural heritage (Gimeno, Portalés, Coma, Fernández, & Martínez; Portalés, Viñals, Alonso-Monasterio, & Morant, 2010), tourism (Kounavis, Kasimati, & Zamani, 2012; Linaza, Gutierrez, & García, 2013), education (Kaufmann & Schmalstieg, 2003; Vargas et al., 2010; Wu et al., 2013), civil engineering (Chi, Kang, & Wang, 2013; Meža, Turk, & Dolenc, 2014), medicine (Birkfellner et al., 2002; Calderita et al., 2015; Kilgus et al., 2015) and, of course, industrial applications, which will be reviewed later on this section.

AR technology is fairly popular because it provides a number of advantages that other technologies cannot offer at the same time (Bimber & Raskar, 2005):

The ability to enhance/augment reality. By using this technology, physical entities can be enhanced with additional elements, such as images, sounds, virtual models, etc. that are not present in the real scene. This feature is the one that provides the name to the AR technology and it is the reason why AR has special interest for supervision tasks, since it is a technology that allows users (in this case, inspectors) to obtain further information than the one that is initially apparent. The ability to integrate virtual and real information could be key for inspecting and signaling a fault condition in a machine, in which for instance, some piece is not working properly. With AR, this fault could be highlighted in such a way that the inspector would not miss the problem. The 3D registration process is key, because, if this process is properly done, the spatial relationship between the virtual and the real object would make the fault signaling unmistakable.

Seamless interaction between virtual and real objects. Interfaces are said to have seams (Ishii, Kobayashi, & Arita, 1994) if there is a functional, temporal, spatial or cognitive element that forces the user to change the way in which he/she interacts with the interface. When using properly designed AR applications, interaction is performed in a natural seamless way, since users can visualize and interact with virtual elements as if they belong to the real environment, without excluding or hiding the real objects, unlike Virtual Reality (VR), in which the real world is purposely hidden to the user. In AR, visual contact is direct and has no restrictions. The ability to transition from real to virtual elements is a very important feature for supervision tasks, since it enables a simple and natural way to monitor real elements and conditions, which is substantially different to performing the supervision task through a model or a depiction of the real element with a static display in a computer.

According to (Milgram & Kishino, 1994) interfaces can be categorized or placed on a continuum depending on the amount of synthetic environment generated by a computer. At the edges of this continuum we can place the real environment (0% virtual) and the purely virtual one (VR, 100% virtual); between these two edges there is a wide variety of interfaces in which both environments are blended, with several degrees of predominance of one with respect to the other. AR applications can use this property to establish gradual transitions between the real world and the virtual one, since the proportion of virtual and real content can be chosen and varied.

It supports a tangible interface. In AR applications, there is a very intimate relationship between virtual objects and the physical world. Physical objects can be augmented through virtual elements, allowing the overlap of these elements dynamically. These objects can be accessible to all users (public use) or may be individualized for each user (private use), so that particular interactions can be defined for different users. Similarly, physical objects can be used to manipulate virtual elements in an intuitive manner. The advantage of this feature is that users do not require any prior knowledge about a particular software application since the interfaces are natural. This could also be very important for supervision tasks, because human supervisors may need to learn to use different protocols and even different software elements to carry out their work. With AR, it is possible to simplify the task by making use of tangible interfaces, and natural interaction metaphors that are quickly identifiable and very easy to use.

The presence of spatial cues for face-to-face and remote collaboration. Augmented Reality allows that the elements generated *in silico* (by a computer) be spatially distributed according to the physical environment. This is one of the features that may be relevant for supervision tasks, since it may be of special relevance for the *in situ* inspection of industrial processes and machines.

Nevertheless, building AR applications is not trivial, as there currently exist a number of technological barriers. In this regards, two main challenges can be reported: the *registration problem* and the *occlusion problem*. On the one hand, the registration problem deals with finding a way to obtain a three-dimensional reference frame to locate the virtual objects with respect to the real objects, in real time. This can be solved in many different ways. Although traditional solutions make use of AR markers or natural features of particular objects inserted in the real world to act as fiducial markers, the last trend is to perform an automatic scan of the real world, eliminating the need to insert artificial objects in the scene. This later solution is less stable, although impressive achievements have been accomplished lately.

The occlusion problem, on the other hand, deals with finding a proper way to render virtual objects when real objects are in front of them in the scene, occluding the virtual elements. Without occlusion handling, users will have the misconception that the real object is further from the viewpoint than the virtual objects when the virtual objects are occluded by the real ones (Tian, Guan, & Wang, 2010). This can lead to motion sickness and to interaction problems. The occlusion problem is one of the most complicated and important problems that can arise in AR applications.

Augmented Reality for Industrial Applications

Regarding the use of AR for industrial processes, various AR applications have been proposed and developed in recent years. The number of works applying AR to industrial applications shows that this technology can be of great benefit to increase productivity, safety and reduce undesired situations due to aforementioned features. AR is mainly used in industrial processes for training, repair, assembly and maintenance, demonstrating in several cases the advantages of using this technology against the usual way of performing these operations (Gavish et al., 2015; Henderson & Feiner, 2011; Sanna, Manuri, Lamberti, Paravati, and Pezzolla, 2015). For instance, in (Henderson and Feiner, 2011), the benefits of AR for maintenance and repair operations are studied, with a case study for military mechanics who require routine maintenance tasks inside a tower of armored vehicles. The study shows that AR allows for faster movements and, in some cases, requires fewer head movements. In addition, the mechanics found the AR intuitive and satisfactory for carrying out the desired tasks.

The use of AR in industrial environments is not new. In fact, it is as old as AR itself, since the paper that would coin the term Augmented Reality (Caudell & Mizell, 1992), describes a system to assist in a manufacturing processes, with a case study in the airline industry. Since then, there have been many scientific contributions in this regard. For example, in (Goose, Sudarsky, Zhang, & Navab, 2003) an AR system with mobile devices is proposed for industrial maintenance and supervision. AR markers are used for real-time 3D registration. As an interesting peculiarity, the system supports both visual and auditory stimuli. In (Alvarez, Aguinaga, & Borro, 2011) an AR tool is presented for assembly and disassembly processes in industrial maintenance operations, offering users visual instructions that are useful for operators, so that they can perform their tasks safer and faster. In (Gay-Bellile, Bourgeois, Tamaazousti, Naudet-Collette, & Knodel, 2012) an application of AR for the motor industry is presented, allowing the assisted maintenance of parts of a vehicle, among others. The system is based on SLAM techniques to allow 3D registration in real time. The visualization is done through a mobile device, a set-up that is recurrently used in many applications for its simplicity, availability and good performance with current state-of-the-art mobile devices (mainly phones and tablets). In (Webel et al., 2013) an AR platform is presented for the training of operators in industrial assembly and maintenance skills. They also use a mobile device for visualization and the system offers haptic feedback using vibra-tactile bracelets that can apply vibration stimuli to the arm, forearm and wrist. On the other hand, in (Ramirez, Mendivil, Flores, & Gonzalez, 2013) an authoring tool is presented in order to develop AR applications for maintenance and training. In (Gimeno, Morillo, Orduña, & Fernández, 2013) an authoring tool is also presented, with the peculiarity that it uses a Kinect sensor for the acquisition of depth, being able to treat the occlusion problem.

More recently, in (Zhu, Ong, & Nee, 2014) a context-dependent AR system is presented to assist in industrial maintenance tasks. In this system, the operator uses a pair of AR glasses for the visualization. For the 3D registration, markers and the ARToolKit libraries are used (HITLabNZ, 2003). In (Fiorentino, Uva, Gattullo, Debernardis, & Monno, 2014) the development of an AR application is provided. The application assists the operator, through instructions, in maintenance works. In this case, the visualization of the augmented environment is done on a large screen, rather than through the use of mobile devices, which is the most common case. On the other hand, in (Funk, Kosch, Kettner, Korn, & Schmidt, 2016) the result of the motionEAP project is presented, which focuses on the use of AR technology to show cognitive help in the work environment. As a result of this project, a series of recommendations are offered when designing AR applications of this type. These include to keep the instructions as simple as possible and show the visual content in the precise place where the operation should be performed.

Finally, it is worth presenting the work published in (Palmarini, Erkoyuncu, Roy, & Torabmostaedi, 2018), where a systematic review of AR applications for maintenance is carried out. They also comment on the next trends of this technology, such as hardware improvement, pattern recognition and man-machine interaction. As it can be seen, most of the works are focused on training or maintenance/ supervision of industrial process. These two pillars represent the foundations of any industrial process, since it is essential that operators are correctly trained in the use of processes, protocols and machines, and these machines are properly supervised and maintained so that they can be used for as much time as possible without fault, increasing productivity and reducing accidents.

Augmented Reality and SCADA for Machine Supervision

Despite the multiple applications of AR found in industrial processes, the use of these in combination with SCADA systems is not common or has not been sufficiently explored, with very few examples of success. For example, in (Soete, Claeys, Hoedt, Mahy, & Cottyn, 2015) the possible opportunities to complement SCADA systems with mixed reality systems are explored. The result provided is based on an improved SCADA visualization system (through enhanced virtuality) and an AR-based supervision tool, visualized through a conventional display. In this case, the same authors point to the need to improve their system through the use of mobile devices.

In (Antonijević, Sučić, & Keserica, 2016) an AR system is described for the automation of substations using the IEC 61850 communication, which shows data provided by a SCADA system in real time. QR codes are used to load the data. However, these codes are also used for real-time 3D registration by means of ARToolKit libraries. Although the authors of that paper prefer this solution due to the fact that QR codes can embed more information than traditional markers, the truth is that the use of these codes also introduces certain limitations in terms of registration. Since QR codes are more detailed than AR markers, for the 3D registration to be properly done, QR codes must be large enough, limiting the distance and field of view of the AR application and, consequently, limiting the movements of the operator in his working environment.

Other works point to the possibility of integrating and designing systems for the *in situ* maintenance or supervision of SCADA devices by means of AR solutions, but without providing a finished tool (Kranzer et al., 2017, Opriş, Costinaş, Ionescu, & Nistoran, 2017). Therefore, we can identify an interesting and productive research avenue in the use of AR for machine maintenance and supervision. This is where the SIMARA project, explained next, is framed.

THE SIMARA PROJECT

The SCADA environment is often managed by a centralized control system that performs all the processes needed to implement the control and supervision tasks of industrial electrical machines. For this reason, it has encouraged the disappearing of visual indicators in machines that run the SCADA protocol. This way, all the information is typically kept centralized and the machines can be spared of incorporating display systems to highlight problems and working conditions. All this information is only shown on the SCADA control panel in a centralized control room, or in SCADA-compatible mobile devices that are expensive. This is an advantage for companies in terms of safety and purchasing costs. It also makes machines more robust since engineers do not need to worry about placing elements to display information, but it poses some disadvantages when one of these machines needs to be repaired or maintenance works are performed: it is usually the case that the machine being repaired is far from where the SCADA-based control room is placed, since this control center/room can be placed in a different location of the same building, in a different building or even in a distant city. In such cases, maintenance personnel needs to be in constant contact with the control center staff in order to manipulate the machine or even to check its status. This results in longer repair times and in lower productivity rates for these tasks. A SCADA-compatible mobile device can be used, but the user still needs to perform several operations to identify the machine among the (potentially large) SCADA list of devices, usually searching by code number.

For this reason, we present the SIMARA project. SIMARA is a Mobile Augmented Reality system that allows workers to visualize and control electrical machines running under the SCADA system, in cases where these machines do not have visual indicators. The SIMARA system allows maintenance personnel to perform reparation tasks as if they were using a real control panel with displays, levers, buttons, etc. Instead of having to go to the control room to check the SCADA signals or use expensive SCADA-compatible devices (which do not make the spatial connection between the machine and the SCADA codes and signals) operators could see the information they need just on top of the electrical machine they are working on, by means of a mobile device and AR technology (see Figure 1). The AR paradigm provides the possibility of blending virtual content (a virtual control panel) with the real world (real electrical machines without visual indicators or local control gauges). This way, workers can do their maintenance/repair at a fraction of the cost that it would take to introduce visual indicators in every machine that is being used.

The SIMARA client application consists of two main elements: a mobile device (usually a tablet because it features a larger screen than a mobile phone) with the AR tool to display the AR content (the virtual panel), and a web-service to connect

Figure 1. The SIMARA AR mobile application in a real scenario

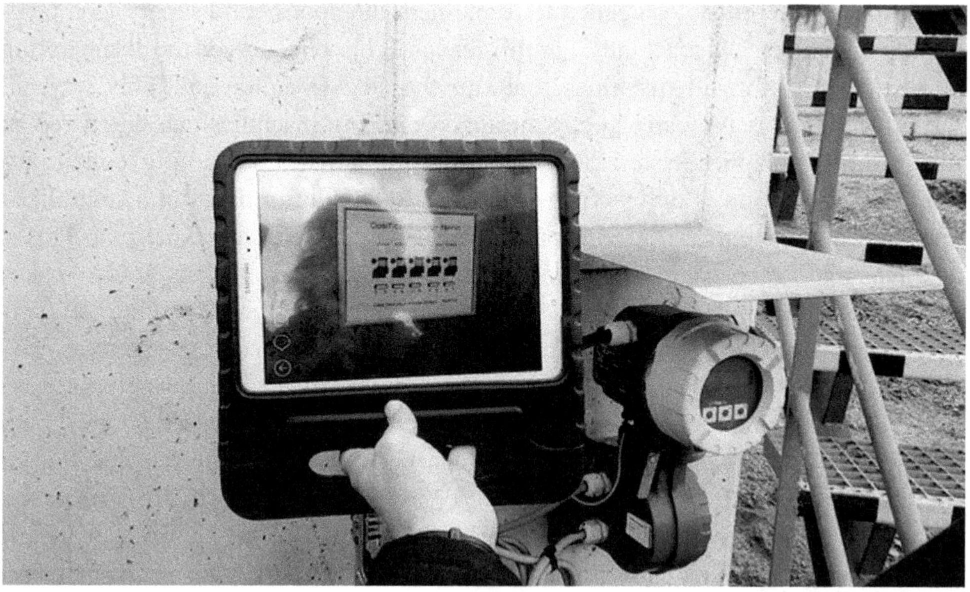

the SCADA control center with the AR application. Therefore, SIMARA acts as a bridge between the physical machines being supervised, repaired or maintained, and the information gathered and visualized on the main control panel of the SCADA control room. The communication between the SCADA system and the web-service is performed by means of OPC (OLE process control) protocol, which is an industry standard.

In addition, SIMARA offers an authoring tool in order to create customized virtual panels and link the information and control of these panels with the OPC signals coming from the SCADA control center. This way, the SIMARA system can be used to control and monitor any kind of SCADA-compatible machine that may be used, not just a closed list of devices. Users can expand the SIMARA system to customize its use for the particular features and requirements of the industrial process in which is used. This is a key advantage, because it provides a configurable system that is not limited to a few predefined set of machines of a particular industry. Universality is one of the initial objectives of the SIMARA project.

The mobile client application uses (simultaneously) two image processing systems: a QR-code-based system and a fiducial marker recognition system. The QR-code is used to provide information about the machine being supervised, whereas the fiducial marker is used for 3D registration processes, so that true AR capabilities are provided. The rationale behind this decision is the following: QR-codes can provide a substantial amount of information but their observation by a computer vision system can be performed only from a near point. On the contrary, AR markers are easier to visualize from more distant locations but the amount of information that they can provide is reduced to a few bytes. The combination of the two vision systems allows SIMARA to provide AR capabilities and at the same time receive the necessary information to properly identify the machine and provide the appropriate virtual panel information.

The benefits of Mobile AR for this case scenario are threefold: (i) it simplifies repair and maintenance tasks with SCADA-based machines, reducing the possibility of errors due to communication misunderstandings between the control room staff and the repair team; (ii) it reduces repair times without the need to purchase additional hardware (except from a simple tablet), therefore increasing productivity; (iii) it provides the possibility to define new virtual control panels to control any electrical machine compatible with SCADA.

DETAILED SYSTEM DESCRIPTION

As it has been already mentioned, the SIMARA project provides both an authoring tool and a mobile client AR application. These are two different software elements

Figure 2. The SIMARA authoring tool

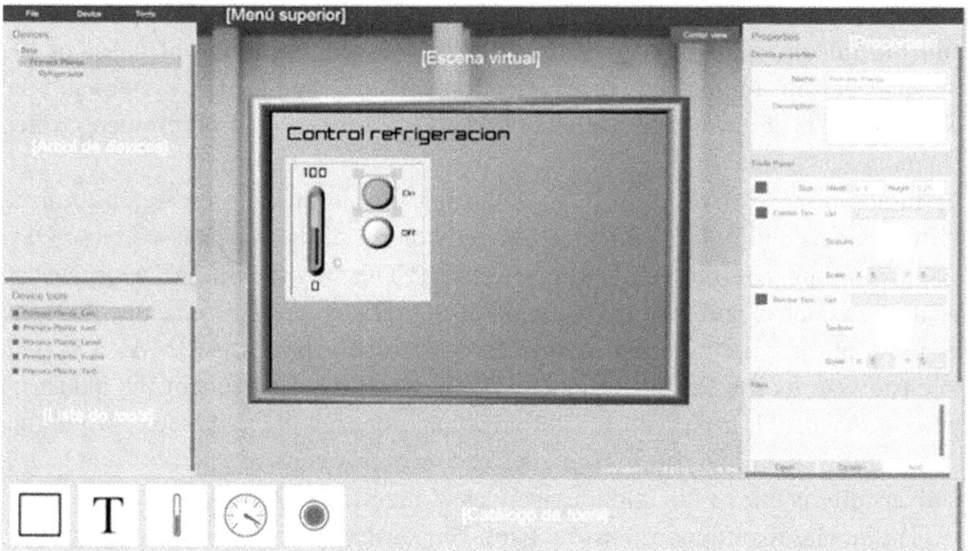

that are, nonetheless, designed to be interoperable and work together. The authoring tool is used to create virtual panels that represent the status of some SCADA-compatible machine. These virtual panels allow also to control and modify the machine's status by means of virtual input controls. On the other hand, the Mobile AR application is used to integrate these virtual panels on top of a real-time video feed of the SCADA machine, by means of the camera of the mobile device in which the client application runs.

One of the differentiating features of the SIMARA system is that it employs a dual computer vision system to implement the AR system (see Figure 3). Although the use of both vision systems is not strictly necessary, it simplifies the AR-application and makes it more robust, since we are aware that some AR systems fail to be successfully used in industrial environments due to a lack of robustness.

Figure 4 shows a diagram of the operation of the SIMARA system. Four elements can be identified in the figure: the SCADA-compatible machine that needs to be supervised, the mobile client device, the SIMARA server and the editor. The editor is in charge of creating the virtual panels and linking their elements to the SCADA signals, by means of the proposed authoring tool. The created panels are stored in the SIMARA server. This server accesses the SCADA system through web services, and manages the information of the virtual panels, so that the mobile client obtains both the definition of the panels and the values generated by the SCADA system. Each SCADA-based machine incorporates OPC devices in charge of generating the SCADA signals. This way, when a user places the mobile application in front

Figure 3. The dual (QR code + AR marker) identification system in a sewage treatment plan

of an actual machine, he/she can observe the virtual panels on top of the view of the machine taken from the camera of the mobile device, simply by pointing to the markers. The SCADA signals will then be translated to various types of visual indicators (needles, levels, lights, etc.) and visualized over the real machine in the appropriate location by means of the AR paradigm.

The web service is implemented in Visual Studio .NET 2014 (C#, ASPX). The configuration data reside in a database. This SIMARA database contains, among other elements, information on the variables that need to be monitored and the polling time of each of the signals (refresh rate of the corresponding SCADA signals' values). When a mobile application makes a request to obtain the values of the panel indicators, the web service creates a single query with all the signals present in the panel (depending on the polling time), so that the number of queries to the SCADA server is reduced. The values of the signals are processed in the web service, updated in the SIMARA database and finally sent to the mobile application. The identifiers of each of the signals are stored in the database based on the configuration created with the authoring tool, which is explained next. This enables to link the virtual panel elements with the SCADA signals.

Figure 4. The four main components of the SIMARA system

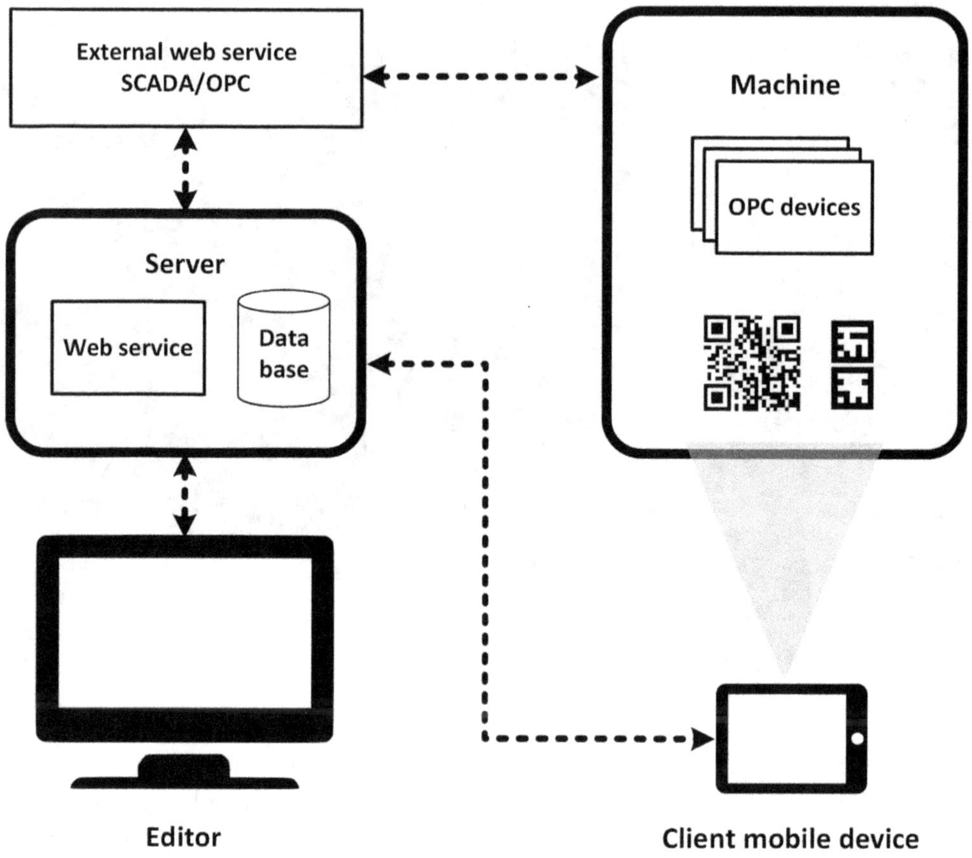

Authoring Tool

The authoring tool consists of a desktop application that allows the creation of virtual panels (called *faceplates*) associated with machinery, in such a way that the user can create panels on which to display the information collected by the SCADA system in real time. The user can choose the way the visualization of the information is performed. Several options are available so that this visualization can be as similar as possible to a real control panel. For instance, for a SCADA signal indicating the temperature from 0 to 100 units, a vertical level can be used to emulate a thermometer (see Figure 2), so that the user quickly understands the state of that particular sensor.

The authoring tool incorporates five types of controls: images, text, levels, needle indicators (gauges) and glowing colored buttons. It also incorporates a graphical interface through which the author is able to obtain a visual representation of the final result of the edition. Therefore, the authoring tool is a WYSIWYG application (What You See Is What You Get).

All the controls are visually configurable and can be associated with different OPC signals, in order to show or modify the value of each particular SCADA signal. Simple formulas can be applied in order to adjust the range of values to the desired one, so that different ranges can be created to generate different outputs for the same or different signals. The OPC signals are associated to the different virtual controls by using the name of each signal in the OPC server. All editing is done online through queries and requests to the web service. Therefore, the authoring tool acts as a manipulator of the information stored in the server. Thanks to this idea, several users can work together in the creation of virtual panels, although concurrent access to the same panel has been limited; that is, two authors are not allowed to modify the same panel at the same time, but they can edit different panels simultaneously. This way, concurrency problems are eliminated.

The authoring tool is implemented using Unity (Unity Technologies, 2018) with the C# programming language for the scripting part of the application. Thus, a sizeable part of the implementation is reusable and can be shared between the editor and the mobile client application, reducing the implementation time and increasing the coherence between the two applications. The authoring tool presents five screens, which can be accessed from the main screen: 1) "User management", where users who have access to machine information (from mobile applications) are managed; 2) "Edit faceplates" where the virtual panels are created; 3) "Project management", where the different projects are managed (a project in the SIMARA context is a set of users and virtual devices); 4) "Other options", where users can configure different options of the web services; 5) "Edit project devices", which is accessed from the "Project management" screen and it is used for allowing users to instantiate the faceplates previously created.

Mobile Augmented Reality Application

The mobile application is designed and implemented with the purpose of allowing the *in situ* control and supervision of SCADA machines in industrial processes. It is important to emphasize this particular feature, because it is not uncommon in SCADA machines to perform their supervision remotely (from the control room) and not *in situ*, since an *in situ* inspection of the machine without the use of mobile devices is typically unfeasible, due to the lack of actual indicators in the machine. Thanks to the AR paradigm, this *in situ* inspection can be performed with a strong spatial connection between the supervising software and the supervised hardware (the electrical machine).

The SIMARA mobile application has the novelty, compared to other works proposed in the literature, of using a dual computer vision system: on the one hand, we use QR codes to identify the machine, so that it is possible to retrieve and later communicate information in real-time through web services (the information is extracted and provided from/to the SCADA network) about the supervised machine; on the other hand, we use traditional AR markers (only once the machine data is loaded) to display the information in an augmented form without severely restricting the freedom of movement of the user. This way, QR codes can be quite small since users are able to approach them to read the information, whereas the 3D registration process is performed by means of traditional AR markers. These AR markers may be of even a smaller size than the QR codes without the loss of registration capabilities that this would imply in case of using QR codes. ARToolkit is used to implement the tracking of the user's point of view.

In addition, during the *in situ* supervision of the machine with the mobile application, when the vision of the AR marker is momentarily lost, the visualization of the information is kept and it is simply shown as in a Virtual Reality (VR) application. This means that, in this case, the application would be showing the information in a virtual way without being aligned with the real world, nevertheless offering a completely understandable visualization to the user. In this display mode the user can modify the size and position of the virtual panel using the tactile interface of the mobile device, as it would do with any image in a conventional viewer (using two fingers to zoom, one to move, etc.).

Figure 5 shows a basic outline of the available screens in SIMARA. The first screen is an identification screen in which the user must enter his/her username and password. Depending on the user credentials, the web service returns a list of projects which the user can access. In each project, the user obtains a hierarchical view of all the virtual devices, so that any of these devices may be chosen and its corresponding virtual panel be visualized. After the authentication process is successfully performed, the user is offered the option of scanning a QR code or directly accessing a particular virtual panel (provided that the user possesses sufficient security credentials to access the information of the panel). In addition, the elements of each panel have a certain level of configurable security restrictions so that users with greater privileges can observe or activate hidden controls that users with fewer privileges cannot do. Once the user is viewing a virtual panel, he/she can return to the previous screens to select a different panel, or they can simply point to a different QR code, since the identification of the QR codes is performed continuously in the background. This continuous detection has a double purpose: to facilitate the change between devices (by simply pointing to a new QR code) and make sure that the right information is always being shown (it is important to avoid

Figure 5. Basic workflow of the SIMARA mobile application

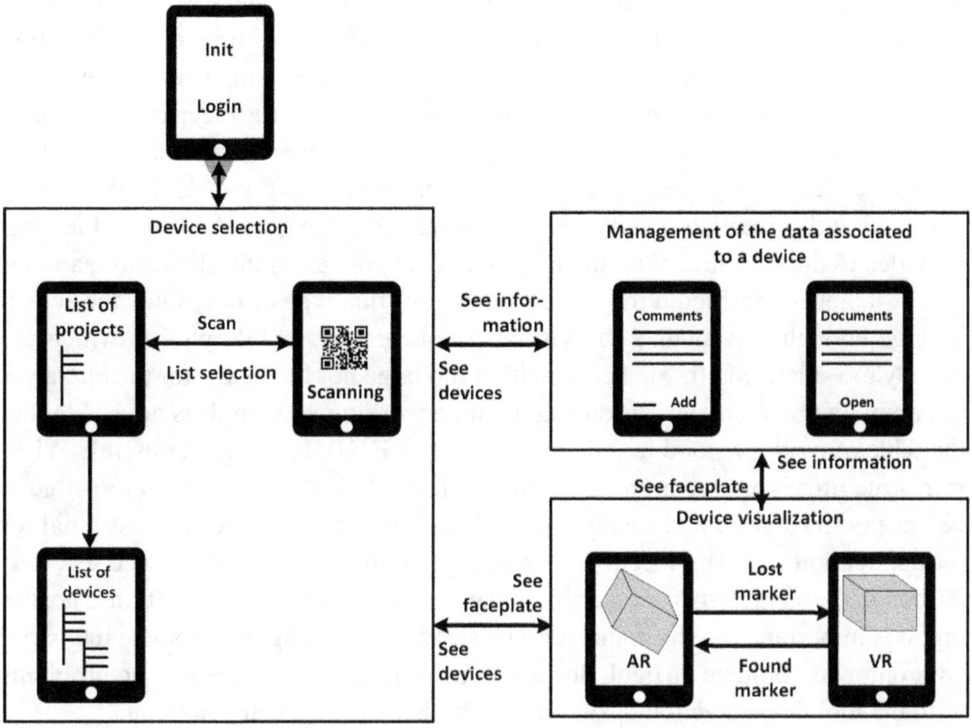

showing the information of a wrong panel because the user has forgotten to scan the QR code or select a new device).

Real Application

SIMARA is an industrial AR system that may be utilized in very different industrial processes. Its design and implementation have been conceived in a general way, so that SIMARA can be adapted to different machines and devices. This universality is one of the main strengths of the system. However, since industrial environments are particularly tough for IT applications, industrial AR applications need to prove that they can be successfully used in real environments. For this reason, it is important that SIMARA could be tested in a real environment within a particular industrial process, to check if its potential users were keen to use. In this regard, SIMARA has been applied to a sewage treatment plant in Spain, in which a SCADA control system is used to supervise the electrical machines (water pumps, collectors, etc.) needed to make the sewage treatment process work. This particular industrial plant was chosen because its control room was placed in a distant location with respect to the electrical machines that were being supervised. For this reason, supervising operators complaint that they would expend a great deal of time communicating with the SCADA control room, and their work could be simplified if they were able to perform an *in situ* supervision and maintenance of the electrical machines.

The electrical machines in the sewage treatment plant were analyzed, AR markers and QR-codes were placed in suitable locations and virtual panels for them were designed and instantiated using the authoring tool. Next, an Android tablet was provided to the operators with the mobile client application installed and ready for use. A tablet is considered the perfect device for this type of use, since the screen is large enough to visualize the virtual panels, unlike mobile phones which are usually too small, and the device is neither too large nor too heavy for the operators to complain about carrying it during a whole working session. It is advisable that the tablet provides a good mobile connection (3G/4G). Relying on possible Wi-Fi infrastructures is not recommended since industrial plants are typically too large to be covered by a good and steady Wi-Fi signal. Internet connection is essential for the application to work, since it is necessary to communicate with the web services. Although the bandwidth necessary to make a smooth use of the application is not too big, it is important that the connection be steady, something that in some industrial environments could be difficult due to electrical noise. Users reported no problems with the mobile connection during its use in the sewage treatment plant.

In the absence of further statistical experiments that quantify precisely the amount of user satisfaction and increase in task performance, the operators of the plant reported no problems and they kept using the application for their work after the test experience was finished. Although further experiments are necessary (using different industrial processes, machines and panels), we can confidently say that SIMARA represents a successful application of AR in industrial environments.

CONCLUSION AND FUTURE WORK

Electrical machine supervision is a very important task. Nowadays, machines are used everywhere, and our daily life depends on machines that need to be running twenty four hours a day. This dependency is likely to grow. For this reason, it is increasingly important to be able to articulate mechanisms to control, supervise and perform proper maintenance of these machines, especially those used in critical industrial process, like power plants, nuclear facilities, sewage treatment plants, etc.

SCADA is one of the technologies that eases the operation and supervision of electrical machines. However, the absence of a spatial connection between the SCADA signals and the machines being supervised, suggests that a different inspection paradigm be proposed or an adaptation of the existing methods could be applied. This is where AR could be of great help and where the SIMARA project is framed. SIMARA features a dual computer-vision system (QR-code + AR marker) that provides a robust client application for the integration of AR and SCADA signals by means of virtual panels shown on top of real SCADA machines. An authoring tool is also provided in order for users to customize the application to their particular needs. To the best of our knowledge, SIMARA is the first complete Mobile AR application to offer these features in combination with SCADA.

Despite AR technology has been used in many diverse application fields and scientific knowledge areas, including various industrial processes, its combination with SCADA systems is still quite novel, with very few cases described by the scientific community. Although the work presented in this chapter shows the advantages of the combination of these two technologies in a specific case with SCADA-based electrical machines, we believe that it can be extended to other industrial processes. We hope that this chapter opens this research avenue so that many more successful AR applications are proposed and effectively used in industrial environments.

Although the presented tools are fully functional, it is possible to continue the work of the SIMARA project and improve it. For instance, the authoring tool could include additional visual indicators and input gauges. In addition, the mobile client AR application could also be improved by means of natural markers instead of using artificial AR markers. From a methodological point of view, it would be very interesting to perform statistical experiments to quantify the performance and value of the system, although user satisfaction seems high. Nevertheless, it would be of interest to test the system in different industrial plants and processes where the requirements and working conditions could be different. Although SIMARA has been designed in a general way, it is not unreasonable to think that particular industries would need particular features that suggest modifications or adaptations of the SIMARA system to fit in some industrial process. Machine supervision is a broad topic. Different machines could have different requirements in terms of safety, time, performance, noise, economic exploitation, etc. that may suggest changes or improvements in the client or authoring tool.

REFERENCES

Bimber, O., & Raskar, R. (2005). *Spatial Augmented Reality: Merging Real and Virtual Worlds*. A. K. Peters, Ltd. doi:10.1201/b10624

Birkfellner, W., Figl, M., Huber, K., Watzinger, F., Wanschitz, F., Hummel, J., ... Ewers, R. (2002). A head-mounted operating binocular for augmented reality visualization in medicine-design and initial evaluation. *IEEE Transactions on Medical Imaging, 21*(8), 991–997. doi:10.1109/TMI.2002.803099 PMID:12472271

Boyer, S. A. (2009). *SCADA: supervisory control and data acquisition*. International Society of Automation.

Calderita, L. V., Bustos, P., Mejías, C. S., Fernández, F., Viciana, R., & Bandera, A. (2015). Asistente Robótico Socialmente Interactivo para Terapias de Rehabilitación Motriz con Pacientes de Pediatría. *Revista Iberoamericana de Automática e Informática Industrial RIAI, 12*(1), 99–110. doi:10.1016/j.riai.2014.09.007

Caudell, T. P., & Mizell, D. W. (1992). Augmented reality: An application of heads-up display technology to manual manufacturing processes. *Proceedings of the Twenty-Fifth Hawaii International Conference on*. 10.1109/HICSS.1992.183317

Chi, H.-L., Kang, S.-C., & Wang, X. (2013). Research trends and opportunities of augmented reality applications in architecture, engineering, and construction. *Automation in Construction, 33*, 116–122. doi:10.1016/j.autcon.2012.12.017

Díaz, D. J., Boj, C., & Portalés, C. (2016). Hybridplay: A new technology to foster outdoors physical activity, verbal communication and teamwork. *Sensors (Basel), 16*(4), 586. doi:10.339016040586 PMID:27120601

Dunleavy, M., & Dede, C. (2014). *Augmented reality teaching and learning. In Handbook of research on educational communications and technology* (pp. 735–745). Springer. doi:10.1007/978-1-4614-3185-5_59

Galatis, P., Gavalas, D., Kasapakis, V., Pantziou, G., & Zaroliagis, C. (2016). Mobile Augmented Reality Guides in Cultural Heritage. *Proceedings of the 8th EAI International Conference on Mobile Computing, Applications and Services*. 10.4108/eai.30-11-2016.2266954

Gimeno, J., Morillo, P., Casas, S., & Fernández, M. (2011). An augmented reality (AR) CAD system at construction sites. In *Augmented Reality-Some Emerging Application Areas*. InTech.

Gimeno, J., Portalés, C., Coma, I., Fernández, M., & Martínez, B. (2017). Combining Traditional and Indirect Augmented Reality for Indoor Crowded Environments. A Case Study on the Casa Batlló Museum. *Computers & Graphics, 69*, 92–103. doi:10.1016/j.cag.2017.09.001

Giner Martínez, F., & Portalés Ricart, C. (2005). *The Augmented User: A Wearable Augmented Reality Interface.* Paper presented at the International Conference on Virtual Systems and Multimedia (VSMM'05), Ghent, Belgium.

Ishii, H., Kobayashi, M., & Arita, K. (1994). Iterative design of seamless collaboration media. *Communications of the ACM*, *37*(8), 83–97. doi:10.1145/179606.179687

Kaufmann, H., & Schmalstieg, D. (2003). Mathematics and geometry education with collaborative augmented reality. *Computers & Graphics*, *27*(3), 339–345. doi:10.1016/S0097-8493(03)00028-1

Kilgus, T., Heim, E., Haase, S., Prüfer, S., Müller, M., Seitel, A., ... Schlemmer, H.-P. (2015). Mobile markerless augmented reality and its application in forensic medicine. *International Journal of Computer Assisted Radiology and Surgery*, *10*(5), 573–586. doi:10.100711548-014-1106-9 PMID:25149272

Kounavis, C. D., Kasimati, A. E., & Zamani, E. D. (2012). Enhancing the tourism experience through mobile augmented reality: Challenges and prospects. *International Journal of Engineering Business Management*, *4*, 10. doi:10.5772/51644

Linaza, M. T., Gutierrez, A., & García, A. (2013). *Pervasive augmented reality games to experience tourism destinations. In Information and Communication Technologies in Tourism 2014* (pp. 497–509). Springer. doi:10.1007/978-3-319-03973-2_36

Meža, S., Turk, Ž., & Dolenc, M. (2014). Component based engineering of a mobile BIM-based augmented reality system. *Automation in Construction*, *42*, 1–12. doi:10.1016/j.autcon.2014.02.011

Milgram, P., & Kishino, F. (1994). A taxonomy of mixed reality visual displays. *IEICE Transactions on Information and Systems*, *77*(12), 1321–1329.

Portalés, C., Gimeno, J., Casas, S., Olanda, R., & Giner, F. (2016). Interacting with augmented reality mirrors. In J. Rodrigues, P. Cardoso, J. Monteiro, & M. Figueiredo (Eds.), *Handbook of Research on Human-Computer Interfaces, Developments, and Applications* (pp. 216–244). IGI-Global. doi:10.4018/978-1-5225-0435-1.ch009

Portalés, C., Viñals, M. J., Alonso-Monasterio, P., & Morant, M. (2010). AR-Immersive Cinema at the Aula Natura Visitors Center. *IEEE MultiMedia*, *17*(4), 8–15. doi:10.1109/MMUL.2010.72

Sheikh, M. A., Nor, N. M., Ibrahim, T., & Bakhsh, S. T. (2017). An Analytical and Experimental Approach to Diagnose Unbalanced Voltage Supply. *Arabian Journal for Science and Engineering*, 1–12.

Sheikh, M. A., Nor, N. M., Ibrahim, T., & Irfan, M. (2017). Unsupervised on-line method to diagnose unbalanced voltage in three-phase induction motor. *Neural Computing & Applications*, 1–16.

Tang, S.-L., Kwoh, C.-K., Teo, M.-Y., Sing, N. W., & Ling, K.-V. (1998). Augmented reality systems for medical applications. *IEEE Engineering in Medicine and Biology Magazine*, *17*(3), 49–58. doi:10.1109/51.677169 PMID:9604701

Tian, Y., Guan, T., & Wang, C. (2010). Real-time occlusion handling in augmented reality based on an object tracking approach. *Sensors (Basel)*, *10*(4), 2885–2900. doi:10.3390100402885 PMID:22319278

Unity Technologies. (2018). *Unity*. Retrieved from https://unity3d.com/es

Vargas, H., Sánchez, J., Jara, C. A., Candelas, F., Reinoso, O., & Díez, J. L. (2010). Docencia en Automática: Aplicación de las TIC a la realización de actividades prácticas a través de Internet. *Revista Iberoamericana de Automática e Informática Industrial RIAI*, *7*(1), 35–45. doi:10.1016/S1697-7912(10)70006-6

Vera, L., Gimeno, J., Coma, I., & Fernández, M. (2011). *Augmented mirror: interactive augmented reality system based on kinect.* Paper presented at the IFIP Conference on Human-Computer Interaction. 10.1007/978-3-642-23768-3_63

Wu, H.-K., Lee, S. W.-Y., Chang, H.-Y., & Liang, J.-C. (2013). Current status, opportunities and challenges of augmented reality in education. *Computers & Education*, *62*, 41–49. doi:10.1016/j.compedu.2012.10.024

KEY TERMS AND DEFINITIONS

3D Registration: The process by which AR applications can obtain a reference spatial framework to place the virtual objects so that they match the expected location with respect to the real ones.

Augmented Reality: A technology that allows seamlessly displaying virtual information blended with the real world with a spatial connection between the virtual and the real objects.

Electrical Machine: A device that uses electrical or electromagnetic power to control or perform any kind of intended action.

Interaction: A kind of action that occurs as two or more objects have an effect upon one another.

Mobile AR: A special form of AR in which mobile devices (smartphones, tablets) are used to visualize real and virtual content. The view of the real word is typically obtained by means of the mobile device's camera.

SCADA: Supervisory Control and Data Acquisition. A protocol and control system architecture that allows performing remote monitoring, control and supervision of electrical machines through a networked data communication system. It is mostly used in industrial environments.

Supervision: The action, process, or occupation of watching a person or activity in order to make certain that everything is performed correctly or safely.

Chapter 6
Invasive Methods to Diagnose Stator Winding and Bearing Defects of an Induction Motors

Muhammad Aman Sheikh
Universiti Teknologi PETRONAS, Malaysia

Nordin B. Saad
Universiti Teknologi PETRONAS, Malaysia

Nursyarizal Mohd Nor
Universiti Teknologi PETRONAS, Malaysia

Sheikh Tahir Bakhsh
King Abdulaziz University, Saudi Arabia

Muhammad Irfan
Najran University, Saudi Arabia

ABSTRACT

Condition monitoring in an induction motor must concentrate on the root causes of the failure modes that exhibit a slow failure sequence. According to published surveys, two-fifths of the faults are due to bearing failures. Inter-turn short circuit faults in stator windings are approximately responsible for one-third of the motor faults. In the last few decades, various methods and alternative techniques have been proposed and implemented to diagnose induction motor faults. In an induction motor, stator winding and bearing faults account the largest percentage of motor failure. Due to the fact these faults can lead the motor to catastrophic failure that are expensive in term of maintenance cost, wastage raw material, and unplanned shutdown. Thus, to diagnose the state of motor and overcome existing problem, the chapter provides detailed invasive methods which are proposed and are currently in practice. Moreover, the chapter also highlights the limitation, scope, and the challenges of existing invasive condition monitoring techniques.

DOI: 10.4018/978-1-5225-6989-3.ch006

INTRODUCTION

Fault diagnosis of the stator winding and bearing defects has been extensively studied. Some of the approaches are based on laboratory tests while other approaches are based on experimental results and analysis through simulation performance (Drif & Cardoso, 2014; Das et al., 2014; Soualhi, Medjaher & Zerhouni, 2015; Se et al., 2015). There have been many papers refereed and analysed by the researchers for these faults. Stator winding and bearing fault diagnosis methods can effectively divided into two main categories: invasive and non-invasive. This chapter presents all the existing invasive methods which are used for condition monitoring and fault diagnosing of an induction motor.

INVASIVE FAULT DIAGNOSIS METHODS

The faults have various effects on physical, electrical, and magnetic properties of the induction motor. Thus, the variation in the properties or behaviour of the motor help to extract the fault index. The fault can be properly identified by detecting and measuring the effects on the properties of the motor. Mostly in invasive fault diagnosis methods, the fault index detection and measurement require special sensors to be installed on the frame of the induction motor. Next subsection introduces the approaches adopted in invasive method to identify the fault.

Acoustic Noise

Acoustic noise is a transient flexible signal that is generated due to permanent deformation of an equipment and these signals can propagate through solid materials. The frequencies associated with acoustic noise are beyond the audible range 20 KHz. Winding and bearing faults can be identified by acoustic noise (Kharche & Kshirsagar, 2014; Germen, Basaran & Fidan, 2014; Mirafzal, Povinelli & Demerdash, 2006). Acoustic noise spectrum of induction motor includes electromagnetic motive (EM) and air conditioning noise. The air conditioning noise relates the distortion of periodic air pressure affected by the rotational parts. While the EM noise refers Maxwell stress on the metal surfaces in the presence of a magnetic field. These stresses in the presences of a magnetic field, induce vibration in the structure of the induction motor that in result emits noise radiation. The mechanical noise and aerodynamic noise are in the range of 12 dB per ester of the speed of induction motor. The higher the speed of induction motor the higher will be EM noise. A limitation of this technique is the requirement of special sensor and acquisition system that should capture acoustic signals with sufficient resolution because the sampling rate is very high typically 2 MHz. This subsequent constraint the diagnosis system to employ more advanced and expensive acquisition systems. Therefore, for long term installations of the senor with limited lifetime in order to manage large volumes of data with extremely fast data rate should be proposed to deal such a huge data rates. Moreover, this technique does not highlight the segregation issue of the stator winding and bearing raceways defects.

Vibration

The vibration can even be observed for entirely healthy motor due to consequences of manufacturing imperfection such as bearing roughness, the slight deviation in the degree of tolerance, poor assembling of motor parts, which can lead to misalignments. Whereas, the frame of vibration also depends on the stator winding faults, single phasing (inter-turn short circuit), supply voltage imbalance, and bearing defects (Seshadrinath, Singh & Panigraphi, 2014; Delgado-Arredondo et al., 2015; Filho, Pederiva & Brito, 2014). Resonance among the stator and the electromagnetic force is the other cause of vibration in electrical machines. Over the past few decades, the detection of a stator and bearing faults through vibration signal has been the subject of large volumes of research. Where most of the work has been conducted to diagnose bearing related faults through continuous monitoring of the bearings. The characteristic fault frequencies in the machine vibration spectrum depending on which bearing surface contains the fault are listed in equations (1) to (3) (Harmouche, Delpha & Diallo, 2015; Leite et al., 2015; Wang, 2007).

Outer raceway localized defect frequency:

$$F_{outR} = \frac{N_b}{2} F_r (1 - \frac{BD cos\beta}{PD}) \tag{1}$$

Inner raceway localized defect frequency:

$$F_{inrR} = \frac{N_b}{2} F_r (1 + \frac{BD cos\beta}{PD}) \tag{2}$$

Rolling ball defect frequency:

$$F_{bll} = \frac{PD}{2BD} F_r (1 - \frac{BD^2 \cos^2 \beta}{PD^2}) \tag{3}$$

where N_b is the number of rolling balls, BD is the diameter of balls, PD is the pitch diameter of the bearing, and F_r is the rotor frequency. The detection of vibration signals is facilitated through accelerometers or vibration sensors which are mostly mounted directly to winding or the bearing of interest. However, this is impractical to mount the sensor for such a place where the access is limited and the signal cannot be routed to acquisition system. More often, the sensors are mounted on the outer casing of the machine. For this, multiple sensors should be employed, mounted near or exactly at the areas of interest (Ali et al., 2015). However, until now there is no significant approach based on vibration is found related to segregation of raceways and stator winding faults.

Temperature

Historically, in electrical machines, the measurement of temperature is widely employed to assess operational stability. Under steady operation, the temperature of winding and bearing is relatively constant but will rise if the factor of winding or bearing operator deviates from a reference frame. This can be determine through the motor performance. In the case of bearing it is linked to failure in the lubricant that reduces boundary lubricated regimes, increased friction and asperity interaction, finally resulting in the rising in temperature (Picazo-Rodenas et al., 2015). While in the case of a stator winding the rise in temperature will damage insulation.

Therefore, thermal protection is the most important features for the fortification of induction motor (Blackburn & Domin, 2015). The stator winding insulation and bearing of an induction motor is too vulnerable against temperature (Garcia-Ramirez et al., 2014; Eftekhari et al., 2013). In order to prolong the longevity of induction motor, its temperature must be monitored accurately. At the incipient stage the temperature index has low reliability and less accuracy to diagnosis the stator winding faults or bearing faults. Environmental temperature is another factor that disturbs cooling system and makes this index less reliable. There may be another factor that rises temperature such as supply voltage imbalance, overloading, and frequent switching of a motor. In order to detect and measure this index, expensive thermocouple sensors or resistive temperature detectors are installed. Due to the size limitation, there is installation difficulty, therefore, the temperature monitoring method is a reliable approach in large machines only.

Partial Discharge

The partial discharge (PD) was known as the corona. The PD is electrical discharge due to the air holes and insulation defects. Excessive heat and weakness in the structure of the motor are two main reasons causing insulation failure and creating air holes (Ojaghi, Sabouri, Faiz, 2014). The PD can corrode the major parts of insulation and lead to exhaustion of insulation. The PD is a primary online method to diagnosis the health of the coil. This method was first introduced in 1976 and was tested on hydro-generators. For online monitoring, special sensors were developed. PD is an electrical discharge of low amplitude, ultrafast pulses that are used for online detection of the stator winding fault (Siddiqui, Sahay & Giri, 2014; Bonaldi et al., 2012). However, PD also has low reliability and less accuracy as compared to non-invasive methods, especially at an incipient stage with weak faults. Installation difficulties and the cost are other drawbacks of this index.

Gas Analysis

Variable and high-frequency creates voltage stresses. These stresses affect insulation endurance and damage winding insulation. The insulation breakdown will generate carbon monoxide. Which penetrates toward the cooling air and causes the insulation failure. The generation and indication of carbon monoxide start demolition of the insulation and appearance of the winding fault. Which can be diagnosed through infrared absorption technique to detect carbon monoxide which states the failure of a stator winding insulation (Pandey, Zope & Suralkar, 2012).

Surge Test

The surge test is also one of the techniques used to detect stator winding faults. In surge test, simultaneously two similarly frequency voltage pulses spectrum are applied to the two winding of a stator, and the reflected pulses are displayed or detected on oscilloscope (Wilson & Instruments, 2003; Xianghui et al., 2007; Warren & Stone, n.d.). The comparison among the pulses can be used to diagnose the faults between stator windings, coils, and the ground. This is a predictive technique and used to detect lack of insulation. An electronic surge test is applicable to locate insulation and asymmetric three-phase winding faults. This technique involves special equipment, it is expensive. In addition, the technique has no feature to perform the online test.

Magnetic Flux

Stator winding fault will create distortion in air-gap flux density and will set up flux. In order to sense the flux, after several inspections, it was found that there is a space in the slots of the stator. Therefore, a new coil can be mounted within the stator that acts as a sensor (Filho, Pederiva & Brito, 2014). In this manner, the new coil was inserted into the stator winding. The winding of newly installed coils were isolated from the main stator windings. It is also used to diagnose bearing faults (Ozeglin, 2008). It is an expensive approach because the flux in the air gap is sensed by the special tools and sensors.

CONCLUSION

Induction motors are work horse component of many industries and are frequently integrated into commercial and industrial available equipment. Although induction motors are reliable electric machines, they are susceptible to many electrical and mechanical faults due to installation issues and environmental factors. In an induction motor, stator winding and bearing faults account the largest percentage of motor failure. The stator winding or bearing fault has various effects on physical, electrical, and magnetic properties of the induction motor. Thus, the variation in the properties or behavior of the motor help to extract the fault index. This chapter provide detail about most of the existing invasive techniques which are used to diagnose stator winding and bearing faults. Due to the fact that there does not exist any fault characteristics frequency model of few fault related to the motor. Therefore, this chapter also highlight the scope and limitation of existing techniques.

ACKNOWLEDGMENT

The authors acknowledge the support of Ministry of Education Malaysia and the Universiti Teknologi PETRONAS Malaysia through the awards of Prototype Research Grant Scheme PRGS/1/2016/TK04/UTP/02/8, YUTP-Fundamental Research Grant and the research facilities provided.

REFERENCES

Ali, J. B., Fnaiech, N., Saidi, L., Chebel-Morello, B., & Fnaiech, F. (2015). Application of empirical mode decomposition and artificial neural network for automatic bearing fault diagnosis based on vibration signals. *Applied Acoustics*, *89*, 16–27. doi:10.1016/j.apacoust.2014.08.016

Blackburn, J. L., & Domin, T. J. (2015). *Protective relaying: principles and applications*. CRC press.

Bonaldi, E. L., Lambert-Torresm, G., da Silva, J. G. B., de Oliveira, L. E. L., & da Silva, L. E. B. (2012). *Predictive Maintenance by Electrical Signature Analysis to Induction Motors*. INTECH Open Access Publisher.

Das, S., Purkait, P., Koley, C., & Chakravorti, S. (2014). Performance of a load-immune classifier for robust identification of minor faults in induction motor stator winding. *Dielectrics and Electrical Insulation. IEEE Transactions on*, *21*, 33–44.

Delgado-Arredondo, P. A., Garcia-Perez, A., Morinigo-Sotelo, D., Osornio-Rios, R. A., Avina-Cervantes, J. G., Rostro-Gonzalez, H., & Romero-Troncoso, R. J. (2015). Comparative study of time-frequency decomposition techniques for fault detection in induction motors using vibration analysis during startup transient. *Shock and Vibration*, *2015*, 1–14. doi:10.1155/2015/708034

Drif, M. H., & Cardoso, A. J. M. (2014). Stator fault diagnostics in squirrel cage three-phase induction motor drives using the instantaneous active and reactive power signature analyses. *IEEE Transactions on Industrial Informatics*, *10*(2), 1348–1360. doi:10.1109/TII.2014.2307013

Eftekhari, M., Moallem, M., Sadri, S., & Hsieh, M. F. (2013). A novel indicator of stator winding inter-turn fault in induction motor using infrared thermal imaging. *Infrared Physics & Technology*, *61*, 330–336. doi:10.1016/j.infrared.2013.10.001

Filho, P. L., Pederiva, R., & Brito, J. (2014). Detection of stator winding faults in induction machines using flux and vibration analysis. *Mechanical Systems and Signal Processing*, *42*(1-2), 377–387. doi:10.1016/j.ymssp.2013.08.033

Garcia-Ramirez, A. G., Morales-Hernandez, L. A., Osornio-Rios, R. A., Benitez-Rangel, J. P., Garcia-Perez, A., & de Jesus Romero-Troncoso, R. (2014). Fault detection in induction motors and the impact on the kinematic chain through thermographic analysis. *Electric Power Systems Research*, *114*, 1–9. doi:10.1016/j.epsr.2014.03.031

Germen, E., Başaran, M., & Fidan, M. (2014). Sound based induction motor fault diagnosis using Kohonen self-organizing map. *Mechanical Systems and Signal Processing*, *46*(1), 45–58. doi:10.1016/j.ymssp.2013.12.002

Harmouche, J., Delpha, C., & Diallo, D. (2015) Improved fault diagnosis of ball bearings based on the global spectrum of vibration signals. *IEEE Transactions on Energy Conversion, 30*, 376-383.

He, W., Zi, Y., Chen, B., Wu, F., & He, Z. (2015). Automatic fault feature extraction of mechanical anomaly on induction motor bearing using ensemble super-wavelet transform. *Mechanical Systems and Signal Processing*, *54*, 457–480. doi:10.1016/j.ymssp.2014.09.007

Kharche, P. P., & Kshirsagar, S. V. (2014). Review of Fault Detection in Rolling Element Bearing. *Int. J. of Innovative Research in Advanced Engineering*, *1*, 169–174.

Leite, V. C., da Silva, J. G. B., Veloso, G. F. C., da Silva, L. E. B., Lambert-Torres, G., & Bonaldi, E. L. (2015). Detection of localized bearing faults in induction machines by spectral kurtosis and envelope analysis of stator current. *IEEE Transactions on Industrial Electronics*, *62*(3), 1855–1865. doi:10.1109/TIE.2014.2345330

Mirafzal, B., Povinelli, R. J., & Demerdash, N. A. (2006). Interturn fault diagnosis in induction motors using the pendulous oscillation phenomenon. *IEEE Transactions on Energy Conversion*, *21*(4), 871–882. doi:10.1109/TEC.2005.853767

Ojaghi, M., Sabouri, M., & Faiz, J. (2014). Diagnosis methods for stator winding faults in three-phase squirrel-cage induction motors. *International Transactions on Electrical Energy Systems*, *24*(6), 891–912. doi:10.1002/etep.1750

Ozelgin, I. (2008). Analysis of magnetic flux density for airgap eccentricity and bearing faults. *International Journal of Systems Applications, Engineering & Development*, *2*, 162–169.

Pandey, K., Zope, P., & Suralkar, S. (2012). Review on fault diagnosis in three-phase induction motor. In MEDHA–2012, Proceedings published by International Journal of Computer Applications. IJCA.

Picazo-Ródenas, M. J., Antonino-Daviu, J., Climente-Alarcon, V., Royo-Pastor, R., & Mota-Villar, A. (2015). Combination of noninvasive approaches for general assessment of induction motors. *IEEE Transactions on Industry Applications*, *51*(3), 2172–2180. doi:10.1109/TIA.2014.2382880

Seshadrinath, J., Singh, B., & Panigrahi, B. K. (2014). Vibration analysis based interturn fault diagnosis in induction machines. *IEEE Transactions on Industrial Informatics*, *10*(1), 340–350. doi:10.1109/TII.2013.2271979

Siddiqui, K. M., Sahay, K., & Giri, V. (2014). Health Monitoring and Fault Diagnosis in Induction Motor-A Review," *International Journal of Advanced Research in Electrical. Electronics and Instrumentation Engineering*, *3*, 6549–6565.

Soualhi, A., Medjaher, K., & Zerhouni, N. (2015). Bearing health monitoring based on Hilbert–Huang transform, support vector machine, and regression. *IEEE Transactions on Instrumentation and Measurement*, *64*(1), 52–62. doi:10.1109/TIM.2014.2330494

Wang, L. (2007). *Induction motor bearing fault detection using a sensorless approach*. Texas A&M University.

Warren, V., & Stone, G. (n.d.). *Review of Electrical Tests to Assess Motor Windings*. I. P. Engineering.

Wilson, J., & Instruments, B. (2003) Current state of surge testing induction machines. *Iris Rotating Machine Conference*, 1-13.

Xianghui, H., Habetler, T. G., Harley, R. G., & Wiedenbrug, E. J. (2007). Using a Surge Tester to Detect Rotor Eccentricity Faults in Induction Motors. *Industry Applications. IEEE Transactions on*, *43*, 1183–1190.

Chapter 7
Types of Faults and Condition Monitoring Methods of Induction Motors

Maheswari M.
Malla Reddy Engineering College, India

Gunasekharan S
Malla Reddy Engineering College, India

ABSTRACT

Induction motors are the electromechanical devices used to convert electrical energy into mechanical energy and work under the principle of mutual inductance. They have stator and rotor as two major parts. They run at constant speed when the supply voltage and frequency are constant and are suited for constant speed drives. They have rugged construction but working environment causes different faults. As per IEEE and EPRI study on induction motor faults, bearing fault and stator faults are 46% and 36%, respectively. The broad categories of the fault are stator winding fault, broken rotor fault, rotor mass unbalance fault, bowed rotor faults, single phasing fault, bearing fault, and crawling. Unbalanced stator voltage and current, oscillations in torque, drop in efficiency and torque, overheating and unwarranted vibration are the effects of these faults. Undetected faults cause complete failure of motor and it is costly in terms of lost production time, maintenance cost, and wasted raw materials.

DOI: 10.4018/978-1-5225-6989-3.ch007

INTRODUCTION

Induction motors are the electro mechanical device used to convert electrical energy into mechanical energy. It works under the principle of mutual inductance such as transformer. So it is also called as Induction motor. It is also known as rotating transformer because of its similar operating principle. The invention of induction motor made most of the industrial operations as automatic process. An induction motor consists of magnetic circuits interlinking two electrical circuits placed in two main parts of the machine. The major two parts of induction motor are stator and rotor as shown in Figure 1. The stationary part of the motor is called as stator and the rotating part is known as rotor. The transmission of power takes place from one part to other through electromagnetic induction principle. In general, stator and rotor consists of two circuit namely electric circuit and magnetic circuit as discussed by Sen (1989). The electric circuit is made up of insulated copper or aluminium which is used to carry the current . The circuit made up of laminated magnetic materials used to carry magnetic flux is known as magnetic circuit is shown in Figure 2.

CONSTRUCTION

The major parts of induction motor are stator and rotor. Stator is the stationary part and rotating part is called as rotor. The parts of induction motor are

Figure 1. Cross Section View of Induction Motor

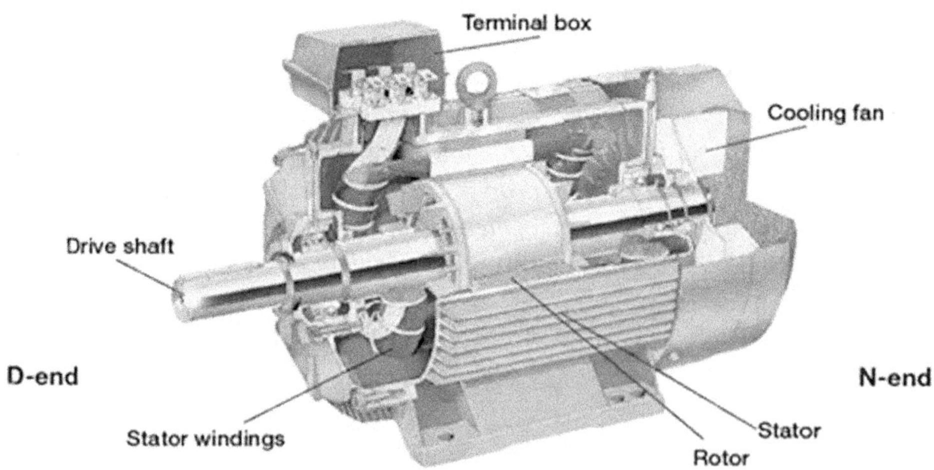

Figure 2. Magnetic Circuit of Induction Motor

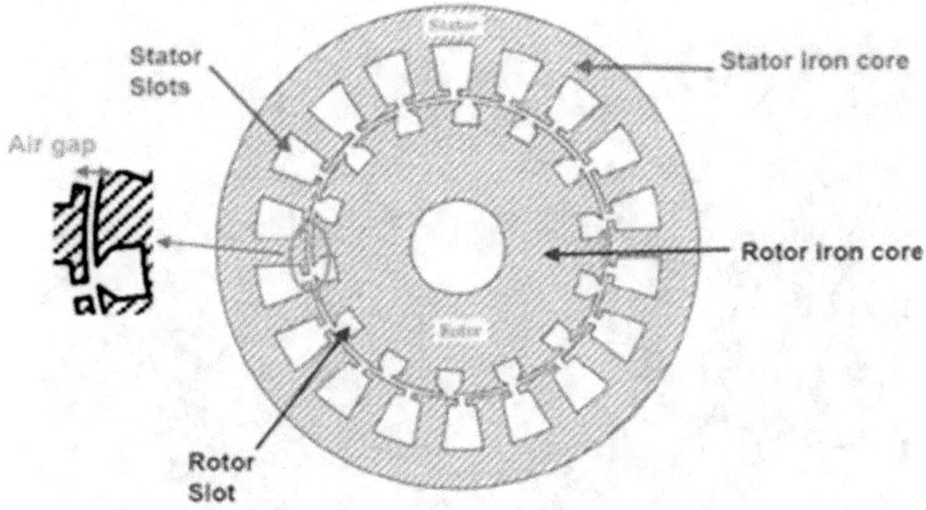

1. Stator
2. Rotor
3. End Flanges
4. Bearings
5. Shaft
6. Cooling Fan
7. Terminal Box

Let us see the structure and functions of each part in detail.

Stator

Stator is a stationary part of the motor and it consists of (a) Outer cylindrical frame (b) Magnetic path and (c) Set of insulated electrical windings as shown in Figure 3.

Outer Cylindrical Frame

Cylindrical frame is made up of cast iron or sheet steel or cast aluminium alloy. It has feet for foot mounting of the motor or a flange for mounting of the motor.

Figure 3. Stator of an Induction Motor

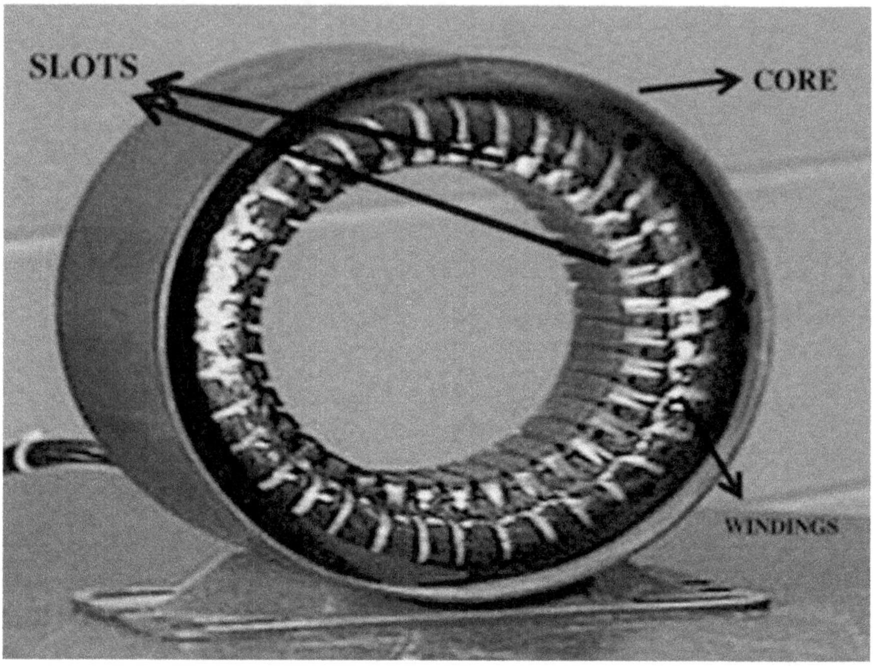

Magnetic Path

It consists of a set of slotted high grade alloy steel laminations which is supported by the stator frame. Magnetic path is laminated to reduce the eddy current loss and heating.

Set of Insulated Electrical Windings

The stator has three sets of coils one for each phase and are separated by 120° for a three phase. These windings are placed inside the slots of the laminated magnetic path.

Rotor

The rotating part of the motor is called as rotor. It is located inside the stator bore and rotates in coaxially with the stator. It is also made up of thin laminated sheets and pressed together in the form of cylinder similar to stator construction. The thin sheets are insulated from each other by means of paper and varnish. The rotor slots consists of electric circuit and cylindrical electromagnetic substance acts as magnetic path. The types of rotor winding are of squirrel cage type or wound rotor type. According to the type of rotor, the induction motors are classified as squirrel cage induction motor and slip ring induction motor as conversed by Kothari and Nagrath (2010).

Squirrel Cage Rotor Type

The rotor consists of set of heavy bars not wires made up of copper, aluminium or alloys which is embedded on the rotor slots. These bars act as rotor conductors. The bars are not parallel to the shaft but there is a skew in construction to reduce the magnetic hum and locking as discussed by Sen (1989). The rotor bars are short circuited at the two ends by means of end rings as shown in Figure 4.

Wound Rotor Type

In this case, the rotors conductors are insulated windings not the bars. The windings are not short circuited by end rings. The end of three windings brought out and connected to the slip rings mounted on the shaft as shown in Figure 5. The external electrical connections are made up of brushes located on the shaft.

Figure 4. Squirrel Cage Rotor Type

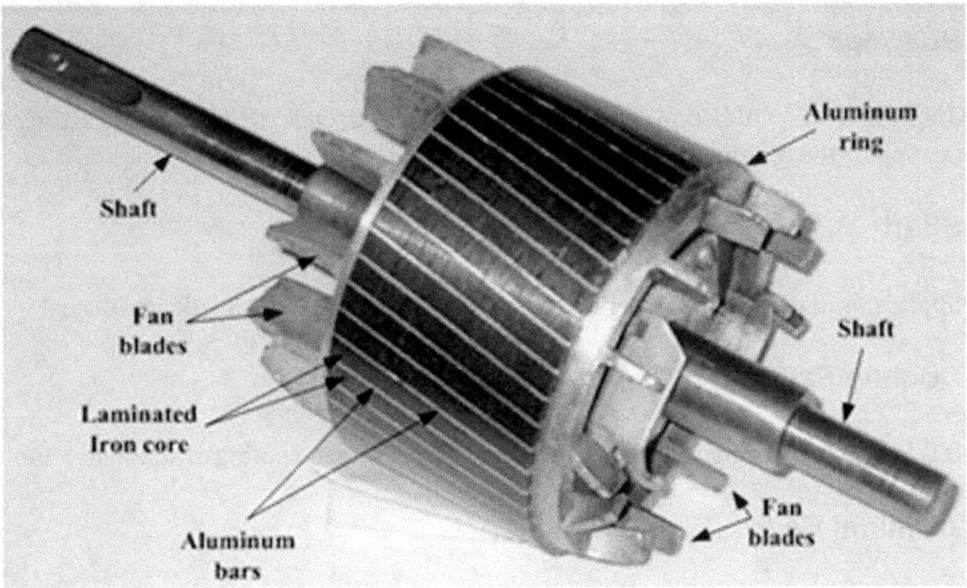

Figure 5. Slip Ring Rotor

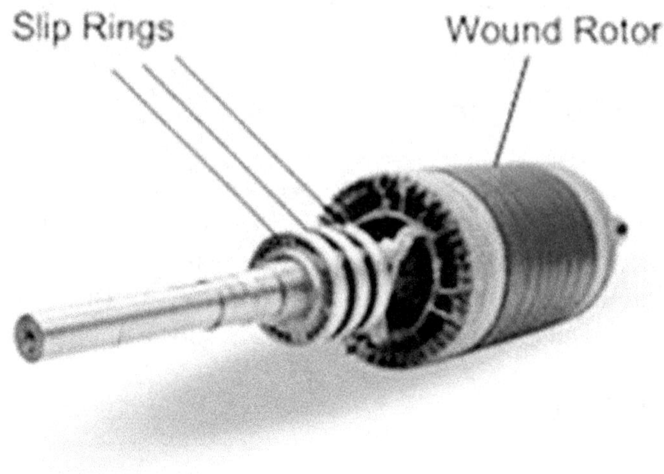

End Flanges

There are two end flanges which are used to support the two bearings on both ends of the motor.

Bearings

There are two set of bearings which are placed at both ends of the rotor and are used to support the rotating shaft.

Shaft

Shaft is used to transmit generated torque into the load and is made up of steel.

Cooling Fan

It is located in the opposite end of the load side for forced cooling of stator and rotor.

Terminal Box

It is on top or on either side of the outer cylindrical frame of stator to receive the external electrical connections.

PRINCIPLE OF OPERATION

When a three phase supply is applied to the stator of an induction motor, an uniform rotating magnetic field is produced in the stator. It will induce an emf in the rotor winding and the current is produced due to the rotor is short circuited. The rotor current will interact with the stator rotating magnetic field to develop a torque in the direction of rotating magnetic field. As the rotor is free to rotate, the torque will cause it to move round in the direction of stator field. Kothari and Nagrath (2010) had said that the above process will make the three phase induction motor as self starting . During the process of conversion of electrical energy into mechanical energy some losses will occur as follows,

- Friction and windage losses 5-15%
- Iron or Core losses 15-25%
- Stator Losses 25-40%
- Rotor Losses 15-20%
- Stray Load Losses 10-20%

Full load efficiency of the induction motor varies from 85 to 97%. Induction motors are simpler, cheaper and efficient. Squirrel cage induction motors are rugged in construction and more efficient than wound rotor type induction motor. Induction motor maintains constant speed when the supply voltage and frequency are constant. Hence it is more suitable for constant speed drives. Several standard designs of induction motor is available in the market to fulfil the requirements of different starting and running conditions of various industrial applications as discussed by Karmakar, et al. (2016) . According to this, the induction motors are classified as Class A, Class B, Class C and Class D. The properties and uses of all classes of Induction motor are listed in Table 1.

Table 1. Various Classes of Induction Motor

	Class A	Class B	Class C	Class D
Properties	Normal starting torque, high starting current and low operating slip	Normal starting torque, low starting current and low operating slip	High starting torque and low starting current	High starting torque, low starting current and high operating slip
Uses	Fan, pump load etc. where torque is low at start	For constant speed drive such as pump, blower	Compressor, conveyors, crashers etc.	For driving intermittent load, e.g. punch press etc.

INDUCTION MOTOR FAULTS, CAUSES AND EFFECTS

Induction Motors are rugged, low cost, low maintenance and reasonably high efficient. They are operating reliably but there are more possibilities for the occurrence of the faults. The most vulnerable parts of the induction motors are bearing, stator winding, rotor bar and the shaft. Also the faults occurs in the non uniform airgap between the stator and the rotor. In IEEE, there were lot of study had been carried out on various motors in industrial applications. According to the IEEE standard 493-1997 the most common fault and their statistical occurrences are listed in Table 2 and the study carried out by General Electric Company under the EPRI sponsorship is given in Table 3 as discussed by Karmakar, et al. (2016).

Faults shown in Table 3 are in broad categories. The faults in Induction motors are listed as follows,

- Broken rotor fault
- Rotor mass unbalance fault
- Bowed rotor fault
- Bearing fault

Table 2. Statistical Data on Motor Faults

Type of Faults	Number of Faults/Failures				
	Induction Motor	**Synchronous Motor**	**Wound – Rotor Motor**	**DC Motor**	**All Total Motors**
Bearing	152	2	10	2	166
Winding	75	16	6	-	97
Rotor	8	1	4	-	13
Shaft	19	-	-	-	19
Brushes or Slip ring	-	6	8	2	16
External Device	10	7	1	-	18
Others	40	9	-	2	51

Table 3. Possibilities of Fault Occurrence on Induction Motor

Studied by	Bearing Fault (%)	Stator Fault (%)	Rotor Fault (%)	Others (%)
IEEE	42	28	8	22
EPRI	41	36	9	14

- Stator winding fault
- Single phasing fault
- Crawling

There is a possibility of occurrence of more than one fault in an Induction motor at a time. Generally, these faults results in as discussed by Thorsen and Dalva (1995)

- Unbalanced stator voltage and current
- Oscillations in torque
- Reduction in efficiency
- Overheating and
- Excessive vibration.

Also the faults increase the magnitude of harmonic current and voltage which will lead to the major effects in the induction motor. In the following paragraph, the causes and effects of faults in Induction motor is going to be discussed.

Broken Rotor Fault

The squirrel cage induction motor has rotor bars and end rings in the rotor, if any one or more bars are broken partially or completely then the faults is known as broken rotor fault. The rotor and broken rotor parts are shown in Figure 6 . There

Figure 6. Rotor and Broken Rotor Bars

are more reasons for the occurrence of rotor fault in the induction more. In squirrel cage induction motor rotor asymmetry occurs due to manufacturing defect. During brazing process, non uniform stress may occurs in cage assembly which causes the failure in rotation of the motor. Also the large centrifugal forces due to heavy end rings may cause more stress on rotor bars. The above said two reasons are the causes of the broken rotor faults. The major effect of this fault is to produce asymmetrical distribution of rotor current. If the motor runs with the asymmetry for a long time, then it causes overheating at the point of crack and it may lead to break the bar. If one bar is broken, then the other side bars carry large current and it may lead to larger electrical and mechanical stresses on the side bars. The rotor continues to rotate in the same condition the rotor bars will be broken and then failure will spread to all the side bars. The crack may occur at bars, end rings or at the joining of bars and end rings. In the most of the cases, the cracks were occurred at the joining of bars and end rings as discussed by Karmakar, et al. (2016).

Causes of Broken Rotor Fault

The major causes of rotor broken bar are as follows,

- Thermal Stress
- Mechanical Stress caused by bearing faults
- Manufacturing defects
- Frequent starting of the motor
- Fatigue of the metal of the rotor bar.

Effects of Rotor Broken Fault

Cracked bars in rotor will causes the side band of frequencies in the stator current given by,

$$f_{brb} = f\left(1 \pm 2ks\right) \tag{1.1}$$

where f is the supply frequency, s is the slip and k is an integer.

From the equation 1.1, it is clear that the band of frequencies will be caused with $f = \left(1 - 2s\right)$ and $f = \left(1 + 2s\right)$ as lower and upper band of frequencies respectively.

The magnitude of the lower side band $f = (1 - 2s)$ over the fundamental will be used as an indicator for the broken rotor fault.

Rotor Mass Unbalance

From the construction of Induction motor, it is clear that the rotor is placed inside the stator bore and rotates coaxially with the stator. In a normal motor, the air gap between the stator inner core and rotor outer core is identical. If the rotor is not aligned centrally with the rotor and its axis of rotation is not same as the stator then the air gap will not be uniform and it is called as air gap eccentricity. The air gap eccentricity may occur in Induction motor due to the rotor faults like, rotor mass unbalance fault and bowed rotor fault. Due to this air gap eccentricity, there will be an unbalanced electromagnetic pull in the induction motor as discussed by Karmakar et al. (2016). The pull will be more where the air gap is small whereas it will be less in large air gap side. Hence the rotor will tend to move in the direction where the pull is more across the air gap. The rotor pull over will be more at the time of starting of an induction motor. In some cases, the rotor may rub the stator which may leads to damage or failure of the stator. Air gap eccentricity also causes vibration and noise.

General Description of Rotor Mass Unbalance

The rotor mass unbalance fault occurs mainly due to the manufacturing defects or otherwise extended period of operation. If any unsymmetrical addition or subtraction of mass in the rotor or due to misalignment or due to bending of shaft causes the change in the axis of the centre of rotation of the rotor. The rotor rubs the stator due to the air gap eccentricity which cause the rotor material body wear out and it is treated as subtraction of mass in rotor. This may leads to unbalanced rotor fault as discussed by Karmakar et al. (2016). It is shown in Figure 7.

Classification of Rotor Mass Unbalance

The rotor mass unbalance is classified into three types according to the position of rotor axis as,

1. Static mass unbalanced rotor
2. Couple unbalanced rotor
3. Dynamic unbalanced rotor

Figure 7. Rotor with Mass Unbalance Fault

Static Mass Unbalanced Rotor

If the shaft rotational axis and weight distribution axis of rotor are parallel but offset then it is called as static mass unbalanced rotor. It is shown in Figure 8 and it cannot be identified without special equipments.

Couple Unbalanced Rotor

If the shaft rotational axis and weight distribution axis of rotor are coincide at the centre of the rotor then it is known as couple unbalance rotor. It is shown in Figure 9.

Dynamic Unbalanced Rotor

If the shaft rotational axis and weight distribution axis of rotor are not coincide then it is referred as dynamic unbalanced rotor and it is a combination of static and couple unbalanced rotor.

Figure 8. Static Mass Unbalanced Rotor

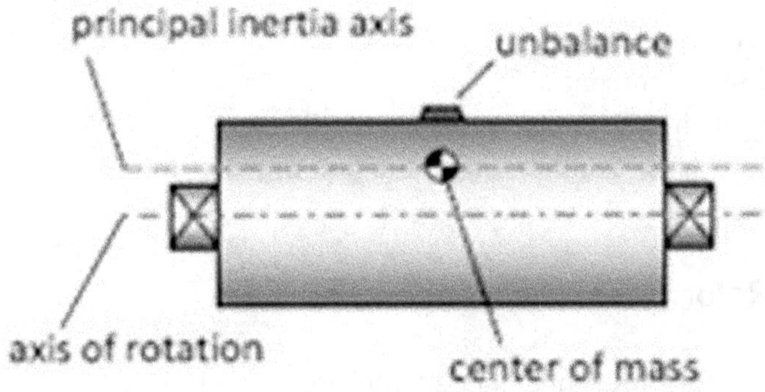

Figure 9. Couple Unbalanced Rotor

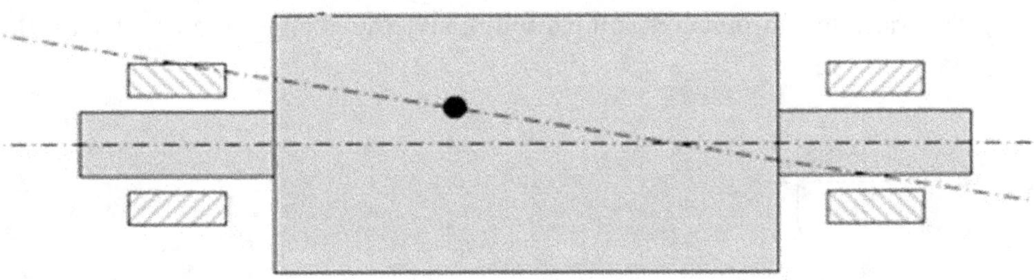

Figure 10. Dynamic Unbalanced Rotor

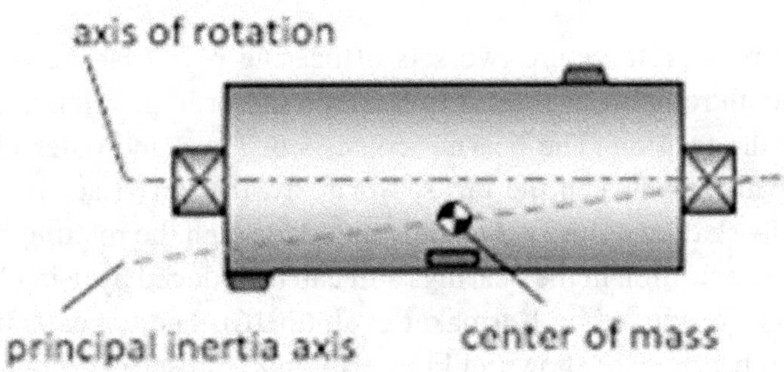

The main causes of rotor mass unbalance in an induction motor are,

- Manufacturing defects
- Internal misalignment
- Bending of shaft
- Non symmetrical addition or subtraction of mass in the rotor
- Extended period of operation

Effects of Rotor Mass Unbalanced Fault

The effects of rotor mass unbalanced faults are,

- It produces dynamic eccentricity which results in the oscillations in the air gap length.
- The oscillations in the air gap length causes variation in the air gap flux density and variation in induced voltage in the stator winding.
- The variation in induced voltage will cause the change in current with the frequency of

$$f_{ubm} = f\left(k\frac{\left(1-s\right)}{p}\right)+1 \tag{1.2}$$

where, f is the supply frequency, s is the slip, p is the number of poles and k is an integer.

Bearing Fault

In Induction motors, there are two sets of bearing placed at the two ends of the rotor for smooth rotation. It is used to held the motor in position and rotate freely by reducing the friction. The bearing consists of inner and outer races and a set of rotating balls in between the races. Normally, the inner race of the bearing is attached to the shaft and the load is transmitted through the rotating balls to reduce the friction. The friction in the bearings still can be reduced by using lubricants like grease or oil as discussed by Karmakar et al. (2016). The typical ball bearing and dissected ball bearing are shown in Figure 11 and 12 respectively.

Any damage in the inner races or outer races or in the ball is termed as bearing faults. Bearing is the weakest component and the largest cause for the fault in the induction motor. There is 41-42% of faults are occurring in the induction motor due to the bearings as per the study of IEEE and EPRI.

Figure 11. Ball Bearings

Figure 12. Ball Bearings (Dissected)

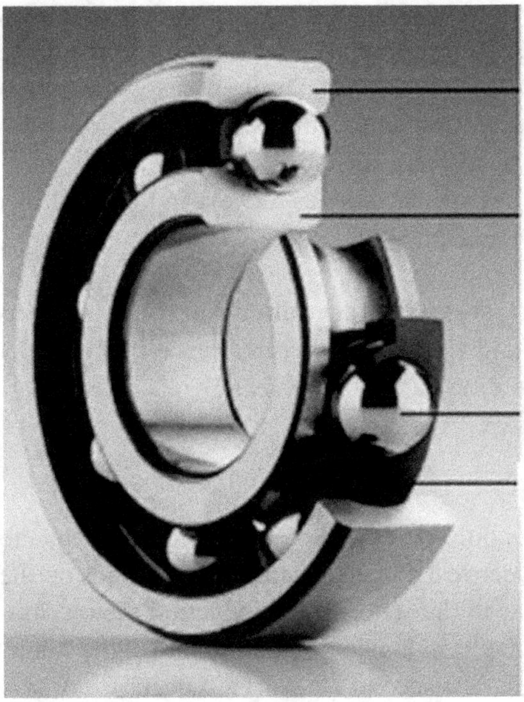

Causes and Effects of Bearings

The causes and effects of bearing fault in induction motor is listed and discussed below,

1. The two races and balls may be annealed due to the excessive load, tight fits and rise in temperature. It may also lead to degrade or destroy the lubricant. Brinelling also occurs in the bearings when the load exceeds the elastic limit of the bearing material.
2. Corrosion also takes place in bearings if it is exposed to the corrosive liquids like acids or corrosive atmosphere. It also occurs when the bearings are handled carelessly during installation. The corrosion leads the motor into early fatigue failure.
3. After a long run of the bearings fatigue failure occurs. It causes fracture and removal of small parts from the two races or balls of the bearings. It is a progressive one and it will spread throughout the bearings for every operation. It will cause more noise and vibration during operation.
4. One of the leading factor for bearing fault is contamination. Lubricants get contaminated by dirt and foreign particles which is mostly present in the industrial environment. It cause more vibration and wear.
5. During restricted flow of lubricant and excessive temperature will leads to lubricant failure. It degrades the property of lubricant and makes excessive wear in races and balls of the bearings. Due to this, temperature will increase and grease will melt and flow out of the bearings. Discoloured of the balls and ball tracks are indication for the lubricant failure.
6. Misalignment of the bearings also leads to wear in the surfaces of balls and races which will rise the temperature of the bearings.

From the above issues, it is evident that any failure in bearings increases the friction which in turn temperature and vibration increases. Hence the bearing temperature and the vibration are the best indicators of the healthy machine.

Stator Fault

During normal operating conditions of the induction motor, it is subjected to various stresses like mechanical, electrical and environmental. Depends on the severity of the stresses, the faults occur in the stator of the induction motor. Stator faults are under two types namely (i) faults in laminations and frame of the stator (ii) faults in stator winding. Among these, the second type of fault occur from 28 – 36% as per the study of IEEE and EPRI. The above said stresses will be maintained if proper operation and maintenance is carried out.

Stator Winding Fault

The fault occurs due to the insulation failure of the stator winding then it is called as stator winding fault or inter turn short circuit fault. Stator winding fault is classified into six types according to the nature of fault occurs between the same phases or others. They are

1. Short circuit between two turns of same phases or turn to turn fault.
2. Short circuit between two coils of same phases or coil to coil fault.
3. Short circuit between two phases or phase to phase fault.
4. Short circuit between turns of all phases.
5. Short circuit between winding conductors and stator core or phase to ground fault.
6. Open circuit fault when any phase gets open.

Different types of stator winding faults are shown in Figure 13, Figure 14 and Figure 15. Short circuit winding fault occurs when partial or total winding gets short circuited and open circuit fault occurs when there is no current flow in the winding or phase or disconnected.

Causes and Effects of Stator Winding Faults

The stator winding faults occurs mainly occurs due to the electrical, mechanical, thermal and environmental stresses due to various operating conditions. The following paragraph will discuss about these stress and its effects in induction motor.

1. **Electrical Stress:** Supply voltage transient causes the electrical stresses. The transient occurs due to line to ground, line to line or three phase faults, due to lightning, opening or closing of circuit breakers or due to variable frequency drives. The transient voltage reduces the life time of the stator windings and cause turn to turn or turn to ground fault.
2. **Mechanical Stresses:** Generally mechanical stresses are caused by movement of stator coil and rotor striking the stator. Stator coil moves due to the force which is proportional to the square of the stator current. This movement will lose the top sticks and damage the insulation of the windings. Rotor may strike the stator due to rotor unbalanced fault and it causes the stator laminations to

Figure 13. Different Types of Fault in Star Connected Winding

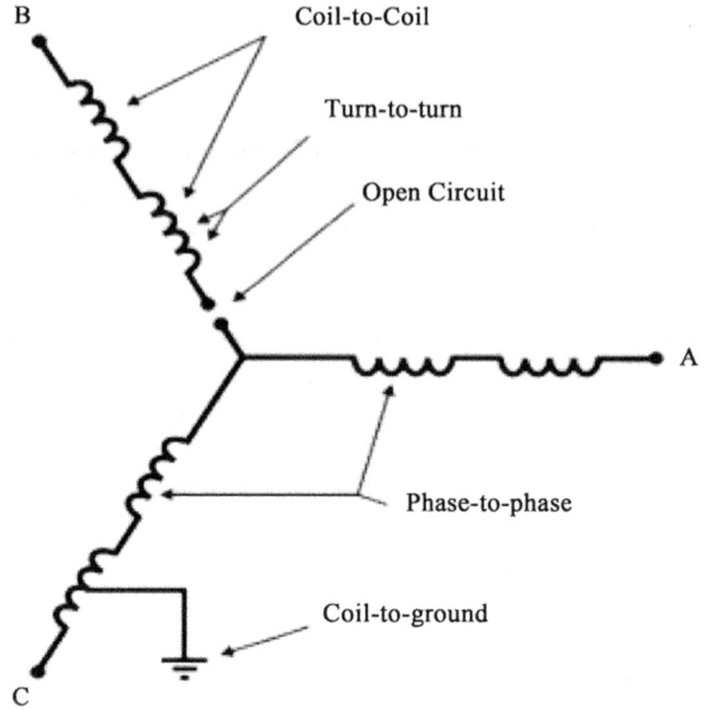

Figure 14. Damaged Stator Winding

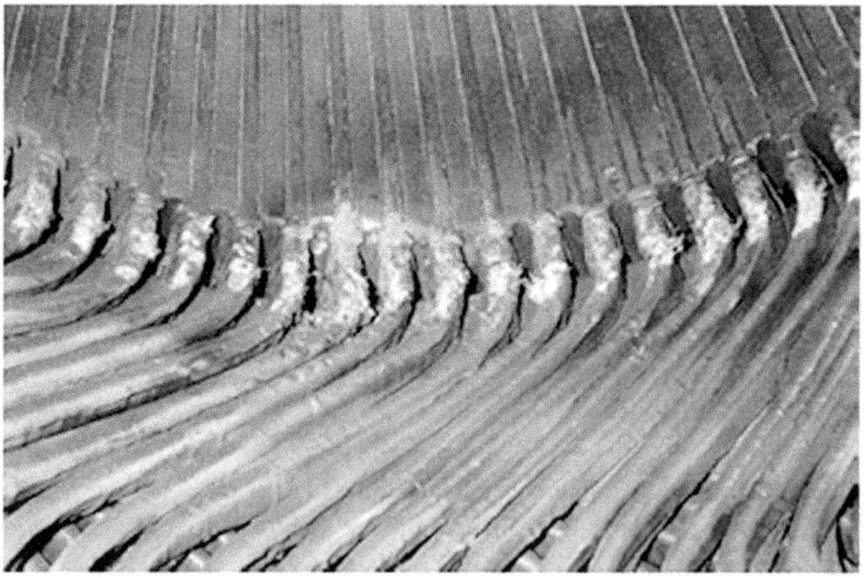

Figure 15. (a) Inter – Turn Short Circuit between turns of the same phase (b) Winding Short Circuited (c) Short circuit between winding and stator core at the end (d) Short circuit between winding and stator core at the middle (e) Short circuit at the leads (f) Short Circuit between phases

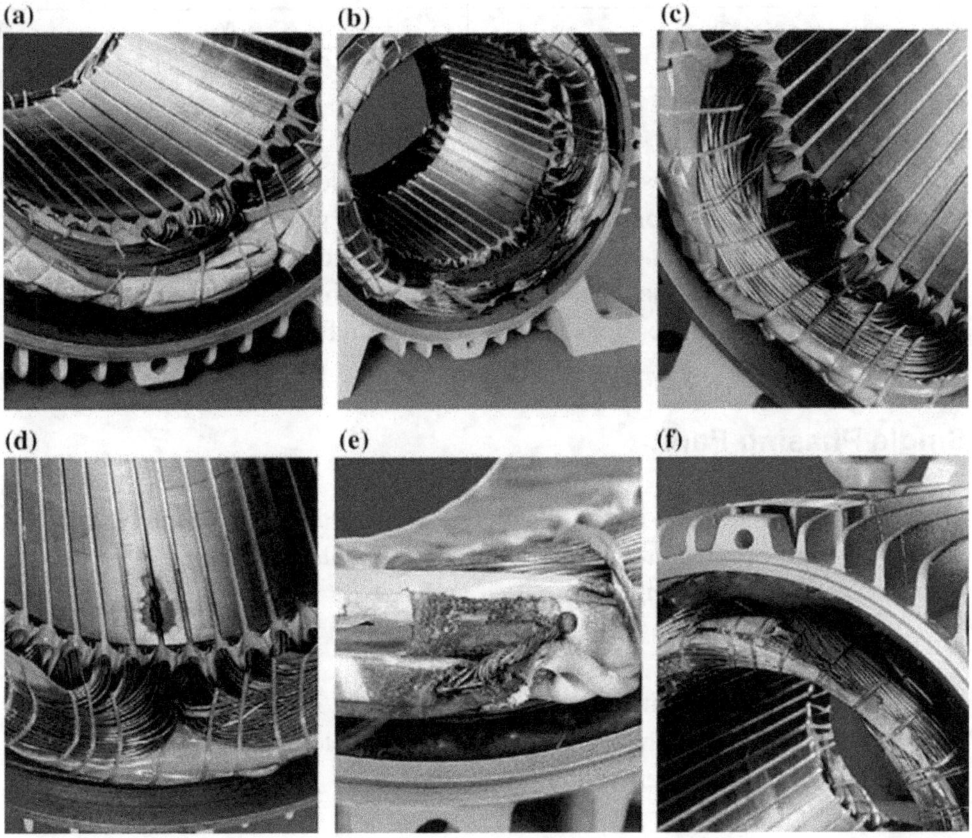

puncture and it will lead to coil to ground fault. Stator windings with open circuit fault will be disconnected in case of high vibration.

3. **Thermal Stresses:** It is the one of the main reason for stator winding insulation failure and it occurs due to thermal overloading. Thermal stress occurs due to over current, obstructed ventilation, higher ambient temperature and unbalanced supply voltage. The stator winding temperature is increased by 25% if there is an increase in voltage imbalance by 3.5% . Also the insulation life is reduced to 50% when the stator winding temperature increases 10% above its rated value. The effect of rise of temperature above the atmosphere on the insulation failure is shown in the Table 4.

Table 4. Effects of Rise of Temperature

Ambient in °C	Insulation life in Hours
30	250,000
40	125,000
50	60,000
60	30,000

4. **Environmental Stresses:** These types of stresses occur when the motor is operating in too cold or too hot or too humid. The presence of foreign particles in the environment may contaminate the insulation of the stator winding and reduce the heat dissipation from the motor to the environment and it will tend to short insulation life. Hence the motors should be situated with free air flow environment.

Single Phasing Fault

In a three phase induction motor, if any one phase is lost in the supply then it is termed as single phasing fault. It is supply related fault of an induction motor.

Causes of Single Phasing Fault:

The major causes of single phasing fault in an induction motor are,

- Fuse blowing in the utility system
- Short circuit in one phase of star or delta connected motor
- Failure of an equipment in the supply system

Effects of Single Phasing Fault

The effects of single phasing fault are

- The three phase motor will not start if the fault persists already in the supply.
- If it happens during running condition, then the motor continues to run. The torque produced by the remaining two phases will manage the load torque. Due to this, the healthy phases will get overheated and leads to damage of the windings.

Crawling

Under some situations, the induction motor does not accelerate even after full load supply is applied but runs at one seventh of its synchronous speed. This phenomenon is known as crawling. It is a kind of electromagnetic fault.

The air gap flux in between the stator and rotor is not purely sinusoidal and it contains odd harmonics. These harmonics will develop unwanted torque. The flux produced by third harmonics and its multiples will have 120° phase shift and hence neutralize themselves.

The air gap flux due to fundamental component rotates at synchronous speed given by $N_s = \frac{120\ f}{p}$ rpm where f is the supply frequency and p is the number of poles. But the harmonic flux rotates at a speed of $\frac{N_s}{k}$ rpm speed where k represents harmonic order, in the same direction expect 5th harmonic. The flux produced by 5th harmonic rotates in the opposite direction to fundamental component. The magnitudes of flux produced by 11th and higher order harmonics are small but flux produced by 5th and 7th harmonics are predominant.

Along with fundamental flux, these harmonics also produce fluxes. Hence the components of torques are (i) torque due to fundamental component rotating with a speed of N_s (ii) torque due to 5th harmonic component rotating at $\frac{N_s}{5}$ in the direction opposite to the fundamental (iii) torque due to 7th harmonic component rotating at $\frac{N_s}{7}$ in the same direction of the fundamental. The torque produced by 5th harmonic is very small and can be neglected. Hence the resultant torque will be the sum of torque due to fundamental component and torque due to 7th harmonic component as shown in Figure 16. The 7th harmonic torque becomes zero at one seventh of the synchronous speed and hence there is a dip in the resultant torque near slip 6/7, which is more significant because the torque decreases when speed increases. Under loaded condition, the motor run at a speed one – seventh of its synchronous speed. This phenomenon is known as crawling. It is predominant in squirrel cage induction motor and it can be avoided by the proper selection number of stator and rotor slots as discussed by Karmakar et al. (2016).

Causes of Crawling

The causes of crawling in induction motor are,

Figure 16. Torque – Slip Characteristics showing resultant of Fundamental and 7th order harmonics

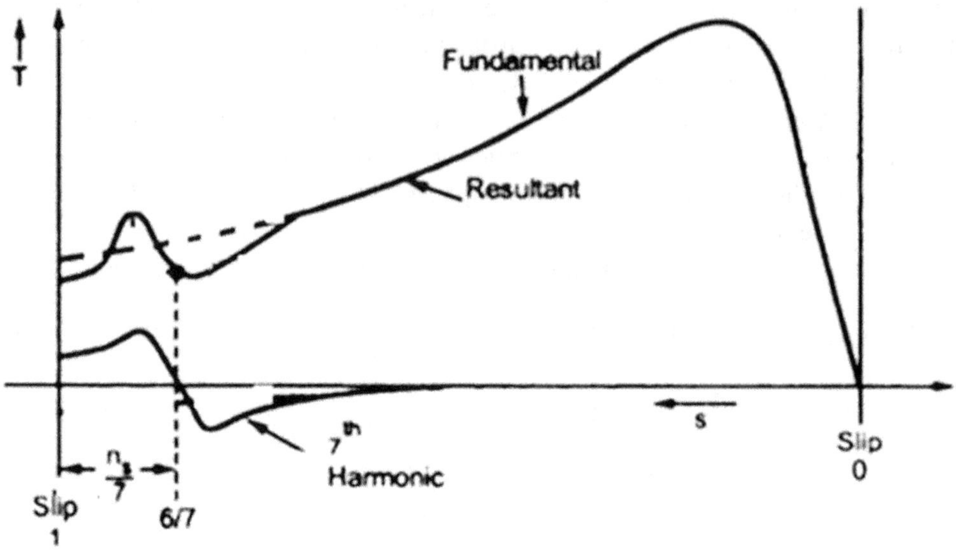

- 7th harmonic components due to abnormal magneto motive force.
- High harmonic components persist in the supply voltage.

Effects of Crawling

The effects of crawling in an induction motor are,

- The motor will not run at rated speed under loaded condition and remain to rotate in one seventh of its synchronous speed.
- It generates enormous heat in the stator winding
- Vibration and noise will be more in the motor.

CONDITION MONITORING OF INDUCTION MOTORS

Condition monitoring is the process of monitoring parameters of the machinery to identify the significant change which is an indicative of a developing fault. Mainly it is used to schedule the maintenance or other necessary actions to be taken to avoid failure and its consequences. The condition monitoring will help the operators to improve the life span of the machines by identifying the conditions.

Condition monitoring of induction motors can be done in three forms such as breakdown maintenance, fixed-time maintenance and condition-based maintenance. In breakdown maintenance, the maintenance takes place only when the motor breaks down. In this case, the motors life span will be comparatively more than others but once the break down occurs then it makes replacement of entire motor and makes more expense. Also it causes loss of productivity due to down time. In case of fixed-time maintenance, motor has to be stopped for a long time to identify the fault type and location. It also requires skilled and trained people to identify and rectify the faults. In case of condition-based monitoring, the parameters are monitored even under normal operation of the motor. Hence it is possible to identify the first symptom of the fault and it can be rectified.

Fault detection and condition monitoring methods can be classified into two categories according to technical point as passive and active methods. Passive methods make use of measured quantities and derive certain fault or condition indicators. These methods are applicable to the machines connected to the supply main directly. It does not have any chance to influence the operation of the motors. But in case of active methods, it requires any one kind of actuators to control the operation of the motor. The actuators may be either inverter or an additional power electronic device which is connected in series with the feeder of the machine.

Another classification of condition monitoring is based on the physical quantities measured as follows,

- Electrical Monitoring estimate only electrical signals like currents, voltages and mechanical angle between rotor and stator by means of encoder. It is useful to identify the electrical origin faults.
- Magnetic Monitoring evaluates the signals determined from measurement coils. These coils may be embedded into the stator coils or it may be located externally.
- Vibration Monitoring assess the vibrational acceleration or velocity through which it is possible to identify the mechanical imbalances. It is difficult to determine the mechanical imbalances using electrical signal.
- Thermal monitoring measures the temperatures either directly or estimated from the state observers or models.
- In mixed physical domain monitoring method, signals from different domains are measured and evaluated.

Structure of Monitoring Methods

The monitoring methods are usually have the processing blocks as shown in Figure 17. The first stage indicates, the measurement of signals. For measurement of voltage and current, the transducers are used. The measurement of rotor speed and rotor angle is measured by means of encoder or found from slotting harmonics of the current either in frequency or time domain. The other parameters like temperatures, vibration and flux in coils are determined by the alternate monitoring methods.

Signal Conditioning is the next stage which does the initial processing of the measured signal. It includes the determination of phase voltages or currents based on the type of connection and in some cases phasor also. In case of evaluation process, the internal machine quantities like stator or rotor temperature or fault signatures from the measured signals. For example, harmonic component of the current indicates the electrical asymmetry of a squirrel cage induction motor. At last the assessment of the condition will be done.

Condition Monitoring Methods for Detection of Stator Fault in Induction Motor

Under normal operating condition, the MMF produced by the stator winding can be synthesized into fundamental and harmonic waves. In this case, the magnitude of MMF fundamental wave will be high and constant. If any fault occurs in the stator winding, then the magnitude of MMF fundamental wave will not be constant. During this situation, the fundamental wave can be composed into forward and reverse travelling waves. The forward travelling wave is represented by positive sequence stator current (i_{sp}) and the reverse travelling wave produced by the stator winding fault is represented by negative sequence current (i_{sn}). The values of sequence current will be determined from phase voltages and currents through transformation matrix. In some situations, the asymmetry in supply voltage also produces negative sequence current. Hence the due care has to be taken care during the evaluation of negative sequence current. In addition to the above quantities, the inherent asymmetry of stator winding has to be considered.

Figure 17. Block Diagram of Condition Monitoring Methods

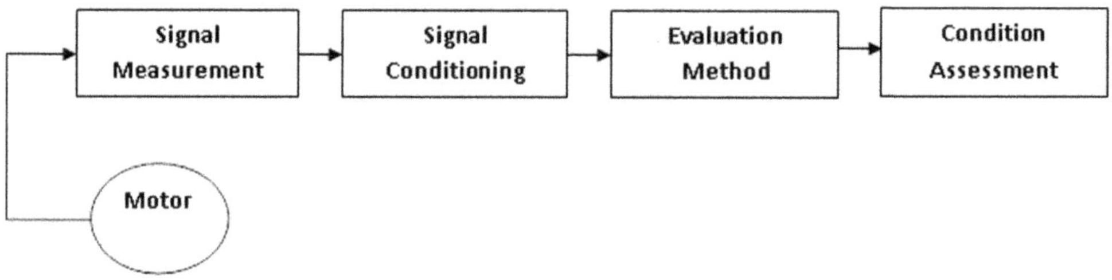

The negative sequence voltage V_{sn} is estimated from the positive and negative sequence currents through sequence impedances (Z_{snp} and Zsnn) by using the following equation,

$$V_{sn} = i_{sp} Z_{snp} + i_{sn} Z_{snn} \tag{1.3}$$

The values of sequence impedance have to be indentified in order to compensate inherent asymmetry. These values are depends upon the operating condition of the motor and it can be performed by means of neural network or functional approximations as discussed by Tallam et al. (2000). Once the sequence impedances are identified then the fault detection technique can be applied under the regular operation of the motor. To estimate the values of positive and negative sequence impedances, it is essential to measure the value of negative sequence voltage. But in case of symmetry supply, it is difficult to measure it. Hence, it can be generated by using a single phase resistor. Then the measured value of negative sequence impedance can be compared with the estimated value. If any difference occurs between the measured and estimated value then it indicates that there is some fault in the stator winding.

In the other way, it can also be done with the measured value of sequence current and estimated value current.

Condition Monitoring Methods for Stator and Rotor Temperature

The temperature of stator winding can be measured by using sensors which are embedded in the winding, windings heads and iron sheet of the stator. The measurement of temperature in the rotor is more difficult since it is a rotating part. The implementation of radio frequency transmission sensors becomes difficult in case of industrial environments because the machines have to be dismounted or modified. Under some situations, the measurement of stator is also not possible. In these cases, the temperature of stator and rotor has to be estimated. The temperature of stator and rotor depends on the operating condition of the motor. Hence homogeneous or averaged temperature can be assumed for the stator slots, each winding head, rotor bar and end rings. By designing a lumped parameter model, these temperatures can be determined as discussed by Kral etal. (2008) in their research. For thermal modelling of the motor, it is required to know all the geometric details of stator, rotor, housing and cooling.

The simplified lumped thermal parameter model is used for practical applications as discussed by Gao et al. (2008) in his research. These models only employ couple thermal time constants. It is determined by identification test or measured in advance before the motor put into operation. Kral et al. (2004b) discussed about a lumped parameter model can also be combined with other resistance estimation method to form a hybrid model in his research. Generally average stator winding temperature can be estimated from the stator resistance by using the following equation,

$$R = R_{ref}\left(1 + \pm_{ref}\left(T - T_{ref}\right)\right) \tag{1.4}$$

where R_{ref} is the resistance at the temperature T_{ref} . $\pm_{ref}$ is the linear temperature coefficient at the reference temperature. T is the actual temperature and R is the actual resistance.

One method of determination of average stator temperature is DC bias method. In this method, direct current does not cause any voltage drop across the stator inductance and it represents only the voltage drop across the stator resistance only. The temperatures in all three phase windings of symmetric machines are assumed as equal. Hence it is enough to determine the temperature of any one winding by applying DC bias. There are lot of ways available for applying DC bias. One such method is, a resistor in parallel with a diode can be used to cause voltage drop across the stator winding as discussed by Lee and Habetler (2003). Otherwise it can be done by using soft starters as discussed by Zhang et al. (2008) or zero sequence voltage by Jacobina et al. (2000). The estimation of stator resistance is required for tuning of inverter drives, such methods are integrated with DSP software of the drive. Also context observers as discussed by Jeon et al. (2002), models as said by Gao et al. (2008) and neural networks by Karanayil et al. (2007) can be used for the determination of stator and rotor resistance.

Condition Monitoring Methods for Detection of Rotor Eccentricity

Rotor eccentricity is of two types, namely static and dynamic eccentricity. The static eccentricity is caused by misalignment of the bearings, end rings or stator ovality and dynamic eccentricity is caused by shaft bent, mechanical resonances, bearing wear and static eccentricity. In practice, both type of eccentricity occurs at the same time in the machine. The two eccentricities give rise to harmonic component with frequency in the stator current as given below,

$$f_i^e\left[m\right] = f_e\left(1 \pm m\frac{1-s}{P}\right) \tag{1.5}$$

where P is the number of pole pairs and subscript e indicates the eccentricity related components and $m \in [1, 2, 3, \ldots]$ is the ordinal number. Due to the interaction of current and flux, power and torque components arise at following frequencies,

$$f_t^e[m] = f_e\left(m(1-s)/P\right) \tag{1.6}$$

Similar to torque, the rotor slots also give rise to additional frequencies in the stator current as follows,

$$f_{iR}^e[m] = f_e\left[(mR \pm c)\frac{1-s}{P} \pm o\right] \tag{1.7}$$

where, $c \in [0, 1]$ indicates static (zero) or dynamic (one) eccentricity and o is an odd integer.

The eccentricity related harmonic components can be determined by means of conventional Fourier analysis as discussed by Thomson and Fenger (2001). But it is not suitable for motors having oscillating loads. An alternate method utilizing negative sequence component of the stator current since it is independent of the actual loading of the machine as discussed by Wu et al. (2007). Eccentricities have mechanical origins and if the fault detection carried out by means of vibration signals then it becomes more accurate results. A fault signature according to the above equations may be caused by mechanical mass imbalance of the rotor. But it is very difficult to detect a mechanical problem by means of current signatures whereas vibration signatures indicate the issue very clearly as discussed by Kral et al. (2004a).

Condition Monitoring Methods for Detection of Faults in Squirrel Cage Rotor

The squirrel cage rotor consists of number of bars and end rings which are placed on the iron sheet of the rotor. In case of small machines, the rotor bars and end rings are aluminium or copper die cast. But in larger machines, the copper bars are insulated and fitted into the iron sheet and end rings are either welded or hard soldered. Under normal operating conditions, the current carried by each ring will be the same and current through the end rings are also equal. In case of any fault in the bar will increase its resistance thereby the current also get diminished. For example, consider there are 40 bars in the squirrel cage rotor. If bar 1 is broken then the current through that bar becomes zero as shown in Figure 18. Then that current will be shared by the nearer bars like 40 and 2. The increase in the magnitude of current will increase the thermal stress of the respective bars. This process will spread the fault from its initial fault location and it will damage the entire machine.

Rotor faults in most of the cases are evaluated from the stator current signatures and are called as current signature analysis (CSA) methods. Under steady state operating conditions, Fourier analysis method is used to determine the rotor asymmetries. For these cases, it is essential to evaluate the magnitudes of current harmonics accurately because low frequency torque imbalances also produce current side band frequencies. Hence it is very essential to identify the exact fault specific components to avoid confusions with load specific harmonics as discussed by Kral et al. (2005).

During transient conditions, load torque and slip are not constant. Hence conventional Fourier analysis is not applicable in this case. It can be solved either by means of compensating modeled load torque or by mathematical transformations and algorithms as discussed by Wu et al. (2005) or multiple signal classification

Figure 18. Rotor bar current when one bar is broken

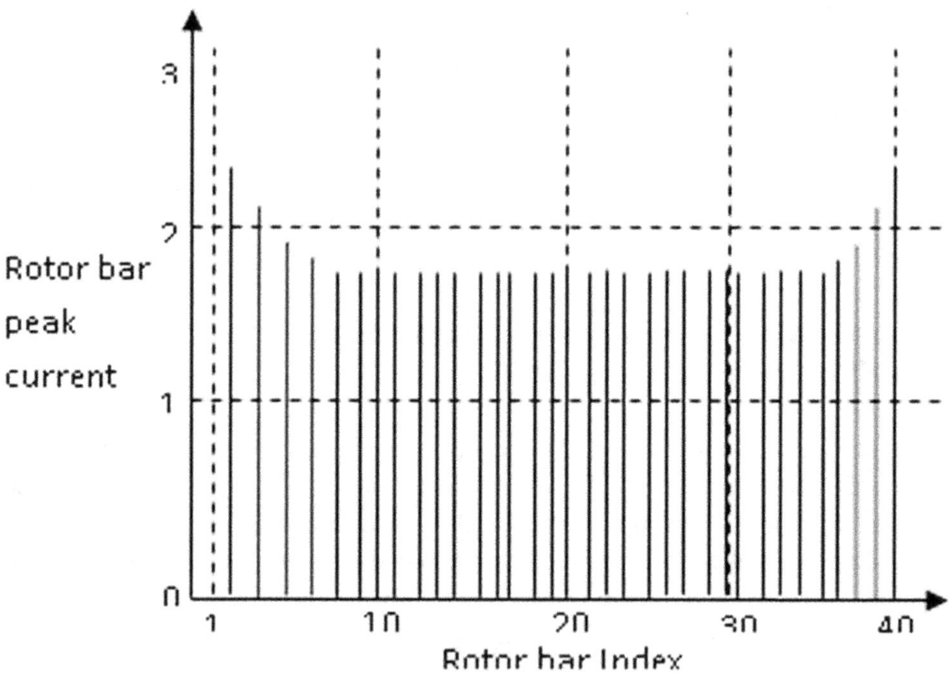

Figure 19. Block Diagram of Vienna Monitoring Method

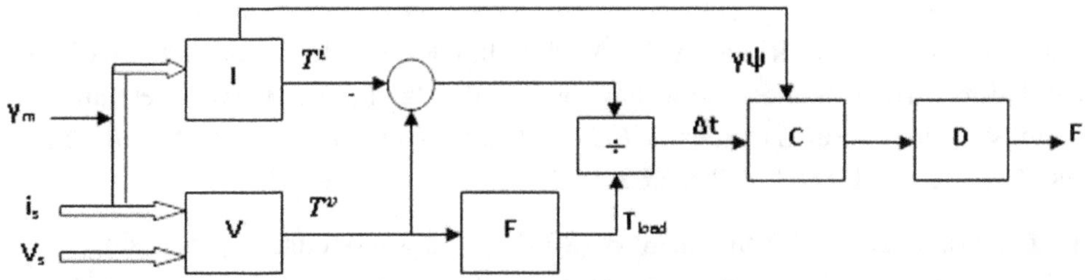

(MUSIC) algorithms to overcome the problem of identification of the non-stationary fault signatures as discussed by Benbouzid and Kliman (2003) or wavelet transforms by Antonino – Daviu et al. (2006).

Apart from all these methods, Vienna Monitoring Method (VMM) is one of the best method to identify the rotor asymmetries. The block diagram of this method is shown in Figure 19. VMM method is a model based approach which compares the torque signals of voltage (T^v) and current (T^i) models of the machine respectively. The voltage model is based on the stator voltage equation in the stator reference frame, whereas the current model relies on the rotor voltage in the rotor reference frame. The input quantities are stator voltage, current phasor and mechanical angle. The major advantages of VMM are that it detects rotor faults under almost any operation and boundary conditions. It works

- Independent of the load torque of the machine.
- Under static and dynamic conditions.
- Under variable load and speed conditions.
- Independent of the inertia of the drive.
- Independent of the applied control technique.

CONCLUSION

This chapter deals about the construction and basic operating principle of the induction motor. In addition to this, the different types of major faults, its sources and effects of induction motors are elaborately discussed. The various types of condition monitoring methods are also conversed.

REFERENCES

Antonino-Daviu, J. A., Riera-Guasp, M., Folch, J. R., & Palomares, M. P. M. (2006). Validation of a new method for the diagnosis of rotor bar failures via wavelet transform in industrial induction machines. *IEEE Transactions on Industry Applications*, *42*(4), 990–996. doi:10.1109/TIA.2006.876082

Benbouzid, M. E. H., & Kliman, G. B. (2003). What stator current processing-based technique to use for induction motor rotor faults diagnosis? *IEEE Transactions on Energy Conversion*, *18*(2), 238–244. doi:10.1109/TEC.2003.811741

Gao, Z., Colby, R. S., Habetler, T. G., & Harley, R. G. (2008). A Model Reduction Perspective on Thermal Models for Induction Machine Overload Relays. *IEEE Transactions on Industrial Electronics*, *55*(10), 3525–3534. doi:10.1109/TIE.2008.926772

Jacobina, C. B., Filho, J. E. C., & Lima, A. M. (2000). On-line estimation of the stator resistance of induction machines based on zero-sequence model. *IEEE Transactions on Power Electronics*, *15*(2), 346–353. doi:10.1109/63.838107

Jeon, S. H., Oh, K. K., & Choi, J. Y. (2002). Flux observer with online tuning of stator and rotor resistances for induction motors. *IEEE Transactions on Industrial Electronics*, *49*(3), 653–664. doi:10.1109/TIE.2002.1005393

Karanayil, B., Rahman, M. F., & Grantham, C. (2007). Online Stator and Rotor Resistance Estimation Scheme Using Artificial Neural Networks for Vector Controlled Speed Sensorless Induction Motor Drive. *IEEE Transactions on Industrial Electronics*, *54*(1), 167–176. doi:10.1109/TIE.2006.888778

Karmakar, S., Chattopadhyay, S., Mitra, M., & Sengupta, S. (2016). *Induction Motor Fault Diagnosis: Approach through Current Signature Analysis*. Springer. doi:10.1007/978-981-10-0624-1

Kothari, D. P., & Nagrath, I. J. (2010). *Electric Machines*. New Delhi: Tata McGraw Hill Education Pvt. Ltd.

Kral, C., Habetler, T. G., & Harley, R. G. (2004). Detection of Mechanical Imbalances of Induction Machines Without Spectral Analysis of Time-Domain Signals. *IEEE Transactions on Industry Applications*, *40*(4), 1101–1105. doi:10.1109/TIA.2004.830762

Kral, C., Habetler, T. G., Harley, R. G., Pirker, F., Pascoli, G. Berguggenberger, H., & Fenz, C. (2004). Rotor Temperature Estimation of Squirrel-Cage Induction Motors by Means of a Combined Scheme of Parameter Estimation and a Thermal Equivalent Model. IEEE Transactions on Industry Applications, 40(4), 1049-1057.

Kral, C., Kapeller, H., Pirker, F., & Pascoli, G. (2005). Discrimination of Rotor Faults and Low Frequency Load Torque Modulations of Squirrel Cage Induction Machines by Means of the Vienna Monitoring Method. *IEEE 36th Annual Power Electronics Specialists Conference,* 2861- 2866.

Kral, C., Pirker, F., & Pascoli, G. (2008). The Impact of Inertia on Rotor Fault Effects – Theoretical Aspects of the Vienna Monitoring Method. *IEEE Transactions on Power Electronics*, *23*(4), 2136–2142. doi:10.1109/TPEL.2008.925183

Lee, S. B., Tallam, R. M., & Habetler, T. G. (2003). A robust, on-line turn- fault detection technique for induction machines based on monitoring the sequence component impedance matrix. *IEEE Transactions on Power Electronics*, *18*(3), 865–872. doi:10.1109/TPEL.2003.810848

Sen, P. C. (1989). *Principles of Electric Machines and Power Electronics*. New York: Willey.

Tallam, R. M., Habetler, T. G., Harley, R. G., Gritter, D. J., & Burton, B. H. (2000). *Neural Network Based On-Line Stator Winding Turn Fault Detection for Induction Motors. In Conference Record of the IEEE Industry Application Conference*. IAS.

Thomson, W. T., & Fenger, M. (2001). Current Signature Analysis to Detect Induction Motor Faults. *IEEE Industry Applications Magazine*, *7*(4), 26–34. doi:10.1109/2943.930988

Thorsen, O. V., & Dalva, M. (1995). A survey of faults on induction motors in offshore oil industry, petrochemical industry, gas terminals, and oil refineries. *IEEE Transactions on Industry Applications*, *31*(5), 1186–1196. doi:10.1109/28.464536

Wu, L., Habetler, T., & Harley, R. (2005). Separating Load Torque Oscillation and Rotor Fault Effects in Stator Current-Based Motor Condition Monitoring. *International Electric Machines and Drives Conference*, 1889-1894.

Wu, L., Habetler, T., & Harley, R. (2007). A Review of Separating Mechanical Load Effects from Rotor Faults Detection in Induction Motors. *IEEE International Symposium on Diagnostics for Electric Machines, Power Electronics and Drives*. 10.1109/DEMPED.2007.4393098

Zhang, P., Lu, B., & Habetler, T. G. (2008). A Remote and Sensorless Stator Winding Resistance Estimation Method for Thermal Protection of Soft- Starter- Connected Induction Machines. *IEEE Transactions on Industrial Electronics*, *55*(10), 3611–3618. doi:10.1109/TIE.2008.927241

Chapter 8
Vibration Monitoring of Rotating Electrical Machines:
Vibration Monitoring

Achintya Choudhury
Sir Padampat Singhania University, India

ABSTRACT

Vibration monitoring is applicable to all rotating machines for defect detection and diagnosis. Measurement and analysis of vibration have also been applied to rotating electrical machines with the objective of fault detection and predictive maintenance. The sources of vibration generation in electrical rotating machines, both electrical and mechanical, have been identified in this chapter. The vibratory characteristics associated with these defects have also been discussed in detail. Analyses of vibratory signatures in time domain, frequency domain, and time frequency domain have been dealt with, and different features and indicators associated with each domain have been described. The details of vibration measurement schemes such as transducers, different signal conditioning elements, as well as characteristics of recording and display devices and their applicability to electrical machines have also been included in the chapter.

INTRODUCTION

Rotating electrical machines, especially electric motors, find extensive applications in Industries. Any unanticipated failure of such electro-mechanical systems may result in serious loss in terms of downtime, consequential damage and safety hazard. A lot of research efforts have therefore been placed on fault detection and diagnosis of electrical machines through different condition monitoring techniques.

DOI: 10.4018/978-1-5225-6989-3.ch008

Among these techniques, vibration monitoring is probably the most widely used means of fault detection. Any rotating machine or machine component is inherently in possession of several probable sources of generating dynamic forces and vibration. With several moving and stationary components of different shapes and relative motion, rotating electrical machines may also generate quite complex vibratory motion. The complexity increases further when excitations are also caused due to magnetic and electromagnetic forces. These mechanical and electrical sources of vibrations are inter-related and complementary to each other.

A detailed vibration analysis is therefore necessary to ascertain the causes of such faults and failures. This chapter includes a detailed discussion on all the causes, both electrical and mechanical, which result in vibratory motion of electrical machines. Analysis of vibratory motions and the suitable mode of analysis based on the complexity of vibration are also discussed in details.

BACKGROUND

Vibration responses of electrical machines have been studied by many researchers in last few decades. Reduction of levels of noise generated from electric motors drew the attentions of researchers quite early (Jordan, 1950; Yang, 1981). In this connection, vibrations caused due to different electrical faults were studied and have been reported in references (Binns & Dye, 1973; Swann, 1963). Most of these researchers also reported the stator responses to these defects. Transverse and torsional responses of rotors were also studied considering the rotor to be either rigid or flexible (Thomson & Dahleh, 2007; Walker et al., 1980; Cudworth et al., 1984; Corbo & Malanoski, 1996). The studies on transverse and torsional responses of rotors of electrical machines have also been reviewed in the book by Tavner et al. (2008).

Vibration monitoring has been included in the reviews made by different researchers on condition monitoring of electrical machines (Vaimann & Kallaste, 2012; Mortazavizadeh & Mousavi, 2014; Negoita et al., 2010). Vibration monitoring has been applied to electrical machines and its components by several researchers and its effectiveness vis-a-vis other monitoring techniques have also been reported (Cameron et al., 1986; Nandi et al., 2001; Payne et al., 2003; Holbert et al., 2006; Tandon et al., 2007; Iorgulescu et al., 2012, Gopinath, 2010; Orman & Pinto, 2013; Gama et al., 2017).

In this chapter, causes of vibration generation in electrical machines, both electrical and mechanical, have been discussed in details. Measurement and analysis of vibration signals in time, frequency and time-frequency domains with important characteristics and indicators have also been described. The suitability of different analyses methods to particular situations with supporting literature has also been discussed.

VIBRATION GENERATION IN ELECTRICAL MACHINES

There are several sources of vibration generation in rotating electrical machines. These sources may be broadly categorized into two types: mechanical and electrical. As stated earlier, these sources are inter-related and complementary. It is therefore very important to understand the reasons and causes for vibration generation before dealing with the monitoring techniques.

Common Mechanical Causes

Following are the common mechanical causes encountered in any rotating machine:

Unbalance

Unbalance is the most common reason for vibrations of rotating machines. The frequency of vibration equals the rotational frequency $\left(1 \times f_s\right)$, f_s being the shaft rotational frequency in Hz and the vibration amplitudes are excessive in the radial direction. Reasons other than unbalance may also result in vibrations at shaft rotational frequency. Vibration due to unbalance may be confirmed by measuring the phase difference between vibrations in two radial directions, horizontal and vertical. For rotary unbalance, this phase difference is 90° (Nakra et al., 1989).

Misalignment

Misalignment is also a frequent cause of vibration generation in rotating electrical machines. Improper orientation of shaft with reference to couplings and bearings often cause misalignment. Vibrations are resulted at multiples of shaft rotational frequency $\left(m \times f_s\right)$. In this case, the amplitudes of vibrations are quite high in axial direction, being 50% or more of that in radial direction (Nakra et al., 1989).

Mechanical Looseness

Mechanical looseness is caused due to loose mounting bolts, excessive bearing clearance, structural cracks etc. Vibrations of high amplitudes are resulted at twice the shaft rotational frequency $\left(2 \times f_s\right)$ or higher multiples (Nakra et al., 1989).

Soft Foot

'Soft foot' refers to the improper contact between machine legs and the supporting baseplate. For easy understanding, it may be compared with a wooden chair with one leg short causing a rocking motion. Causes of soft foot include twisted or warped machinery foot or foundation, dirt and other unwanted materials and improper shims under machine foot. The rocking motion caused by uneven foot is not the only vibratory problem caused due to soft foot. The soft foot tends to bend with tightening of base bolts. This bending, in turn, causes a disturbance in the alignment of the shaft axis. This also results in airgap eccentricity. Relative motion between baseplate and machine foot can be detected by phase measurement (Nakra et al., 1989).

Bearing Defects

Both sleeve bearings and rolling bearings are used in electrical machines. Vibrations caused due to defects in these components are discussed in the following paragraphs:

Rolling Bearings

Rolling bearing defects are probably one of the most important causes of vibrations in electrical machines, especially in induction motors. Some studies have shown that 44% of AC motor failures are due to bearing defects (Holbert et al., 2006). Rolling bearing defects may be categorized into two types: localized and distributed. Localized defects which include pits, spalls and cracks, are primarily caused due to fatigue failure. Distributed defects, on the other hand, include surface roughness and waviness and are caused due to manufacturing error, improper installation and abrasive wear (Tandon and Choudhury, 2000). Defects on bearing elements normally excite resonances in high frequency zone (> 5 kHz). These resonances are amplitude modulated by characteristic defect frequencies which are specific to the location of the defect and depend only on kinematic considerations. For normal speeds, these frequencies lie in the low frequency range and are usually less than 500 Hz. The expressions for these characteristic frequencies have been included in Table 1 (Tandon and Choudhury, 1999).

Table 1. Characteristic defect frequencies for a rolling element bearing (f_s: shaft frequency; d: rolling element diameter; D: Pitch diameter; β: Nominal contact angle; Z: Number of rolling elements)

Defect Location	Characteristic Frequency
Cage	$\left(\frac{f_s}{2}\right)\left[1-\left(\frac{d}{D}\right)\cos\beta\right]$
Outer race	$\left(\frac{Zf_s}{2}\right)\left[1-\left(\frac{d}{D}\right)\cos\beta\right]$
Inner race	$\left(\frac{Zf_s}{2}\right)\left[1+\left(\frac{d}{D}\right)\cos\beta\right]$
Rolling element	$\left(\frac{Df_s}{d}\right)\left[1-\left(\frac{d}{D}\right)^2\cos\beta\right]$

Sleeve Bearings

In sleeve or journal bearings, a converging fluid film, pumped by the motion of the shaft, separates the shaft and bearing liner. However, as the applied load on the bearing increases, the oil film thickness decreases and metal to metal contact may occur. In sleeve bearings, vibration is mainly caused due to 'oil whirl' or 'oil whip'. The oil between shaft and bearing is subjected to shear and tends to rotate at average speed of shaft and bearing. The rotating oil film thus pushes the shaft around the internal clearance of the bearing setting oil whirl (Nakra et al., 1989). Oil whirl vibration is nearly half the speed of the shaft (0.43 to 0.48 f_s) (Tavner et al., 2008).

The shaft whip occurs due to dry frictional forces caused due to metal to metal contact between shaft and bearing. The friction force and an opposing force at the center cause a moment opposing the shaft rotation. This moment causes a whirling or whipping motion at high frequencies which may correspond to critical rotor speeds ((Nakra et al., 1989).

Broken Rotor Bars

Faults such as cracks in rotor bars are among the common causes of vibration generation in electrical machines. Poor manufacturing, such as defective casting or joining, is a primary cause of such fault. Cracks and faults in rotor bars are also caused due to the following reasons (Vaimann and Kallaste, 2012): thermal stresses due to over-load or non-uniform heat distribution, magnetic stresses due to electromagnetic forces, residual stresses due to fabrication or other mechanical stresses due to different dynamic forces. A broken rotor bar causes an increase in resistance resulting in loss of current flow in the broken bar and increased current flow in other rotor bars. This uneven distribution of current and resulting thermal stresses propagate the defect causing further damage of the rotor bars and cage (Vaimann and Kallaste, 2012).

Airgap Eccentricity

Airgap eccentricity may be of two types: static and dynamic. Static eccentricity is caused mainly due to imperfections in manufacturing and installation such as stator core ovality or incorrect positioning of stator core. This results in a fixed radial air gap eccentricity. Dynamic air gap eccentricity is resulted when the center of the rotor is not at the center of rotation. This causes a varying air gap that rotates with the rotor. Air gap eccentricity causes an unbalanced magnetic pull which results in vibration (Nandi et al., 2001). This also enhances the risk of stator-rotor rub, which may seriously damage different components of the electrical system (Vaimann and Kallaste, 2012).

Stator Faults

Interactions between airgap flux and currents flowing in slot windings result in forces acting on stator core. Similar forces are also produced on the end windings due to interaction between end leakage flux and winding current (Tavner et al., 2008). These forces will obviously depend on the current distribution and geometries of airgap. Disturbances in either will cause a change in harmonics of the force wave acting on stator core. The resulting vibration becomes very serious when the frequency of force wave matches with any of the natural frequencies of the system.

Stator winding fault also starts with a short circuit between the adjacent turns in the stator windings. This in turn results in production of high current and winding overheating. Early detection of such faults is necessary because this may cause serious damage to other components of the system. However, vibration is not the most suitable diagnostic tool for early detection of such faults. Rather the resulting damages may be detected by vibration.

MONITORING METHODS AND INDICATORS

Vibration produced by the causes as explained above are measured and analyzed to detect the fault that caused the vibration. Depending on the complexity of the machine and the signal produced, different monitoring techniques are applied. The general principle of vibration monitoring is depicted in the block diagram shown in Figure 1. The available vibration monitoring techniques may be broadly classified into three categories: (i) Time domain analysis, (ii) Frequency domain analysis, and (iii) Time-frequency domain analysis. The details of these methods and their applicability to different components of rotating electrical machines have been explained in the following subsections.

Time Domain Analysis

In time domain analysis, vibration measured at selected points on the machine is expressed in terms of a single value. The method is comparatively simple. However, its applicability to diagnosis of specific faults in electrical machines is limited because it is difficult to identify the source of vibration by this method. An indication of the general machine health condition can be obtained from changing trend of the concerned value. The measurement parameters include some indices, moments of probability density and shock pulse method.

Figure 1. A block diagram representing the process of vibration monitoring

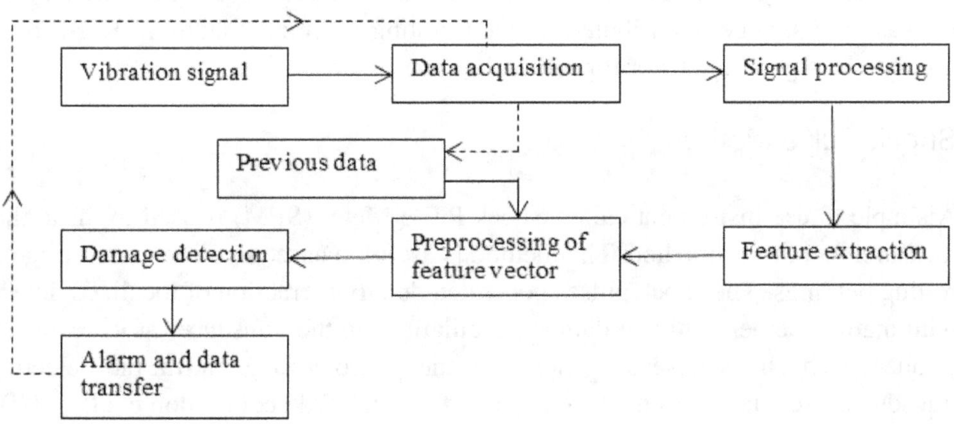

Indices

Different indices which are used to measure the overall vibration level include peak value, peak-to-peak value and root mean square (rms) value of the vibration signal. Out of these, rms value is the best indicator because it considers a little history of the signal concerned. Sometimes the crest factor, defined as the ratio of the peak to rms value is used as the indicator. The measurand is normally the vibration velocity measured in the bandwidth of 0.01 – 1 kHz or 0.01 – 10 kHz. Vibration measured on the bearing cap is compared with machine vibration standard to check the acceptability of these vibrations. One such standard for 0.01 – 1 kHz bandwidth and different power ranges based on the German Vibration Standard VDI 2056, is illustrated in Figure 2.

Moments

Some moments of the probability density distribution of the vibration signal are also used for vibration monitoring. The first four moments are mean, standard deviation, skewness and kurtosis. Out of these, the fourth order moment, kurtosis, finds maximum application. The kurtosis is defined as (Rao, 2004)

$$\kappa = \frac{1}{\sigma^4} \int_{-\infty}^{\infty} \left(x - \overline{x}\right)^4 f\left(x\right) dx$$

where $f(x)$ is the probability density of vibration, $x(t)$. $\overline{x}$ is the mean value and σ is the standard deviation of $x(t)$. Kurtosis indicates impulsiveness of the signal and increase in its value is attributed to deteriorating machine condition. Kurtosis is found to be a good indicator for rolling bearing failure.

Shock Pulse Method

A simple to use instrument called Shock Pulse Meter (SPM) is used to make the measurement in this method. The method is useful in case vibration monitoring of rolling bearings. The shock pulses generated due to interaction of localized defect with mating element, initiate damped oscillations in the transducer at its resonant frequency. The shock pulses are generated in the ultrasonic range and the piezoelectric transducer used has a resonant frequency of around 32 kHz (Tandon et al., 2007). Maximum normalized amplitude of the damped oscillation is measured by the SPM and is an indicator of the bearing condition. The normalized shock pulse value is given by the following expression (Tavner et al., 2008)

Figure 2. Vibration standard VDI 20156 depicting acceptability of vibration limits

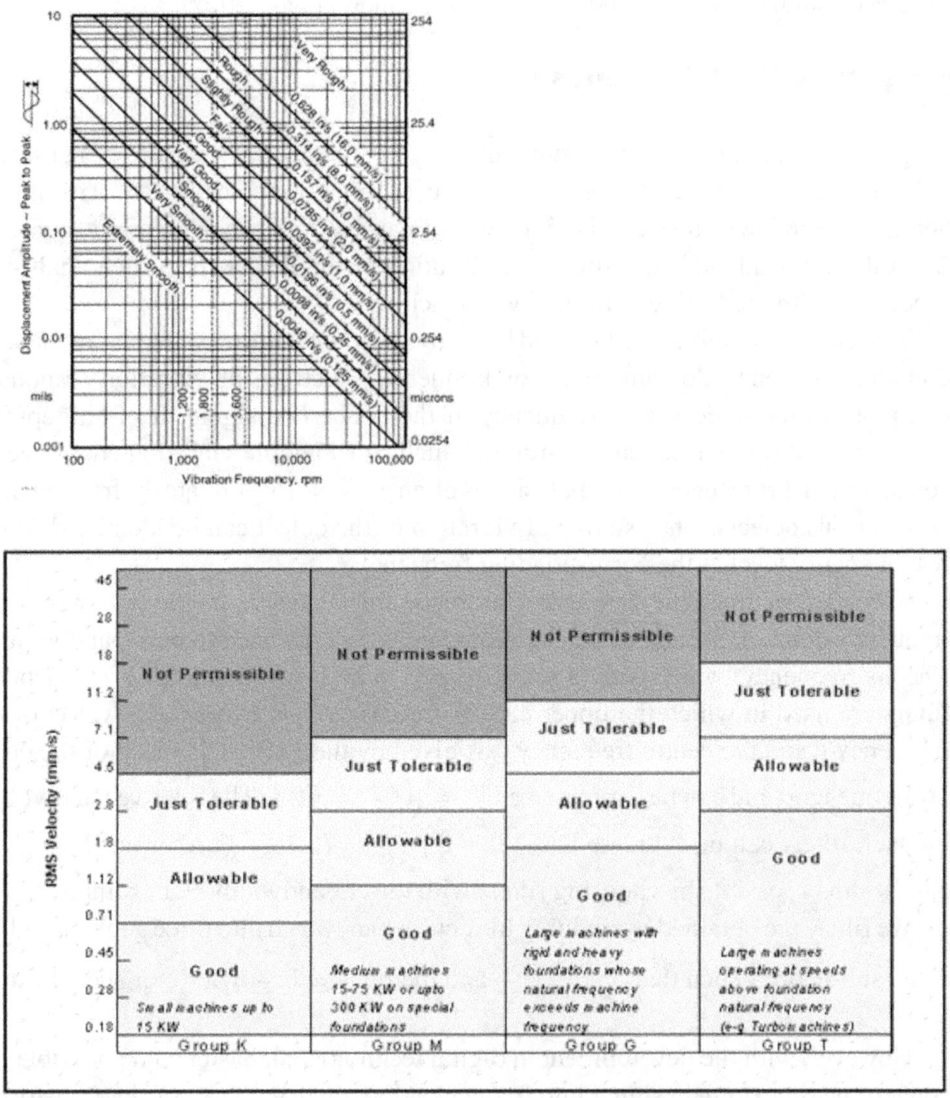

$$SPV = \frac{R}{N^2 F^2}$$

Where R is the SPM reading, N is the RPM of the shaft and F is a factor related to bearing geometry. Effectiveness of the method vis-à-vis other vibration and acoustic monitoring techniques as well as motor current signature analysis method have been studied (Tandon and Nakra, 1992; Tandon et al., 2007).

Frequency Domain Analysis

In spite of its simplicity, the time domain analysis does not provide sufficient useful information for machine diagnosis. Only basic information about the state of machine health, acceptable or unacceptable, may be obtained from time domain responses. It is difficult to identify the source of vibration or the defect from such analysis, especially with increasing complexity of machines.

This concern is suitably addressed by analyzing the vibration signal in frequency domain. Frequency domain signal or frequency spectrum of vibration response is a plot of amplitude versus frequency of the concerned signal. It often depicts one or more discrete frequencies around which considerable vibration energy gets concentrated. From theoretical analysis it is often possible to correlate the frequencies with various defects. Thus source of vibration or the defect can be identified from rise in energy level at the corresponding frequency.

Spectrum or frequency analyzers are used for analysis of vibration signal in frequency domain. In their earlier versions, signal energy used to be separated into various frequency bands using a set of filters. In an octave band analyzer, octave filters are used in which the upper cut-off frequency f_u is twice the lower cut-off frequency f_l and the centre frequency f_c is given by the geometric mean of the cut-off frequencies and can be expressed as $f_c = \sqrt{f_u f_l} = \sqrt{2}\, f_l$ The relative bandwidth for such filters can be expressed as. $\left(f_l - f_u\right) / f_c = 70.7\%$. Further narrowing of bandwidth is possible by choosing filters with lower bandwidth. For example, third octave filters are obtained by dividing the octave bandwidth into three geometrically equal subsections such that $f_u = 2^{1/3} f_l$ and the relative bandwidth for such a filter is 23%.

However, with the development in digital technology, it has become possible to convert the time domain signal into frequency domain using fast Fourier transform (FFT) algorithm. Spectrum analyzers are available for last couple of decades with built-in FFT software. Such analyzers can be used for real-time frequency analysis of the signal. Vibration signals may also be stored in data acquisition systems and analyzed off-line in computers using appropriate software.

As explained earlier, vibration may occur in electrical machines due to electrical as well as mechanical faults. Frequencies associated with most of these faults have also been identified through continued research. The expressions for frequencies associated with these faults have been compiled in Table 2 (Nakra et al. 1989, Trutt et al., 2001, Tavner et al. 2008). These expressions have been obtained in terms of shaft frequency, f_s and electrical supply frequency, f_{se} which are related by the following equations

For synchronous machine,

$$f_s = \frac{f_{se}}{p}$$

For asynchronous machine,

$$f_s = \frac{\left(1-s\right)f_{se}}{p}$$

Where s is the fractional slip for asynchronous or induction machine and p is the number of pole pairs.

Measures to Improve Signal to Noise Ratio (SNR)

Energy contents of the vibrations produced by different components of an electrical machine are different. The transducer mounted on the machine does not only receive vibration signals from different components of the machine, but from the load attached to the machine and other macrostructures. As such, the weaker signals often get buried under the stronger signals and significant components at the frequencies of interest may not be obtained from the spectral analysis. Pertinent examples are the vibrations received from defective rolling element bearings and broken rotor bars. It is therefore very important to retrieve the components related to the weaker signal, which may remain buried under noise.

The measure of the level of desired signal relative to the background noise is measured in terms of signal to noise ratio (SNR). SNR is often measured in terms of decibel (dB) and is given by the following expression

$$SNR = 20\log_{10}\frac{V_S}{V_N}$$

Where V_S and V_N denote voltages of desired signal and background noise in millivolts. A positive value of SNR indicates the signal level to be higher than the noise. Zero or negative value indicates the level to be equal to or less than the noise level. Different techniques have been applied to improve the signal to noise ratio of vibration signals obtained from complex machinery. Improvement of SNR basically depends on elimination of noise using suitable filters. Use of a bandpass filter to retain the bandwidth consistent with the signal is preferred over high-pass or low-pass filtering. However, simple filtering always does not provide the desired results, for example, in case of rolling element bearing vibration. Some additional processing of the signal is required to obtain the desired results.

Adaptive Noise Cancelling Technique

Adaptive noise cancelling (ANC) technique has been proposed to improve the SNR of a corrupted signal. The technique involves two inputs – a primary input consisting of corrupted signal and a secondary or reference input containing noise correlated with the primary noise. The technique has been found to enhance the effectiveness of spectral analysis (Chaturvedi and Thomas, 1982).

Envelope Detection Technique

When vibration is generated due to interaction of a physical defect with another element, resonances are excited. In the process, the natural frequencies of the system, component or structure get modulated by the periodicity of defect interaction. The process basically involves amplitude modulation of the carrier signal which is one of the natural frequencies. Envelope detection technique involves the process of demodulation in which the modulating signal is isolated from the carrier signal. The modulating signal has the frequencies associated with the defect and thus information related to defect can be obtained from the modulating signal. During the process of demodulation, noise also gets eliminated. The process of envelope detection technique is illustrated in Figure 3 (Tandon and Choudhury, 1999). The time-domain signal is bandpass-filtered about one of the natural frequencies and thus eliminating unwanted vibration and noise. Envelope detection of bandpass-filtered signal involves elimination of the carrier signal through rectification and low-pass-filtering of the signal. Hilbert transform can also be applied on the signal to separate the envelope from the carrier signal (Govardhan et al., 2017) and the same has been found to be quite effective. A spectral analysis of the envelope signal in the low-frequency range reveals the frequencies associated with the defect.

The technique has been found to be especially useful for detection of defect in rolling element bearings. Several researchers (McFadden and Smith, 1984; Tandon and Nakra, 1992) have reported success in detecting relevant spectral components by this method. Even for dynamic load on bearing, the technique has been found to be effective (Govardhan et al., 2017).

Table 2. Expressions for frequencies associated with different faults in electrical machines

Type of Defect	Expression	Remarks
Unbalance or dynamic eccentricity	f_s	90° phase difference between vibrations in two perpendicular directions, e.g. vertical and horizontal
Misalignment	$n \times f_s$	High amplitude vibration in axial direction
Mechanical looseness/ soft foot	$2 \times f_s$ and multiples	Soft foot may also cause misalignment
Damaged rolling element bearing	$\left(\frac{f_s}{2}\right)\left[1-\left(\frac{d}{D}\right)\cos\beta\right]$	Refer Table 1 for fault location on particular element; Vibrations also occur in high frequency range (> 5 kHz)
	$\left(\frac{Zf_s}{2}\right)\left[1-\left(\frac{d}{D}\right)\cos\beta\right]$	
	$\left(\frac{Zf_s}{2}\right)\left[1+\left(\frac{d}{D}\right)\cos\beta\right]$	
	$\left(\frac{Df_s}{d}\right)\left[1-\left(\frac{d}{D}\right)^2\cos\beta\right]$	
Oil whirl in sleeve bearings	$(0.43-0.48)f_s$	
Airgap eccentricity and unbalanced magnetic pull for synchronous machine	f_s	
Airgap eccentricity and unbalanced magnetic pull for induction machine	$f_s\left[(nN_r \pm k_e)(1-s) \pm pk\right]$	
Commutator faults in lap wound DC machine	$2pk_cf_s$	
Commutator faults in wave wound DC machine	$2k_cf_s$	
Broken rotor bars in induction machine	$f_s \pm \frac{2nf_{se}}{p}$	Small amplitude vibrations
Stator winding faults in induction and synchronous machine	$pf_s, 2pf_s, 4pf_s$	Disappears when electric supply is removed

$n =$ Integer; $N_r =$ number of rotor slots; $k_e =$ eccentricity order; $s =$ fractional slip in induction machine; $p =$ number of pole pairs; $k =$ space harmonics of stator windings MMF, 1,3,5,7,...; k_c = number of faulty commuter segments

Figure 3. The process of extraction of demodulated spectra by envelope detection technique

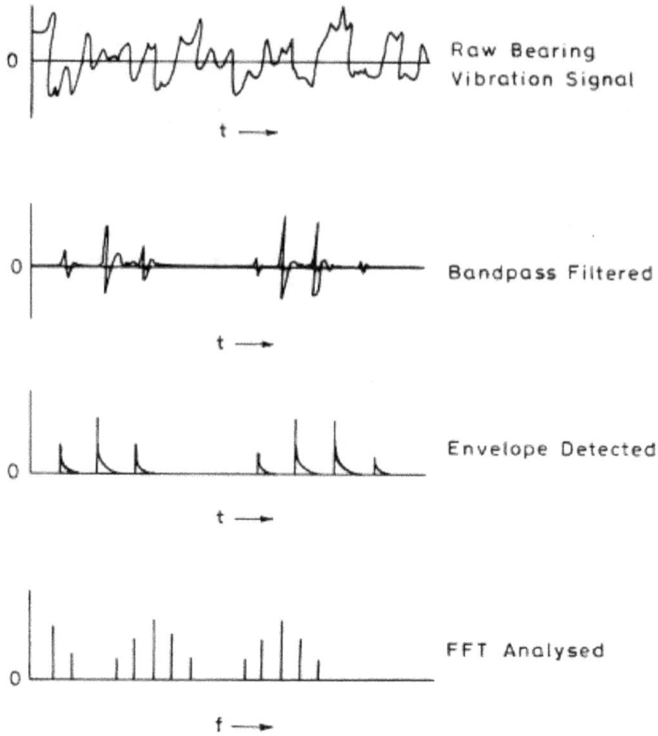

From the above discussion, it is evident that spectral analysis is a very powerful tool for vibration monitoring of any rotating machine, in general, and for rotating electrical machines, in particular. Even for broken rotor bars, for which vibration energy is in the lower range, spectral analysis could be used successfully to detect spectral components at fault frequencies as stated in Table 2 (Gama et al., 2017). The technique, however, has some weaknesses also. As discussed above, the effectiveness of the technique gets impaired in the presence of high value of background noise. The effectiveness gets further reduced when the signal is not perfectly stationary in nature.

Time - Frequency Domain Analysis

For any rotating electrical machine, the stationarity of the vibration signal cannot be guaranteed. For example, the shaft frequency of an induction machine depends on electrical supply frequency and slip. Both these factors may cause a fluctuation in the shaft frequency resulting in loss of stationarity. Possibility of slip exists wherever two components in contact have relative motion between them. In all such cases, stationarity of the signal will be adversely affected.

The search for a vibration monitoring technique that can retain the strength of spectral analysis even in a noisy and quasi-periodic environment, has led to the application of several techniques in time-frequency domain. These techniques can simultaneously generate time and frequency information of the signals and are more effective when the vibration signal lacks stationarity.

Short Time Fourier Transform

Short time Fourier transform or short-term Fourier transform (STFT) is used to determine the frequency content of a signal over a localized section as it changes with time. In practice, the signal is multiplied by a window function which is non-zero for a very small interval of time. Fourier transform of the resultant signal is taken as the time progresses, resulting in a two-dimensional representation of the signal. Mathematically, continuous-time STFT of a signal $x(t)$ is given by the following expression:

$$X(\tau, f) = \int_{-\infty}^{\infty} x(t) W(t-\tau) e^{-j2\pi ft} dt$$

where $W(t-\tau)$ is the window function of fixed width centered on time τ.

Wigner-Ville Distribution

Wigner-Ville distribution (WVD) is also used as transform for time-frequency analysis of a signal. WVD also uses a window to analyze the spectral density of a signal. The distribution was originally proposed by Wigner for applications in quantum mechanics. Subsequently, its suitability for analysis of nonstationary or quasi-periodic signals was detected. For a continuous complex signal $x(t)$, the distribution is given by the following equation (McFadden and Wang, 1990):

$$W_x(t, f) = \int_{-\infty}^{\infty} x\left(t + \frac{\tau}{2}\right) x^*\left(t - \frac{\tau}{2}\right) e^{-j2\pi f\tau} d\tau$$

Where * denotes complex conjugate. An equivalent definition of the distribution is also given by the following expression in terms of $X(f)$ being the Fourier transform of $x(t)$

$$W_X(f,t) = \int_{-\infty}^{\infty} X\left(f + \frac{\eta}{2}\right) X^*\left(f - \frac{\eta}{2}\right) e^{-j2\pi t\eta} d\eta$$

Thus, a symmetric role exists in time and frequency domains and $W_x(t,f) = W_X(f,t)$

Wavelet Analysis

The time-frequency domain analyses described above, viz., STFT and WVD, are handicapped by the stationarity of the windows used in these techniques. This disadvantage could be overcome in the multi-resolution analysis technique called wavelet transform (Mortazavizadeh & Mousavi, 2014; Paliwal, 2015). Wavelets are small wave-like functions of limited length with finite energy and satisfy the following equations.

$$\int_{-\infty}^{\infty} \psi(t)\,dt = 0$$

$$\int_{-\infty}^{\infty} \psi^2(t)\,dt < \infty$$

Some of the wavelets, which have found widespread applications in analysis of vibration signal, are shown in Figure 4 (Burrus et al., 1998).

Wavelet transform provides multi-resolution analysis of the signal using an appropriate wavelet which can be scaled and translated. Continuous wavelet transform (CWT) of a signal $f(t)$ is given by the following expression (Paliwal, 2015):

$$CWT(a,b) = \frac{1}{\sqrt{|a|}} \int_{-\infty}^{\infty} f(t)\,\psi^*\left(\frac{t-b}{a}\right) dt$$

Figure 4. Some of the wavelets used in analysis of vibration signal

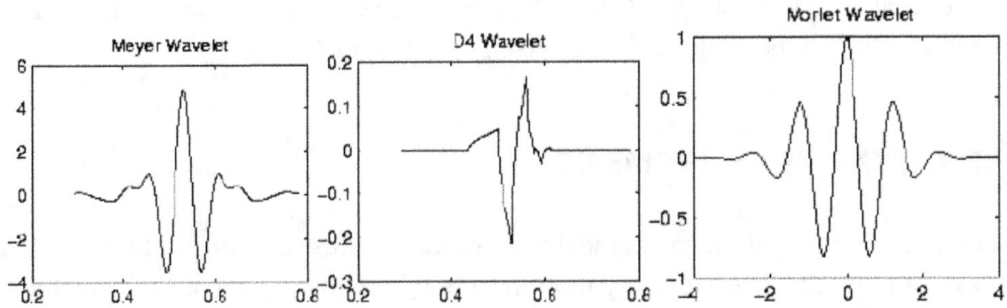

Parameters a and b represent scale and translation indices respectively. Scale index is related to frequency and translation index results in time shifting. Thus wavelet transform is not restricted by a fixed window. Discrete wavelet transform is also used in signal analysis and is given by the following expression (Paliwal, 2015):

$$DWT\left(c,d\right)=\frac{1}{\sqrt{2^c}}\int_{-\infty}^{\infty}f\left(t\right)\psi^*\left(\frac{t-2^c d}{2^c}\right)dt$$

Scale and translation indices in the discrete transform are represented by 2^c and $2^c d$ respectively.

Because of its multi-resolution properties, wavelet transform is being widely used in analysis of non-stationary and transient vibration signals. A typical plot showing the variation of wavelet coefficients against scale and translation is depicted in Figure 5.

Figure 5. A typical plot of wavelet transform of a vibration signal

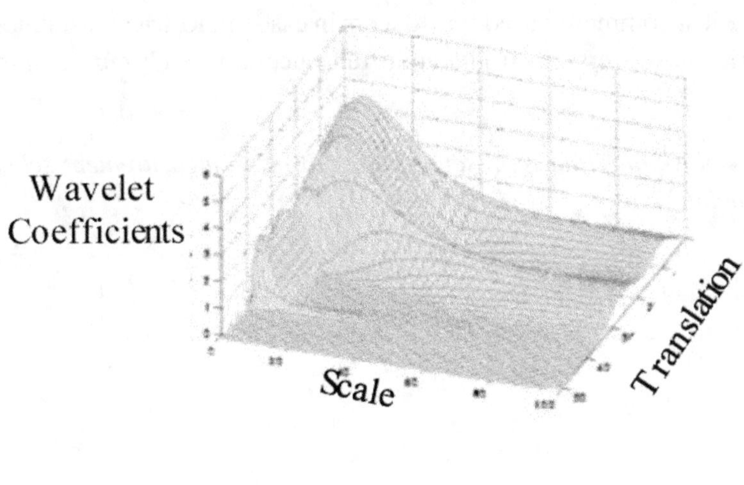

Several researchers (Wang and McFadden, 1996; Prabhakar et al., 2002; Holm-Hansen et al., 2004; Paliwal et. al, 2014) have reported success in vibration monitoring of rolling element bearings and gears using wavelet analysis.

VIBRATION MEASUREMENT

The steps involved in the measurement of vibration are illustrated in the block diagram shown in Figure 6. The vibratory motion (or the dynamic force) of the machine is sensed by an element called transducer which converts it into an electrical signal. The signal produced by the transducer normally requires further processing or conversion to bring it to the required level. The processed signal is then displayed and analyzed.

The vibration measurement scheme often depends on the following considerations: (i) expected ranges of frequencies and amplitudes; (ii) sizes and location of the machines involved; and (iii) type of data processing desired. Different elements used in the vibration measurement are elaborated in the following paragraphs.

Transducer Element

In vibration measurement, a transducer is a device that transforms the physical variable involved (in this case, vibratory motion) into an equivalent electrical signal. Vibration transducers can be broadly classified into two types: (i) seismic type that is directly attached to the vibrating body; and (ii) fixed reference or proximity type in which the vibratory motion is measured against a fixed datum. In case of electrical machines, both these types are used depending on the class of machines, type of bearings and stiffness of casing. Seismic transducers are basically spring-mass-damper systems and are further classified into (a) vibrometers for which the transducer natural frequency is low; and (b) accelerometers with high natural frequency. Most commonly used transducers in case of electrical machines include piezoelectric transducers, electrodynamic transducers and eddy current transducers.

Figure 6. A block diagram representing the vibration measurement scheme for a rotating machine

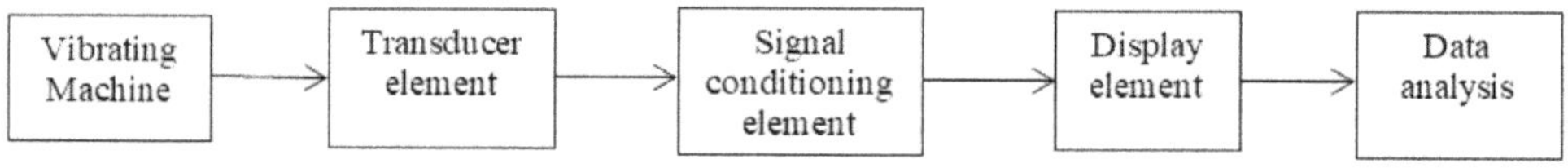

Piezoelectric Transducer

Piezoelectric transducers are of self-generating type and use piezoelectric crystals such as quartz, lithium sulfate and Rochelle salt which generate electrical charge when subjected to deformation or mechanical stress. For vibration measuring transducers (Figure 7), piezoelectric crystals are placed between the mass and the frame which is rigidly attached to the vibrating body. The relative motion of the mass with respect to the frame, gives an electrical output. Since the crystal acts like a spring of very high stiffness, the natural frequency of the transducer is very high and is therefore used as an accelerometer.

Piezoelectric transducers are used as casing mounted transducers to measure vibration in horizontal, vertical and axial direction. Advantages of piezoelectric accelerometer include compactness, ruggedness, high sensitivity and high frequency range. Because of these advantages, this type of transducer finds the widest applications in vibration measurement.

Electrodynamic Transducer

Electrodynamic transducers are based on the principle that when an electrical conductor, in the form of a coil, moves in a magnetic field, a voltage is generated in the conductor. Induced voltage is proportional to the rate of change of flux. The output voltage is proportional to the velocity of the vibrating object. Electrodynamic transducers can be of both seismic type and fixed reference type as illustrated in Figure 8.

In case of a seismic type electrodynamic transducer, the magnet acts as the seismic mass and is placed in a stationary coil. The mass of the magnet being quite high, the natural frequency of the instrument is low. The transducer, thus, can be

Figure 7. A piezoelectric transducer

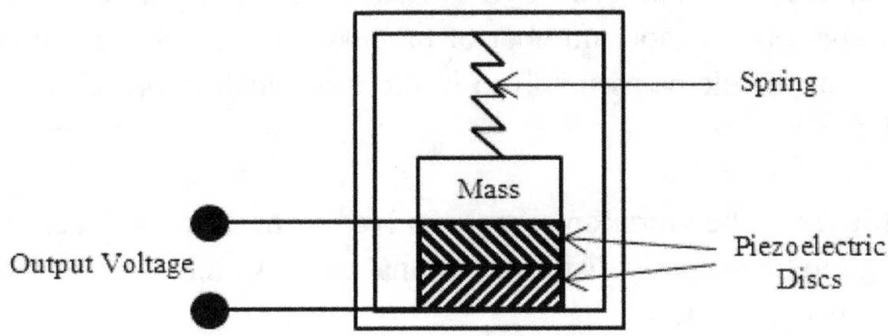

Figure 8. Electrodynamic transducers (a) seismic type; and (b) fixed reference type

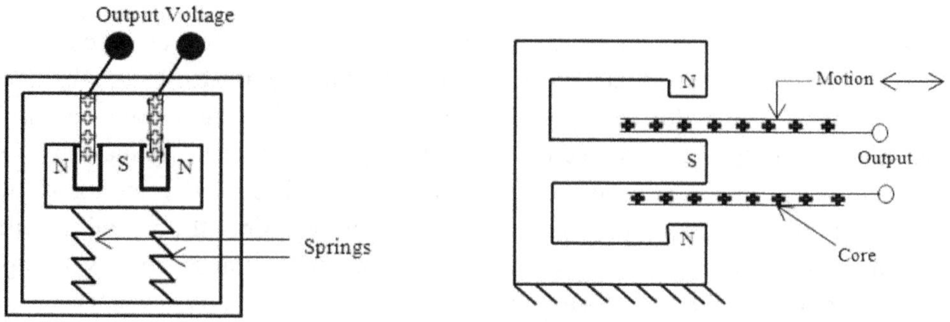

categorized as a vibrometer and is used as a velocity pick up. Electrodynamic type of fixed reference transducer has the magnet fixed and the coil is attached to the vibrating object.

Eddy-Current Transducer

Eddy current transducers are of non-contacting displacement pick up types of sensors. A coil is excited by an A.C. supply inducing a primary alternating magnetic field around the coil. This magnetic field induces small current in the metallic surface of the vibrating body, called eddy current. The eddy currents create an opposing magnetic field that interferes with the primary field, resulting in beats. The interaction of the magnetic fields is dependent on the distance between the transducer and the metallic vibrating surface. As the vibration causes a change in this distance, the system produces a voltage output proportional to the displacement of the metallic surface.

Signal Conditioning Element

The output signal from the transducer often requires modification before the same can be displayed, recorded or analyzed. Based on the requirements, the signal is passed through one or more number of processes during the signal conditioning stage. Some of the elements involved in the modification processes in this stage are as follows:

1. **Amplifiers:** The vibration signal received from the transducer is often weak enough to be properly displayed or analyzed. An amplifier is therefore used to amplify the signal.

2. **Filters:** Depending on the frequency range of importance, filters are used to remove unwanted signals. These filters may be of low-pass, high-pass or band-pass types. In case of rotating electrical machines, vibrations produced by some of the defects, e.g., unbalance, misalignment etc. lie in the low frequency ranges requiring low-pass filtering. On the other hand, defects in rolling element bearings excite resonances in high frequency range. For appropriate analysis, the vibration signals are often processed through high-pass as well as band-pass filters.
3. **Differentiating/ Integrating Elements:** Vibration signals received from a piezoelectric accelerometer and an electrodynamic transducer are proportional to acceleration and velocity of the vibrating body respectively. Parameters to be studied are often different from those obtained from the output of the transducer. Acceleration is the preferred parameter for high speed machines, velocity for medium speed and displacement for low speed machines. The output from the transducer is therefore passed through differentiating/ integrating circuits to obtain the desired parameter. For example, if output proportional to velocity of the vibrating object is required, output signal from a piezoelectric accelerometer will have to be processed through an integrating element.
4. **A-D (Analog To Digital) Converters:** The outputs of transducers are normally analog in nature. However, with the development in digital technology, it becomes necessary to convert the signals into digital forms before the same is displayed, recorded or further analyzed.
5. **Compensating Elements:** Frequency responses of signals sometimes require improvement to match the characteristics of the display device. Compensating elements are used for this purpose. For example, output from a piezoelectric accelerometer having high impedance, is processed through an impedance converter, viz., a cathode follower to convert it into a signal with low impedance as required by the display device. Cathode followers often form part of the charge amplifiers used in conjunction with piezoelectric accelerometers (Nakra et al., 1989).

Display/ Recording Element

These elements form the final stages of the measurement system. They may be of analog or digital type, depending on whether the display/ recording are of continuous or discrete type respectively. The devices included in this group of elements may be broadly classified into three types: basic system, portable system and computer-based system. The basic system normally consists of a vibration meter to measure overall vibration levels (RMS or peak values of acceleration or velocity) and a stroboscope to indicate the speed of machine.

Portable devices are used for display/ recording of results in fields, remote locations and for multiple uses in industries. With advancement in technology, portable devices these days can also perform some analyses of the signal. There are many portable devices available, which have built-in software to perform several analyses ranging from determination of time domain indicators to FFT analysis with SNR improvement. These devices are often equipped with sufficient memory to store/ record data for comparison and analysis in future.

Computer-based devices are useful when the number of machines and monitoring points are high and the complexities of faults are severe. A centralized database can also be maintained to enhance diagnostic capabilities. A class of powerful software is available for detailed analyses of the signal. This class includes both generic and customized varieties. Some of the software (e.g. MATLAB) have toolboxes meant for different analyses of the same signal. Graphical displays of output are also possible for easier understanding of the parameters.

CONCLUSION

Attention has been focused on condition monitoring of electrical machines for last couple of decades. Vibration measurement and analysis is one of the widely used techniques applied for the purpose. Vibration is produced in electrical machines due to different causes. These causes include both electrical and mechanical varieties and have specific features associated with them. For extraction of these features, vibration analysis is carried out in time, frequency and time frequency domains. Some indicators associated with time domain analysis, in general, provide overall information about machine health without specifying the cause. Spectral analysis, however, can provide information indicative of specific causes. Sometimes additional steps are required along with spectral analysis to improve SNR. When the signal is of non-stationary nature, time frequency domain analyses become effective, wavelet analysis being the most practiced among them. Vibration measurement and analysis also involves proper choice of transducer, appropriate signal conditioning elements and effective recording/ storage and display devices.

Vibration monitoring, though quite effective in general, is not the best suited technique for some of the electrical machine defects for which other monitoring techniques are found to be more effective. However, there are many components in electrical machines for which vibration monitoring is the most effective technique. For example, bearing defects which account for majority of failures in electrical machines, are best detected by vibration measurement and analysis.

REFERENCES

Binns, K. J., & Dye, M. (1973). Identification of principal factors affecting unbalanced magnetic pull in cage induction motors. *Proceedings of IEE – Pt. B, 120*, 349 – 354.

Burrus, C. S., Gopinath, R. A., & Guo, H. (1998). *Introduction to wavelets and wavelet transforms*. Prentice Hall.

Cameron, J., Thomson, W., & Dow, A. (1986). Vibration and current monitoring for detecting air gap eccentricity in large induction motors. *Proceedings of IEE – Pt. B, 133(3)*, 155 – 163.

Chaturvedi, G. K., & Thomas, D. W. (1982). Bearing fault detection using adaptive noise cancelling. *ASME Journal of Mechanical Design*, *104*(2), 280–289. doi:10.1115/1.3256337

Corbo, M. A., & Malanoski, S. B. (1996). Practical design against torsional vibration. In *Proceedings of the Twenty fifth Turbomachinery Symposium* (pp. 189 – 222). Texas A&M University.

Cudworth, C. J., Smith, J. R., & Mykura, F. (1984). Mechanical damping of torsional vibrations in turbogenerators due to network disturbances. In *Proceedings of the Third International Conference on Vibration in Rotating Machinery* (pp. 139 – 145). Institution of Mechanical Engineers.

Gama, R. S., Silva, C. C. J., Nascimento, C. E., Silva, A. M., & Cesar, D. C. (2017). Comparing MCSA with vibration analysis on rotating electrical machines to detect broken bars – a case study. *Journal of Multidisciplinary Engineering Science and Technology*, *4*(8), 7918–7913.

Gopinath, S. (2010). Study on electric motor mass unbalance based on vibration monitoring analysis technique. In *2nd International Conference on Mechanical and Electrical Technology*. Singapore: IEEE. 10.1109/ICMET.2010.5598417

Govardhan, T., Choudhury, A., & Paliwal, D. (2017). Vibration analysis of a rolling element bearing with localized defect under dynamic radial load. *Journal of Vibration Engineering & Technologies*, *5*(2), 165–175.

Holm-Hansen, B. T., Gao, R. X., & Zhang, L. (2004). Customized wavelet for bearing defect detection. *ASME Journal of Dynamic Systems. Measurement and Control*, *126*(4), 740–745. doi:10.1115/1.1850534

Holbert, K. E., Lin, K., & Karady, G. G. (2006). Enhancement of electrical motor reliability through condition monitoring. In *IFAC Symposium on Power Plants and Power Systems Control* (pp. 255 – 260). Kananaskis, Canada: Academic Press.

Iorgulescu, M., Alexandru, M., & Beloiu, R. (2012). Noise and vibration monitoring for diagnosis of DC motor's faults. In *13th International Conference on Optimization of Electrical and Electronic Equipmen*. Brasov, Romania: IEEE. 10.1109/OPTIM.2012.6231919

Jordan, H. (1950). *The low noise electrical motor*. Essen: Springer Verlag.

McFadden, P. D., & Wang, W. (1990). *Time-frequency domain analysis of vibration signals for machinery diagnostics (I) Introduction to Wigner-Ville distribution.* Report no. 1859/90, University of Oxford

McFadden, P. D., & Smith, J. D. (1984). Model for the vibration produced by a single point defect in a rolling element bearing. *Journal of Sound and Vibration, 96*(1), 69–82. doi:10.1016/0022-460X(84)90595-9

Mortazavizadeh, S. A., & Mousavi, S. M. G. (2014). A review on condition monitoring and diagnostic techniques of rotating electrical machines. *Physical Science International Journal, 4*(3), 310–338. doi:10.9734/PSIJ/2014/4837

Nakra, B. C., Yadava, G. S., & Thuestad, L. (1989). *Vibration measurement and analysis*. New Delhi, India: National Productivity Council.

Nandi, S., Ahmed, S., & Toliyat, H. (2001). Detection of rotor slot and other eccentricity related harmonics in a three phase induction motor with different rotor cages. *IEEE Transactions on Energy Conservation, 16*(3), 253–260. doi:10.1109/60.937205

Negoita, A., Scutaru, G., & Ionescu, R. M. (2010). A brief review of monitoring techniques for for rotating electrical machines. *Bulletin of the Transalvania University of Brasov, 3*(2), 277–284.

Orman, M., & Pinto, C. (2013). Vibration, sound and thermal analysis as a condition monitoring methods for electric motors. *Applied Mechanics and Materials, 239-240*, 462–467. doi:10.4028/www.scientific.net/AMM.239-240.462

Paliwal, D., Choudhury, A., & Govardhan, T. (2014). Identification of faults through wavelet transform vis-à-vis fast Fourier transform of noisy vibration signals emanated from defective rolling element bearings. *Frontiers of Mechanical Engineering, 9*(2), 130–141. doi:10.100711465-014-0298-6

Paliwal, D. (2015). *Identification of single and multiple defects in rolling element bearing through wavelet transform* (Unpublished doctoral dissertation). Sir Padampat Singhania University, Udaipur, India.

Prabhakar, S., Mohanty, A. R., & Sekhar, A. S. (2002). Application of discrete wavelet transform for detection of ball bearing race faults. *Tribology International, 35*(12), 793–800. doi:10.1016/S0301-679X(02)00063-4

Rao, S. S. (2004). *Mechanical vibrations*. Delhi, India: Pearson Education Inc.

Swann, S. A. (1963). Effect of rotor eccentricity on the magnetic fields of a non-salient pole machine. *Proceedings of IEE, 110*, 903–915.

Tandon, N., & Choudhury, A. (1999). A review of vibration and acoustic measurement methods for the detection of defects in rolling element bearings. *Tribology International, 32*(8), 469–480. doi:10.1016/S0301-679X(99)00077-8

Tandon, N., & Choudhury, A. (2000). A theoretical model to predict vibration response of rolling bearings in a rotor bearing system due to distributed defects under radial load. *ASME Journal of Tribology, 122*(3), 609–615. doi:10.1115/1.555409

Tandon, N., Yadava, G. S., & Ramakrishna, K. M. (2007). A comparison of some condition monitoring techniques for the detection of defects in induction motor ball bearings. *Mechanical Systems and Signal Processing, 21*(1), 244–256. doi:10.1016/j.ymssp.2005.08.005

Tavner, P., Ran, L., Penman, J. & Sedding, H. (2008). *Condition monitoring of rotating electrical machines*. London: The Institution of Engineering and Technology.

Thomson, W. T., & Dahleh, M. D. (2007). *Theory of vibration with applications*. Delhi, India: Pearson Education Inc.

Trutt, F. C., Sottile, J., & Kohler, J. L. (2001). Detection of AC machine winding deterioration using electrically excited vibrations. *IEEE Transactions on Industry Applications, 37*(1), 10–14. doi:10.1109/28.903120

Tse, P. W., Peng, Y. H., & Yam, R. (2001). Wavelet analysis and envelope detection for rolling element bearing fault diagnosis – their effectiveness and flexibilities. *ASME Journal of Vibration and Acoustics, 123*(3), 303–310. doi:10.1115/1.1379745

Vaimann, T., & Kallaste, A. (2012). Condition monitoring of electrical machines. In *11th International Symposium "Topical Problems in the Field of Electrical and Power Engineering" and "Doctoral School of Energy and Geotechnology II"* (pp. 209-212). Academic Press.

Walker, D. N., Adams, S. L., & Placek, R. J. (1981). Torsional vibration and fatigue of turbine-generator shafts. *IEEE Transactions on Power Apparatus and Systems*, *PAS-100*(11), 4373–4380. doi:10.1109/TPAS.1981.316848

Wang, W. J., & McFadden, P. D. (1996). Applications of wavelets to gearbox vibration signals for fault detection. *Journal of Sound and Vibration*, *192*(5), 927–939. doi:10.1006/jsvi.1996.0226

Yang, S. J. (1981). *Low noise electric motors*. Oxford, UK: Clarendon Press.

Chapter 9
A State of Art Critique Review of Role of Optimization in NDTS for Effective Condition Monitoring

Ashish Khaira
Samrat Ashok Technological Institute, India

Ravi K. Dwivedi
Maulana Azad National Institute of Technology, India

ABSTRACT

Nondestructive testing (NDT) techniques play a pivotal role in a condition-based monitoring system. Generally, using an optimization technique for optimizing the available solution gives added advantage. By keeping this fact in mind, this chapter is an attempt to provide a detailed review of research work from 2000-2018 that covered the role of optimization for effective condition monitoring using NDTs. It was observed from literature that little work found related to NDT process optimization. The researcher and practitioner can use the outcome of this work as a beacon and improve it further.

INTRODUCTION

As per Escaler et al. (2006) and Garnier et al. (2011), the early stage and precise fault detection are vital for machine life expectancy as it becomes more trouble-free to plan necessary corrective measures accordingly and it is also essential for optimum performance of machine/components. Non destructive testing (NDT)

DOI: 10.4018/978-1-5225-6989-3.ch009

as the name suggests, is a technique in which component tested for flaw without disintegrating or altering its present structure. It was highlighted by Willcox et al. (2003), D'Orazioa et al. (2008) and Krautkrämer et al. (2013) that NDT is an efficient fault detection technique which helps in reduction of Mean Time to Repair (MTTR) by identifying the fault cause with its location and increases Mean Time between Failure (MTBF) due to proactiveness. As per Khaira et al. (2015), the application of these techniques is in material testing, defect detection in welding, defect detection in casting, forging etc.

The word "optimization" means selecting the best available option from a wide range of possible alternatives. There are several sources of complexity, such as the computational difficulties in modeling, the potentially high number of free variables, or a high number of objectives and constraints. While, decision makers, identifies the most preferred one among the solutions. The decision makers choose one reasonable alternative from a limited set of available ones.

The organization of this research paper is as follows: In section 2, review of optimization techniques in NDTs covered. In section 3, the result and discussion part covered and then, the conclusions drawn in the last section. The contribution and significance of this research study are twofold; firstly, the research work from year 2000-2018 by different researchers, that includes the combinations of NDT using optimization reviewed; secondly, the untouched techniques were highlighted to provide research gaps for further research.

REVIEW OF OPTIMIZATION TECHNIQUES IN NDTS

The following section covers more than a decade research and the contribution of different researchers in the field of NDTs by using optimization techniques.

Zhongqing Su et al. (2006) comprehensively reviewed lamb wave-based damage identification methodology for composite structures, addressing the advances and achievements in these techniques in the past decades using forward (direct) algorithm, Inverse algorithm (General inverse approaches, Artificial intelligence, Tomography) etc. Major importance is given to the unique characteristics and mechanisms of Lamb waves in laminated composites; approaches in wave mode selection, generation and collection. Furthermore, it was needed to select appropriate lamb modes in terms of identification of individual types of damage. The excitation frequency, waveform, cycle number and other signal characteristics should be comprehensively considered during mode selection and diagnostic waveform design. Undesired modes can only be suppressed rather than completely eliminated. The multi-mode purification is an intriguing but challenging issue. Damage identification is normally a typical yet complex inverse pattern recognition problem and cannot be effectively solved by traditional logic approaches. Detecting multiple damages is another issue worthy of investigation. The use of Lamb waves seems more viable than other tools for its exclusive responses to different damage status, but more complex scattering phenomena complicate the lamb wave signal interpretation, and distinguishing such signals remains problematic.

Z. Hameeda et al. (2009) reviewed different techniques, methodologies and algorithms, including Global Fault Detection System, Artificial intelligence, FDI (fault detection and isolation), Recursive isolation developed to monitor the wind turbine performance as well as an early fault detection to keep away the wind turbines from catastrophic conditions due to sudden breakdowns. The bicoherency or the skewness function, or simply the bicoherence is:

$$b_x(\omega_1, \omega_2) = \frac{B_x(\omega_1, \omega_2)}{\sqrt{S_x(\omega_1) S_x(\omega_2) S_x(\omega_1 + \omega_2)}} \quad (1)$$

Sometimes a slightly different denominator used where,

$$S_x(\omega_1) S_x(\omega_2) = E[X(\omega_1) X^*(\omega_1)] E[X(\omega_2) X^*(\omega_2)] \quad (2)$$

Fausto Pedro García Márquez et al. (2012) reviewed state-of-the-art in the condition monitoring (CM) of wind turbines, describing the different maintenance strategies, CM algorithms, including Statistical methods, Trend analysis, Filtering methods, Time-domain analysis, Cepstrum analysis, Time synchronous averaging (TSA), Fast-Fourier transform (FFT), Amplitude demodulation, Order analysis, Wavelet transforms, Hidden Markov models etc.

Zoubir Mehdi Sbartaï et al. (2012) showed statistical fusion of multiple NDTs viz. Ground-penetrating radar (GPR), electrical resistivity and ultrasonic pulse velocity using two techniques, namely response surface method (RSM) and artificial neural networks (ANNs). In this paper, NDT techniques GPR, electrical resistivity and ultrasonic pulse velocity combined to predict more accurately concrete properties such as strength and water content. The results obtained show the effectiveness of the statistical modeling to predict the properties of concretes by fusion of NDT measurements. In the context of this study, the performances of the two techniques of fusion appear relevant in terms of water content and concrete strength prediction. ANN models exhibit better predictive ability than RSM ones. In Response Surface Method (RSM) Input variables, x_i ($i = 1 \ldots n$) also called basic variables of the phenomenon, characterized by a set of recorded statistical information θ_j (j = 1 . . . p) as distribution functions (which could be correlated or not), standardized moments, etc. To create a response surface model, one must provide:

1. Representative and ordered basic variables $\{x\} = \{x_1 . . ., x_n\}$. Their representativeness is to use only explanatory variables and one need also to order the measured values of each variable from its lowest value of its greatest.
2. The statistical information set of the vector {x} (distribution functions $\{\theta\} = \{\theta_1 . . ., \theta_n\}$).
3. The function which is an approximation of the response ''y'', explicitly formulated in terms of {x} knowing the variable statistics.
4. A metric (distance function) for the basic variables and the response space. This metric measures how good the fit of the approximation function is to the response ''y''.

M.R. Nasiri et al. (2011) presented a method to automate the cavitations fault detection in centrifugal pumps using vibration signature analysis. The author used the neural network approach to identify three healthy/faulty conditions, namely normal, moderate and fully developed cavitations; data obtained from vibration analysis were used as input for neural analysis. Experiments were also conducted to evaluate the presented model. Also the number of sensors and the best sensor positions studied. If just one sensor is available then, radial position should be preferred, and, if two sensors are available, then they should be used in radial and back positions and if three sensors are available then, the cavitations would be detected without any error by placing the sensors in radial, back and front positions.

Jose Vicente Abellan-Nebot (2010) reviewed six key issues involved in the intelligent machining systems development viz. the different sensor systems applied to monitor machining processes, the most effective signal processing techniques, the most frequent sensory features applied in modeling machining processes, the sensory feature selection & extraction methods for using the relevant sensory information, the design of experiments required to model a machining operation with the minimum amount of experimental data and the main characteristics of several artificial intelligence techniques to facilitate their application/selection. The paper outlined the possible methods to employ in selecting or extracting the most meaningful features for the machine process modeling. This paper also covers briefly about different sensors and their limitations. The criteria for sensor fusion explained effectively Feature selection/extraction part discussed in this paper is worth reading. A generic methodology for design of experiment (DOE) in machining monitoring systems is also covered in this paper.

Bernhard Sick (2002) investigated 138 publications dealing with on-line and indirect tool wear monitoring in turning by means of artificial neural networks. This article investigated the methods used at the digital pre-processing level, the feature extraction level, the wear model level, and the decision level. The article compares the methods applied in these publications as well as the methodologies used to select certain methods, to carry out simulation experiments, to evaluate and to present results, etc. Engage in evolutionary algorithms for structure optimization. Focus on the development of procedures for an automatic determination of the optimal length of a reference (observation) time interval for local temporal information (short-term or long-term information).

Dhirendr Nath Thatoi et al. (2012) covered different techniques and methodologies for fault diagnosis and condition monitoring of damaged structures subjected to varying dynamic loading. The methods used are classical, wavelets transform, and finite element methods, artificial intelligence methods, and numerical experimental methods. Furthermore, research on the dynamics of cracked structures to monitor a structure and diagnose faults at the earliest possible stage but using artificial intelligence techniques; it was observed that the forecasting time for fault diagnosis improves a lot in comparison to other methodologies. The classical and finite element method can be used as direct method for getting the vibration signatures (e.g., natural frequencies and mode shapes) where as the artificial intelligence methods can be used as an inverse technique for condition monitoring of vibrating systems. In future the scope of hybrid artificial intelligence techniques can be studied for analyzing the behavior of dynamic structures.

Zhigang Tian et al. (2012) covered a physical programming based approach to deal with proportional hazards model, physical programming (Multi-Objective Optimization Approach) along with review of multiple papers which use this technique in some industrial application. The maintenance cost and reliability objectives evaluated, based on proportional hazards model and a control limit condition based maintenance replacement policy. An example of shear pump bearings in a food processing plant used to illustrate the proposed approach. The basic model of proportional hazard model (PHM) combines a Weibull baseline hazard function with a component, including all the covariates which affect the time to failure, as follows:

$$(t, Z(t)) = \frac{\beta}{\eta}\left(\frac{t}{\eta}\right)^{\beta-1} \exp\{\sum_{i=1}^{m} y_i z_i(t)\} \tag{3}$$

Where h (t, Z (t)) is the hazard value, or failure rate value, at time t, given the values of z_1 (t), z_2 (t) . . . z_m (t). The first part of this model is the baseline hazard function $\beta/\eta\ (t/\eta)^{\beta-1}$, which takes into account the age of the equipment at the time of inspection, given the values of parameters β and η. The second part exp $\{\gamma_1 z_1 (t) + \gamma_2 z_2 (t) + \ldots + \gamma_m z_m (t)\}$ takes into account the covariates to consider them as the key factors reflecting the health condition of equipment.

Po-Chun Lin et al. (2011) stated previous fault patterns, results of detection analysis, historical records and data, and expert's experiences to ascertain fuzzy principles and calculate approximately the failure probability index of the components of a power transformer. This work used the standard data of IEC, IEEE, and CIGRE as condition parameters. According to the characteristics of each condition parameter, comparative degradation was introduced to reflect the degree of influence of the factors on the transformer condition. The method of fuzzy mathematics was adopted to find out the subordinate function of the transformer condition.

Linilson Padovese Silmara et al. (2004) analyzed the make use of the probabilistic neural network (PNN) as an automatic diagnostic technique for defect detection in rolling bearings using vibration signals. The effect of the metric and that of the mono and multi sigma in the PNN performance analyzed and discussed. Two forms of training sets were constructed, one using the power spectral density of the signals and another using a blend of scalar parameters. The results permit obtaining excellent classification rate as well as analyzing the influence of the main PNN parameters in its performance. A further step will be to adapt the Genetic algorithm (GA) to optimize a third case of sigma, one for each category or class of classifying space.

Jason Y.L. Goh et al. (2004) described micro deformation analysis using correlation (MDAC) and uses usual optical images to work out both in and out-of-plane sub-pixel displacements devoid of interpolation or curve-fitting. This paper reveals MDAC as an option to scanning acoustic microscopy (SAM) and dye penetrate testing for defect detection in IC chip possessing advantages over both techniques. A non-contact displacement measurement technique (MDAC) applied to the problem of measuring thermally induced strain in IC devices. More purposely, two stacks of images with dissimilar axial positions (through focus, Z) are taken; one before (cold) and one after deformation (hot). Also covers displacement and strain map consequences of IC devices that emphasize the advantages of using MDAC over/ in combination with SAM and dye penetrate testing. Specimen preparation for MDAC is minimal, required only cross-sectioning and finishing of the sample such that a few surface features remain (residual roughness). MDAC has also provided significant advantages over comparable systems such as SAM, dye penetrate testing and interferometer. Hill-Climbing Algorithm also used which is effective method, but rarely seen in other papers. For low light level conditions in which regions of the image contain only a few grey levels (<4 bits), the correlation algorithm has fewer features to lock-on, leading to reduced accuracy. To overcome this problem, a camera with a larger number of grey levels may be used. Alternatively, if this is not possible, a scheme may be implemented that combines a number of images with different exposures into a single extended dynamic range image.

A. Carvalho et al. (2014) presented pattern nonlinear classifiers using artificial neural networks, to check the possibility to detect and classify defects in pipelines inspected through magnetic pigs (MFL technique). Several tests were performed on specimens with defects artificially inserted (internal and external corrosion and lack of penetration). Furthermore, this work may be continued fulfilling more tests in new pipelines, to increase the classifier robustness as well as increasing the scope of the defects to be classified such as excessive penetration, root bead misalignment, and hollow bead. Besides, it is possible to apply new filters and methods to reduce the signal magnitude to turn the process still faster and more efficient.

Herna´n D.Benı´tez et al. (2009) demonstrated training of 3 learning machines (MLP, RBF and SVM) and validated with a reference-free thermal contrast. The study of the thermal diffusivity impact on the modified DAC curves and the MLP classification error shows that affects more severely the DAC curves and the MLP classification errors than variations. Dynamic principal components analysis (DPCA) and k-nearest neighbor algorithm used as an optimization technique. The disadvantage of using learning machines in thermal NDT (TNDT) is that this approach requires at least one sample with a known set of defects to obtain the training dataset and that once the system trained, then this can only be used for the same material with the same experimental platform.

H. Kasban et al. (2011) covered welding defect detection using radiography and neural approach. The image is lexicographically ordered into a 1D signal. The Mel frequency Cepstral coefficient (MFCCs) and the polynomial coefficients are extracted from either the 1D signal, one of its transforms, or both of them. The discrete cosine transform (DCT), discrete sine transform (DST), and discrete wavelet transform (DWT) can be used for this purpose. An artificial neural network (ANN) trained with the extracted features to create a database. This paper clearly explained through experiment that the proposed approach can be used in a reliable way for automatic defect detection from radiography images with noise and blurring presence.

Anna Runnemalm et al. (2014) suggested a single-sided setup of a thermography system with a flash lamp as excitation source. This set-up prepared to mount on an industrial robot to achieve a fully automatic inspection system. With the suggested setup of an automated thermography system, the inspection can be operator-independent, and with a stationary inspection cell and the time, test place and situation can be controlled. The analysis algorithm intends to find out the spatial region in the obtained images analogous to the successfully welded area, i.e., the nugget size. Experiments illustrate that the arrangement is able to distinguish spot welds, compute the nugget diameter, and based on the information also separate a spot weld from a stick weld. The results were compared to the results of a peel test for each piece. In order to avoid uncertainties in the comparative measurement, the peel tests were performed by a skilled person with experience of measuring spot weld diameters. Even if the extracted characteristics of the welds most of the times are correct, the analysis is sometimes confused by the physical deformations of the surface created by the welding process. The most common method for inspection of spot welds is ultrasonic testing, but experience shows that the test reliability is often depending on the time, place, testing situation, and skill of the operator. General system for thermal non-destructive testing explained properly. Therefore, the absolute temperature, which requires exact knowledge about the emissivity, is not needed.

Q. Liu et al. (2007) presented a forward formulation to enable ultrasonic measurement of anisotropy and the directivity of dendrites in a weld, also describes the forward problem of a project that targets in developing a technique to calculate approximately the way of dendrites in an austenitic weld. The ray tracing technique, as an approximate method for simulating the behavior of ultrasound energy in regions that considered large compared with the ultrasound wavelength, based on geometrical optics. In simulations, the numerical method applied in the calculation from an incident ray to a transmitted ray. Only the transmitted ray with the same mode followed while the reflected ray is not under consideration. In order to further investigate the ray distribution from the probe through the weld, two additional rays included to each ray on both sides.

Iraklis Lazakis et al. (2016) discussed innovative solutions for ship inspection and maintenance. The paper presented the latest literature review on Hull Condition Assessment (HCA), image processing, Condition Based Maintenance (CBM) and the state-of-the-art in condition monitoring techniques. Maria Del Pilar et al. (2017) proposed a model that provides a structured solution for a complex problem. It was based on the analytic hierarchy process and pair wise comparison used for the assessment. Susilo et al. (2018) covered radiography process optimization using voltage, current, and exposure time. The radiography images were also analyzed using Contrast to Noise Ratio (CNR).

Result of Review

This research work based on in-depth and orderly investigation of the literature related to non destructive testing. Table 1 shows details about domains & optimization techniques which already applied in different-2 fields. This table is very useful for upcoming researchers to identify research gaps in this field. The Table 2 summarizes the description about various key NDT techniques along with their limitations.

Table 1. Details of domains and optimization technique

Author	Year	Domain	Optimization Technique
Zhongqing Su et al.	2006	Guided Lamb waves Detection Technique	Forward (direct) algorithm, Inverse algorithm (General inverse approaches, Artificial intelligence, Tomography)
Z. Hameeda et al.	2009	Vibration Analysis, Acoustic Emission, Thermography, Radiography	Global Fault Detection System, Artificial intelligence, FDI (fault detection and isolation), Recursive isolation
Fausto Pedro García Márquez et al.	2012	Ultrasonic testing, Vibration Analysis, Acoustic Emission, Thermography, Radiography	Statistical methods, Trend analysis, Filtering methods, Time-domain analysis, Cepstrum analysis, Time synchronous averaging (TSA), Fast-Fourier transform (FFT), Amplitude demodulation, Order analysis, Wavelet transforms, Hidden Markov models
Zoubir-Mehdi Sbartaï et al.	2012	Electrical resistivity methods (Wenner and four-probe square array),	Principle component analysis, ''PCA'' using MATLAB
Jose Vicente Abellan-Nebot ·	2010	Intelligent Machine Monitoring System	Artificial Intelligence
Z. Hameeda et al.	2009	CBM and fault detection	Power engineering-based algorithm, FFT Algorithm, Strain monitoring-based algorithm, Optical fibers-based algorithm
Bernhard Sick	2002	CBM	Neural Network
Dhirendr a Nath Thatoi et al.	2012	Classical method,	Artificial Intelligence
Zhigang Tian et al.	2009	CBM optimization	Proportional hazards model, Physical Programming (Multi-Objective Optimization Approach)
Po-Chun Lin et al.	2011	Machine condition monitoring and fault diagnostics	Fuzzy Comprehensive Evaluation
Jose Vicente Abellan-Nebot	2010	Acoustic Emission	Artificial Intelligence
Linilson Padovese Silmara et al.	2004	Automated Detection	Probabilistic Neural Network (PNN)
Jason Y.L. Goh et al.	2004	Scanning Acoustic Microscopy (SAM), Dye Penetrant Testing	Micro Deformation Analysis using Correlation(MDAC), Hill-Climbing Algorithm
A. A. Carvalho et al.	2014	Magnetic Testing	Neural networks
Herna´n D.Benı´tez et al.	2009	Infrared non-destructive testing	Dynamic principal components analysis (DPCA) and k-nearest neighbor algorithm.
H. Kasban et al.	2011	Radiography images	Neural Approach
Anna Runnemalm et al.	2014	Thermography	Analysis Algorithm.
Q. Liu et al.	2007	Ultrasonic Testing	Ray Tracing Algorithm

Table 2. Comparison of NDT techniques

NDT Techniques	Limitations
Visual Inspection (VT) – In this the inspector inspects the surface using his/her naked eyes therefore it does not require sophisticated apparatus.	• Not suitable for sub- surface defects • Prone to low reliability of fault detection as success rate solely depends upon inspectors skills.
Liquid Penetrant Testing In this the inspector inspects the surface using colored or fluorescent dye to increase the visibility of defects.	• Multiple stages therefore, difficult to automate • Proper lighting and clean surface is essential
Ultrasonic Testing In this the inspector inspects the surface by transmitting ultrasonic waves inside through specimen. Only single sided access needed and equipments are portable.	• Coupling medium needed • Highly skilled person needed • Surface should be free from rust &and excessive paints. • Thin parts difficult to inspect
Eddy Current In this the inspector inspects the surface using the principle of electromagnetic induction. In this method both coated and uncoated surface can be tested and there is no need of surface preparation.	• Selection of probe size is critical. • Not suitable for complex parts.

It is necessary to know the distribution of year wise paper in order to know the actual position of usage of optimization techniques. Figure 1 shows a crescent tendency in the articles time distribution. The graph clearly indicates that most of the work done in 2010-14.

For a researcher it is essential to know the areas in which these techniques used and areas in which are remains untouched. The figure 2 shows graph between number of research papers and domain. This graph shown in figure 2 clearly indicates that most of the work done in generalized condition monitoring and least work are on thermal/infrared, electrical resistivity and radiography. The Figure 3 shows plots between optimization techniques v/s number of research paper. From this plot it is evident that neural networks & artificial intelligence are the most commonly used techniques while other techniques are less frequently used. The Figure 4 shows plots between areas of research v/s number of research papers. It clearly indicates that most of the time optimization techniques are mostly for review as compared to real life application. The Figure 5 shows name of journal/conference v/s number of research papers. It is evident from graph that most of the time researcher preferred NDT & E International journal for publishing optimization in NDTs based papers.

Figure 1. Year of publication v/s number of papers

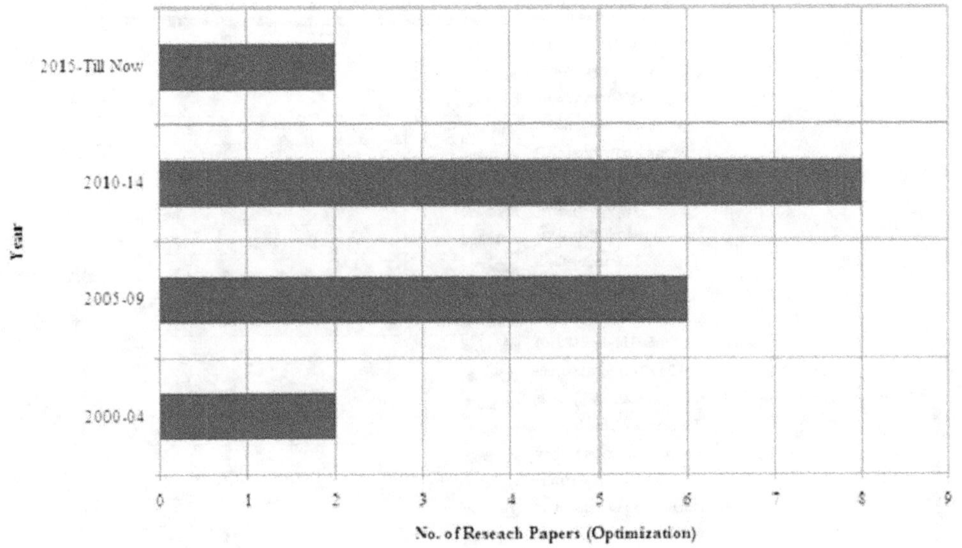

Figure 2. Domain v/s number of research papers

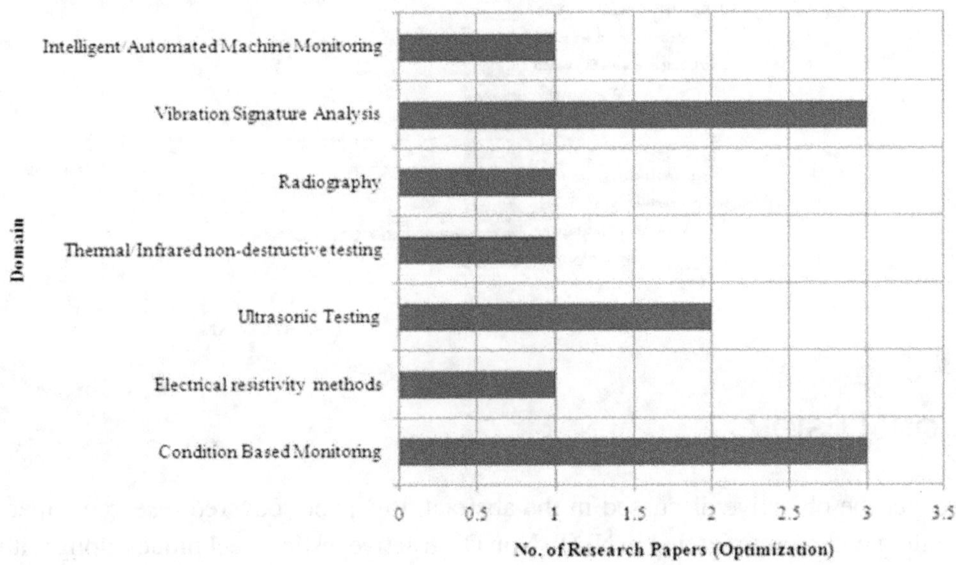

Figure 3. Optimization techniques v/s number of research papers

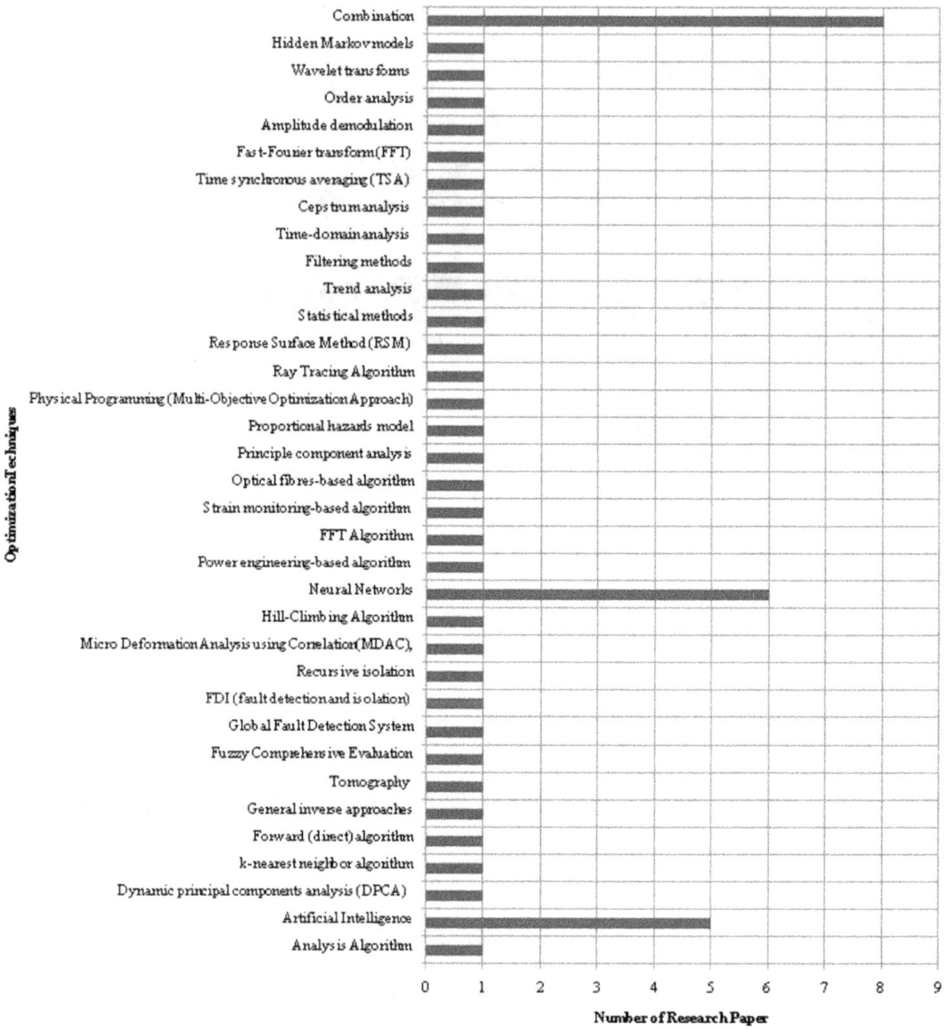

CONCLUSION

As per the objective discussed in the abstract, this paper covered research papers dealing with a optimization in NDT (Non Destructive testing) techniques along with some practical examples. This paper outlined the benefits of optimization technique in NDT. The paper highlighted that parameter optimization is necessary to ensure least erroneous operation. Optimization techniques can apply with lubrication, magnetic testing too. Furthermore, fuzzy logic, genetic algorithms, etc. can explore to use with NDTs.

Figure 4. Area of research v/s number of research papers

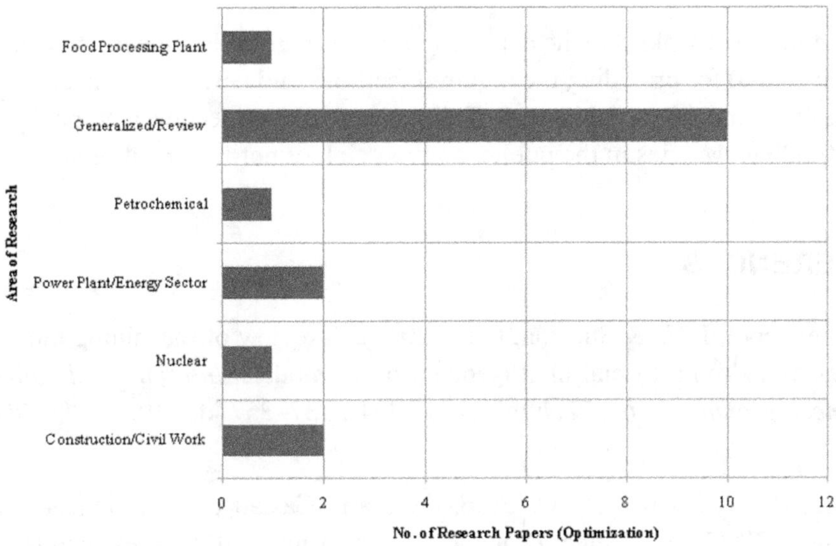

Figure 5. Name of journal/conference v/s number of research papers

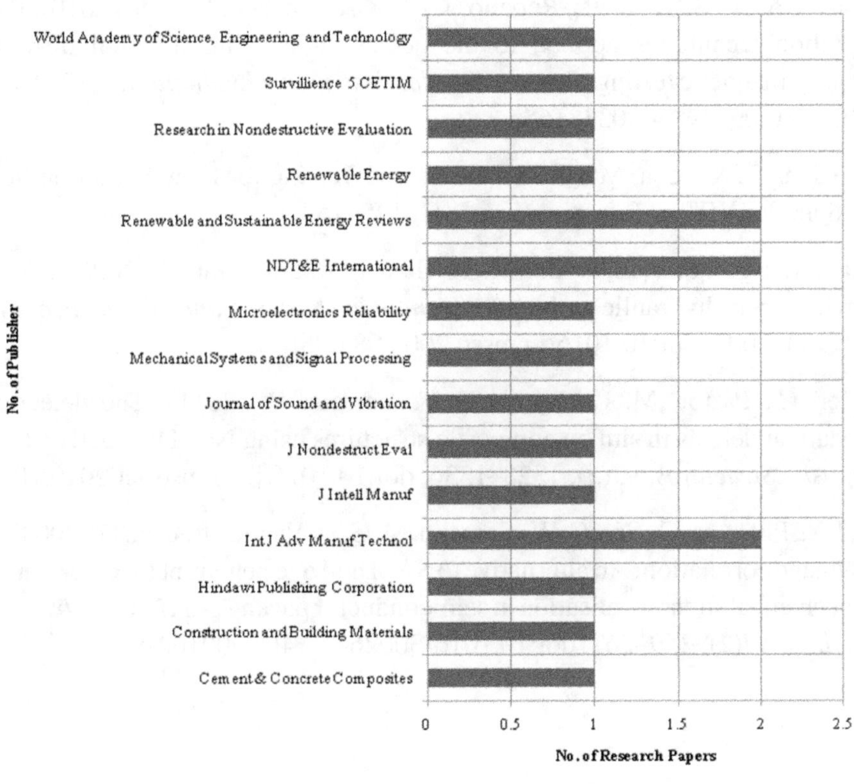

ACKNOWLEDGMENT

The authors are thankful to the department of Mechanical Engineering of MANIT-Bhopal (India) for providing the required facilities and permissions needed for the successful completion of this work. This research did not receive any specific grant from funding agencies in the public, commercial, or not-for-profit sectors.

REFERENCES

Abellan-Nebot, J. V., & Subirón, F. R. (2010). A review of machining monitoring systems based on artificial intelligence process models. *International Journal of Advanced Manufacturing Technology*, *47*(1-4), 237–257. doi:10.100700170-009-2191-8

Benítez, H. D., Loaiza, H., Caicedo, E., Ibarra-Castanedo, C., Bendada, A., & Maldague, X. (2009). Defect characterization in infrared non-destructive testing with learning machines. *NDT & E International*, *42*(7), 630–643. doi:10.1016/j.ndteint.2009.05.004

Carvalho, A. A., Silva, R. R., Rebello, J. M. A., & Sagrilo, L. V. S. (2010). Pattern recognition techniques applied to the detection and classification of welding defects by magnetic testing. *Research in Nondestructive Evaluation*, *21*(2), 91–111. doi:10.1080/09349840903381655

D'Orazioa, T., & Leoa, M. (2008). Ultrasonic Testing for Flaw Detection in Mild Steel Sample. *NDT & E International*, *41*, 145–154.

Escaler, X., Egusquiza, E., Farhat, M., Avellan, F., & Coussirat, M. (2006). Detection of cavitation in hydraulic turbines. *Mechanical Systems and Signal Processing*, *20*(4), 983–1007. doi:10.1016/j.ymssp.2004.08.006

Garnier, C., Pastor, M. L., Eyma, F., & Lorrain, B. (2011). The detection of aeronautical defects in situ on composite structures using Non Destructive Testing. *Composite Structures*, *93*(5), 1328–1336. doi:10.1016/j.compstruct.2010.10.017

Goh, J. Y., Pitter, M. C., See, C. W., Somekh, M. G., & Vanderstraeten, D. (2004). Sub-pixel image correlation: An alternative to SAM and dye penetrant for crack detection and mechanical stress localisation in semiconductor packages. *Microelectronics and Reliability*, *44*(2), 259–267. doi:10.1016/S0026-2714(03)00162-8

Hameed, Z., Hong, Y. S., Cho, Y. M., Ahn, S. H., & Song, C. K. (2009). Condition monitoring and fault detection of wind turbines and related algorithms: A review. *Renewable & Sustainable Energy Reviews*, *13*(1), 1–39. doi:10.1016/j.rser.2007.05.008

Hernandez, M. D. P. C., & Labib, A. (2017). Selecting a condition monitoring system for enhancing effectiveness of power transformer maintenance. *Journal of Quality in Maintenance Engineering*, *23*(4), 400–414. doi:10.1108/JQME-07-2015-0027

Kasban, H., Zahran, O., Arafa, H., El-Kordy, M., Elaraby, S. M., & El-Samie, F. A. (2011). Welding defect detection from radiography images with a cepstral approach. *NDT & E International*, *44*(2), 226–231. doi:10.1016/j.ndteint.2010.10.005

Khaira, A., Srivastava, S., & Suhane, A. (2015). Analysis of relation between ultrasonic testing and microstructure: A step towards highly reliable fault detection. *Engineering Review*, *35*(2), 87–96.

Krautkrämer, J., & Krautkrämer, H. (2013). *Ultrasonic testing of materials*. Springer Science & Business Media.

Lazakis, I., Dikis, K., Michala, A. L., & Theotokatos, G. (2016). Advanced ship systems condition monitoring for enhanced inspection, maintenance and decision making in ship operations. *Transportation Research Procedia*, *14*, 1679–1688. doi:10.1016/j.trpro.2016.05.133

Lin, P., & Gu, J. (2011). Research on Transformer Condition-based Maintenance System using the Method of Fuzzy Comprehensive Evaluation. *World Academy of Science, Engineering and Technology*, *5*(8), 452–457.

Liu, Q., & Wirdelius, H. (2007). A 2D model of ultrasonic wave propagation in an anisotropic weld. *NDT & E International*, *40*(3), 229–238. doi:10.1016/j.ndteint.2006.10.004

Márquez, F. P. G., Tobias, A. M., Pérez, J. M. P., & Papaelias, M. (2012). Condition monitoring of wind turbines: Techniques and methods. *Renewable Energy*, *46*, 169–178. doi:10.1016/j.renene.2012.03.003

Nasiri, M. R., Mahjoob, M. J., & Vahid-Alizadeh, H. (2011, April). Vibration signature analysis for detecting cavitation in centrifugal pumps using neural networks. In *Mechatronics (ICM), 2011 IEEE International Conference on* (pp. 632-635). IEEE. 10.1109/ICMECH.2011.5971192

Padovese, L., & Vicente, S. A. (2004). Optimizing probabilistic neural networks by the use of genetic algorithms for rolling bearing fault diagnosis. *CEP*, *5508*, 900.

Runnemalm, A., Ahlberg, J., Appelgren, A., & Sjökvist, S. (2014). Automatic inspection of spot welds by thermography. *Journal of Nondestructive Evaluation, 33*(3), 398-406.

Sbartaï, Z. M., Laurens, S., Elachachi, S. M., & Payan, C. (2012). Concrete properties evaluation by statistical fusion of NDT techniques. *Construction & Building Materials, 37*, 943–950. doi:10.1016/j.conbuildmat.2012.09.064

Sick, B. (2002). On-line and indirect tool wear monitoring in turning with artificial neural networks: A review of more than a decade of research. *Mechanical Systems and Signal Processing*, *16*(4), 487–546. doi:10.1006/mssp.2001.1460

Su, Z., Ye, L., & Lu, Y. (2006). Guided Lamb waves for identification of damage in composite structures: A review. *Journal of Sound and Vibration*, *295*(3-5), 753–780. doi:10.1016/j.jsv.2006.01.020

Thatoi, D. N., Das, H. C., & Parhi, D. R. (2012). Review of techniques for fault diagnosis in damaged structure and engineering system. *Advances in Mechanical Engineering*, *4*, 327569. doi:10.1155/2012/327569

Tian, Z., Lin, D., & Wu, B. (2012). Condition based maintenance optimization considering multiple objectives. *Journal of Intelligent Manufacturing*, *23*(2), 333–340. doi:10.100710845-009-0358-7

Willcox, M., & Downes, G. (2003). *A brief description of NDT techniques*. Toronto: NDT Equipment Limited.

Yulianti, I., Addawiyah, A., & Setiawan, R. (2018, March). Optimization of exposure factors for X-ray radiography non-destructive testing of pearl oyster. *Journal of Physics: Conference Series*, *983*(1), 012004.

Chapter 10
Induction Machine Rotor and Stator Faults Detection by Applying the N-F Network

Souad Saadi Laribi
Ibn Khaldoun University, Algeria

ABSTRACT

This chapter focuses on the monitoring and diagnosis of induction machine faults, particularly the broken rotor bars. The design of a system for monitoring, detecting, and locating incipient faults for different loads of the machine is achieved by the use of advanced intelligent techniques based on ANFIS-based neuro-fuzzy network. The knowledge base is based on indicators derived from the stator current spectral analysis of the machine which in addition has to detect and assess the number of faulty bars.

INTRODUCTION

Real-time monitoring of electrical machines is recommended to detect faults and overloads in machines, especially those related to electricity generation and those of high power drives. The measured signals containing the significant information due to the fault are usually voltages, currents, speed, vibrations and noise emissions (Benouzza, 2006; Meo, Gentile & Ometto, 2003; Petrovic & Terzic, 2009; Belhamdi, 2014). The development of modern diagnostic methods is increasing due to advances in microelectronics (microprocessor) and signal processing (Petrovic & Terzic, 2009; Belhamdi, 2014; Laribi & Bendiabdellah, 2010; Benouzza et al., 2001).

DOI: 10.4018/978-1-5225-6989-3.ch010

Through studies in the literature, the ability of the method of spectral analysis of the stator current to detect faults affecting the cage motor has been demonstrated. The spectral contents obtained made it possible to highlight the phenomena foreseen by the theory (the presence of characteristic harmonics of the fault) and thus to detect the presence of the fault. However, the use of this method for identifying the nature of the fault and the estimation of its severity requires the knowledge of the expert who is not always available. Today, thanks to the techniques of artificial intelligence, especially Neuro-Fuzzy Networks (NFN), this obstacle is overcome. These networks are able to manage a monitoring procedure of which they represent the expertise side and thus take the role designed for the expert (Laribi et al., 2006; Laribi, 2005; Benouzza, Benyettou & Bendiabdellah, 2004).

In this chapter, the implementation of a monitoring system based on the ANFIS-NFN that can predict and diagnose faults affecting the cage motor is introduced. The envisaged system can predict bar faults in cage motors and identify the nature of each of them, as well as the evaluation of its severity.

The NFN used are trained to determine the degree of severity of the fault according to a set of data. In this respect, it is envisaged to develop a deterministic data base of the learning inputs / outputs from the results obtained from the simulation of the motor behavior with and without faults. Once the learning task or the network training procedure is complete, their performance is validated by recognition tests as well as the overall performance of the system.

MODELISATION OF THE MACHINE

A three-phase cage motor is considered; its rotor consists of N_b insulated bars uniformly distributed on the surface of the rotor and short-circuited by two rings. In order to study its performances, a model of the motor is used, where the cage is considered as a mesh circuit as depicted in Figure 1. The number of differential equations obtained is equal to the number of bars plus one in order to take into consideration one of the two rings (Benouzza, 2006; Meo, Gentile & Ometto, 2003; Laribi, 2016; Laribi et al., 2018).

Under classical assumptions, the mathematical model of the machine is given by the equations of the tensions below:

$$\left[V\right]=\left[R+\frac{dL}{dt}\right]\left[I\right]+\left[L\right]\frac{d}{dt}\left[I\right] \tag{1}$$

Where

Figure 1. Mesh circuit of the cage rotor

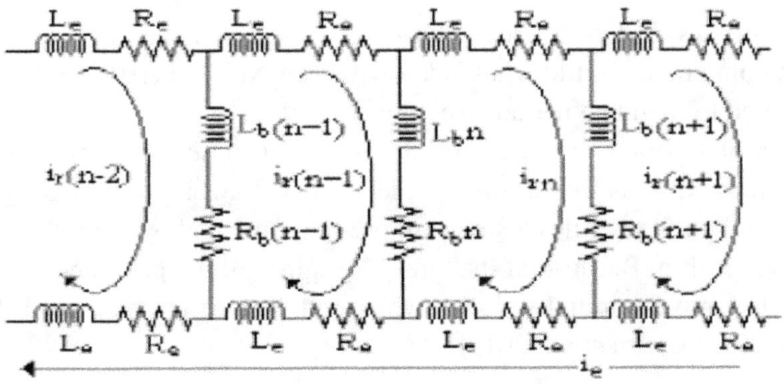

$$\frac{d}{dt}\left[I\right] = -\left[L\right]^{-1}\left[R + \frac{dL}{dt}\right]\left[I\right] + \left[L\right]^{-1}\left[V\right] \tag{2}$$

STATOR CURRENT SPECTRAL CONTENT WITH AND WITHOUT FAULTS

Healthy Motor

In the case of the healthy operation of the induction cage motor, there appears on the stator current spectrum a single harmonic at the frequency 50 Hz which is the fundamental one (Asfani et al., 2012; Costa, Kashiwagi & Mathias, 2015; Chen et al., 2007).

Faulty Motor

The imbalance caused by the broken rotor bars fault influences the current of the supply by the generation of new harmonics on both sides of the fundamental, their frequencies are given by the following expression (Asfani et al., 2012; Costa, Kashiwagi & Mathias, 2015):

$$\begin{aligned} f_{b1,2} &= f_s\left(1 \pm 2kg\right) \\ k &= 1,2.. \end{aligned} \tag{3}$$

NEURO-FLOU NETWORKS (NFN)

Neural networks and fuzzy logic have distinct advantages and disadvantages. They are combined to build what is known as an NFN where the fuzzy system is implemented following a neural structure.

The starting point of the work on the fusion of these two approaches, namely the NFN technique, is the desire to reduce their drawbacks by highlighting in a system based on fuzzy rules a structure pre-neuronal and by the introduction of a learning algorithm. Because of its "fuzzy" origins, all the prior knowledge can be incorporated into the initial weights, which reduces the learning time (Bessam et al., 2016, 2017; Laribi et al., 2017).

Different models of NFN

Given their modes of learning, their structures, and also their applications, several types of NFN can be presented (Bessam et al., 2016, 2017; Laribi et al., 2017):

- NFN of Horicawa
- NFN of Glorennec
- **ANFIS:** Adaptive Network Fuzzy Inference System
- **NNDFR:** Neural Network Driven Fuzzy Reasoning
- **NEFCLASS:** Neuro-Fuzzy Classification
- **NEFPROX:** Neuro-Fuzzy Function Approximation
- **FRN:** Fuzzy Rule Net
- **ARIC:** Approximation Reasoning-Based Intelligent Control
- **NEFCON:** Neuro-Fuzzy Controller.

Among these models, ANFIS is the most used for the detection of faults (Laribi et al., 2017).

Structure of the Used NFN

The NFN used in this chapter work is of the ANFIS type. Figure 2 depicts the classical structure of this network.

Given the fault to be studied, an ANFIS network is used that is designed to determine the severity of a specific fault. The case of the healthy motor is usually introduced into the network as a reference with zero severity. The network has two inputs, representing the ratio of the amplitude of the first and second harmonic characteristic of the fault by the amplitude of the fundamental harmonic, and an

Figure 2. Structure of the NFN

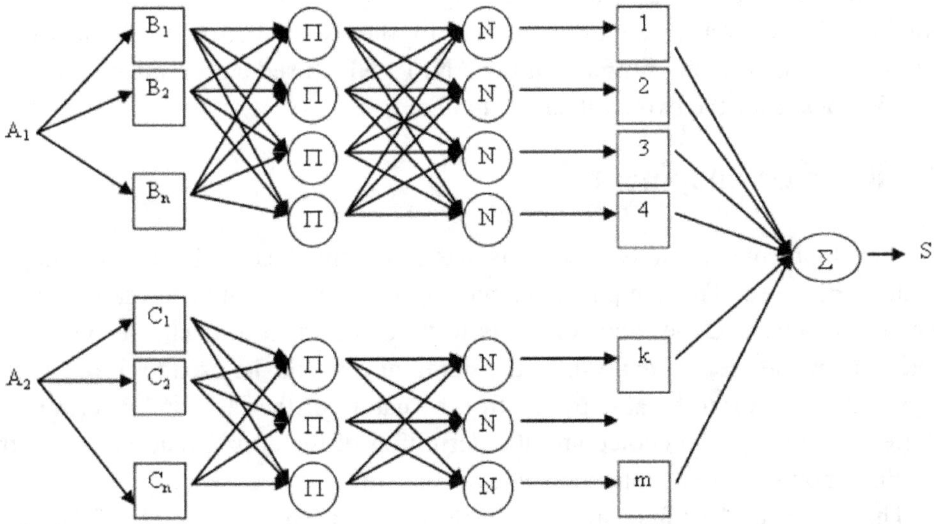

output representing the severity of the fault. Since the ANFIS structure is based on the Suggeno's reasoning, the fuzzy rules are then of the following type:

Rule 1: If A1 is B1 and A2 is C1, then S = p1.A1 + q1.A2 + r1
Rule I: If A1 is Bi and A2 is Ci, then S = pi.A1 + qi.A2 + ri

Where Bi represents the membership functions of the A1 entry; Ci represents the membership functions of the input A2, S represents the output of the neuron weighted by the degrees of membership of the input variables, and pi, qi and ri are the constant parameters of the consequence part (i = 1, ..., number of fuzzy rules).

There is no law to determine the table of the rules, the type and the number of membership functions assigned to each input variable. Generally, the choice is made empirically by examining its influence on the reaction of the network. As one builds on the test and error criterion, the membership functions are changed and the rules are automatically obtained.

Once the choice and the definition of the appropriate values of the parameters for the learning have been made, the results are checked as well as the general performances of the network by a test of recognition by presenting validation data (which are also derived from the results of simulation in our case). By parameters, it is meant the input / output data, the size of the set of these data, the type and number of membership functions, the step and the number of learning iterations. If the results obtained are not satisfactory, the network is again adjusted by applying different values of the appropriate parameters, to return to the previous steps and repeat the same procedures until the outputs and / or desired performances are obtained. Under Matlab environment and by the use of its toolbox dedicated to the fuzzy logic, the subprograms of the ANFIS-NFN used are in order to start the learning phase for the two fault cases studied.

Database Development

The development of a database consists in determining the learning inputs / outputs of the NFN used. These inputs / outputs are chosen according to the diagnostic method applied and the types of faults to be detected. In this chapter work, the detection by the spectral analysis of the stator current is to be used for broken rotor bars fault. In order to be able to identify the nature of the fault and determine its degree of severity, the choice and the definition of the appropriate values of the learning inputs / outputs must be made for this aim.

The presence of the fault studied is reflected in the spectral content of the stator current of the motor by the appearance of the characteristic harmonics of the fault. These harmonics are characterized by their frequency of occurrence and their amplitudes and they are related to the slip (the state of the motor load). It remains to know now if the variation of the load as well as the gravity of the fault influence on their amplitudes. To do this, several simulations are carried out for the various cases of motor operation (healthy motor and faulty motor with broken rotor bars fault) at different load torque values and different degrees of severity.

Healthy Motor

The stator current spectrum obtained from the simulation is illustrated by Figure 3. It remains to know now if the variation of the load as well as the severity of the fault influence on their amplitudes

Faulty Motor With Broken Bars Fault

As an example, two cases of the bars faults are simulated to show the influence of the variation of the load and the degree of severity on the amplitudes of the harmonics characteristic of these faults. These cases show the variation of the load for the same severity (loads of 3.5Nm and 7 Nm) and the variation of the severity for the same load. They are presented as follows:

- One broken bar fault for a load torque of 3.5 Nm.
- One broken bar fault for a load torque of 7 Nm.

Figure 3. Spectral analysis of the stator current (healthy motor)

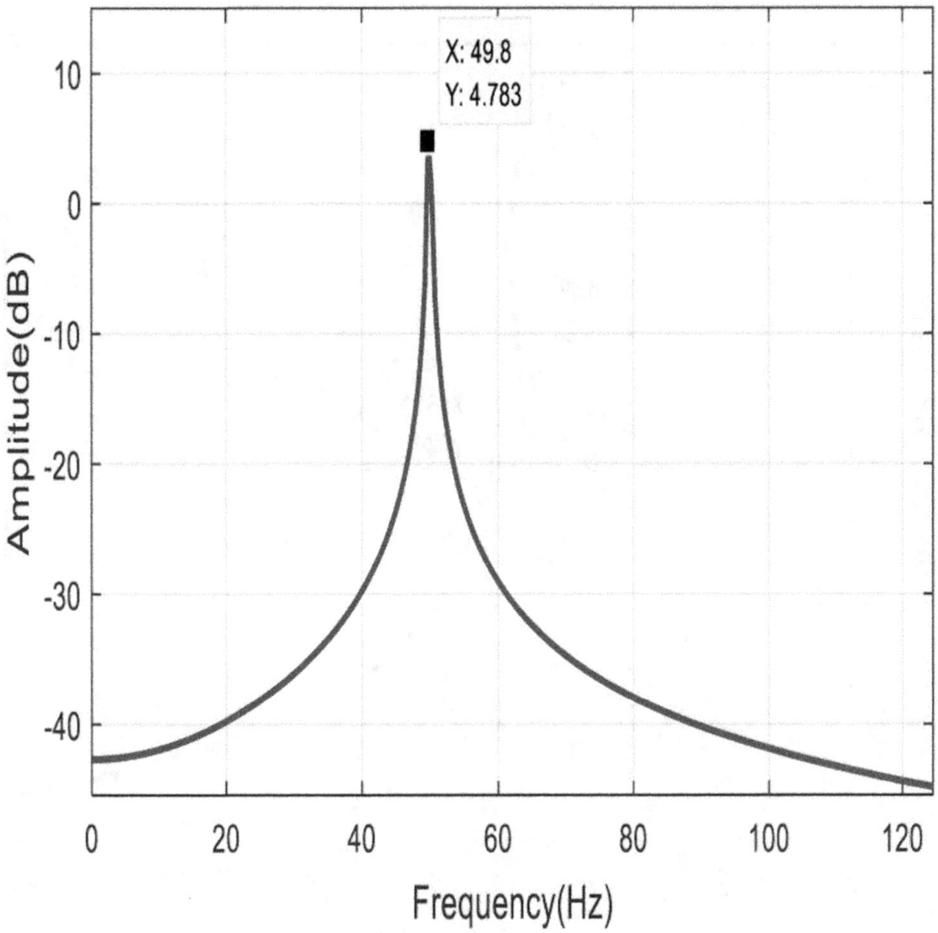

- Two broken bars fault for a load torque of 3.5 Nm.
- Two broken bars fault for a load torque of 7 Nm.

Figsures 4, 5, 6 and 7 respectively show the four spectra of the stator current obtained for the four cases. Note that the amplitudes of the two characteristic harmonics of the fault increase with the increase of the value of the load torque.

As a result, it is found that the amplitudes of the characteristic harmonics are used to evaluate the severity of the fault. This is a reason to serve as learning inputs of the used NFN, while the outputs correspond to the degrees of severity of the fault. However, for more credible learning, one needs to have a fairly rich database, i.e. sufficient data (input / output) pairs. In order to achieve this, simulations of both

Figure 4. Spectrum of the motor current with one broken bar fault under a load torque 3.5N.m

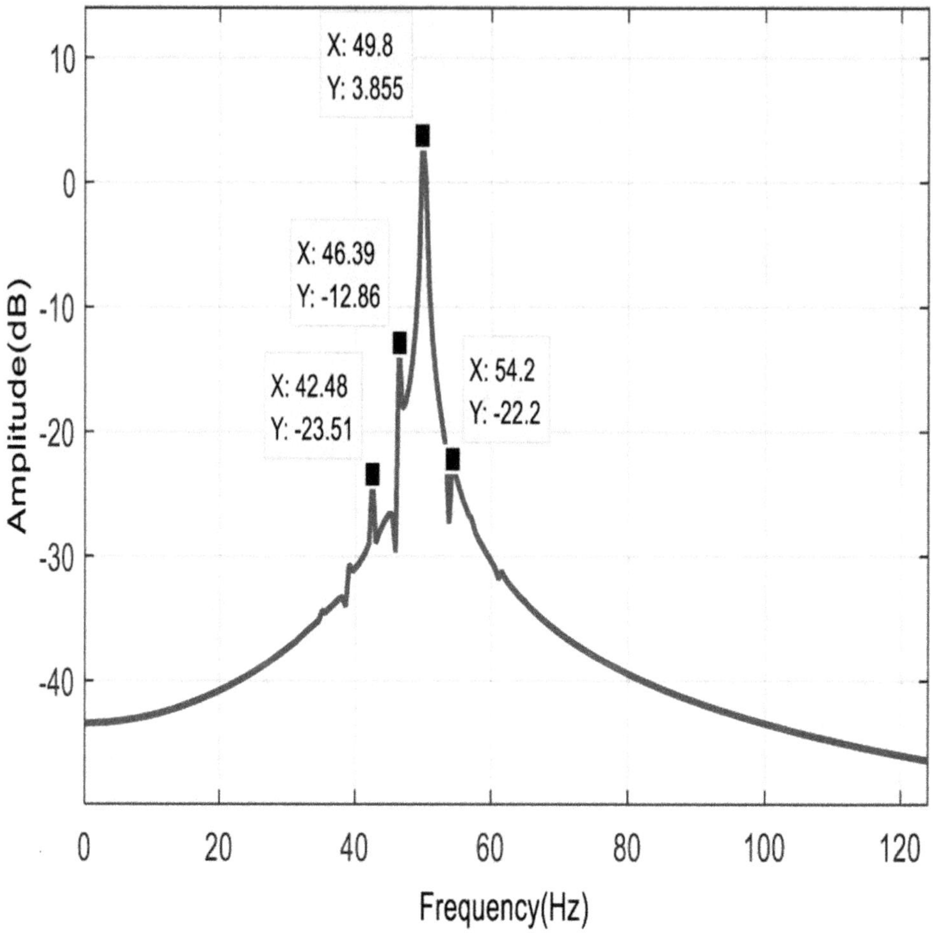

the healthy motor and the faulty motor behaviors are to be conducted for different degrees of severity by varying the stepwise load torque from its zero value to the nominal load. For each severity and each load, a fault case is simulated and so on.

The simulation is performed for the following operation cases:

- Healthy motor for different loads.
- Faulty motor for one and two bars faults for different loads.

The results obtained are saved as data files for later exploitation under Matlab environment.

Figure 5. Spectrum of the motor current with one broken bar fault under a load torque of 7N.m

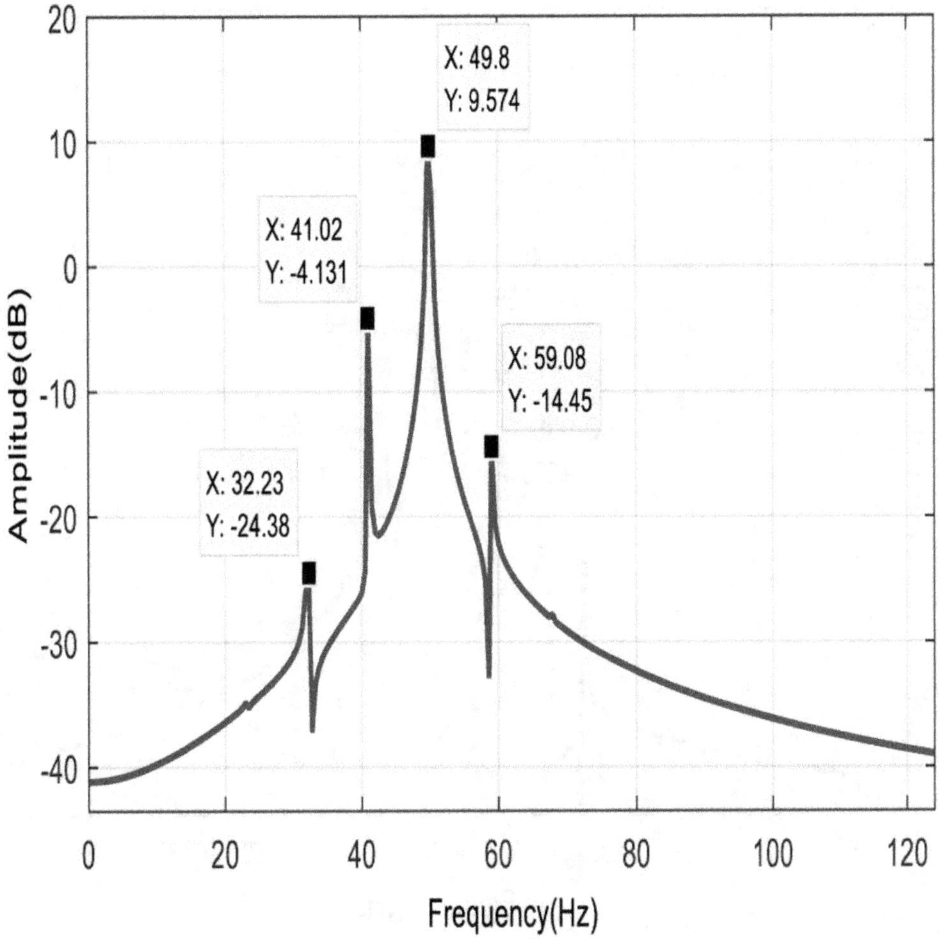

Once the spectral analysis of the stator current has been carried out, by using the fast Fourier transform under Matlab environment, the amplitudes of the fundamental harmonic for the case of the healthy motor and those of the characteristic harmonics of the faults are noted and then recorded in data files (one file for each case of motor operation).

On the basis of the results obtained and in order to eliminate the load effect, it is proven more judicious to take as inputs of the used NFN the ratios of the amplitudes of the two harmonics characteristic of the fault by the amplitude of the fundamental for the bars fault. The outputs of the used NFN are the severities of the faults.

Figure 6. Spectrum of the motor current with two broken bars fault under a load torque 3.5N.m

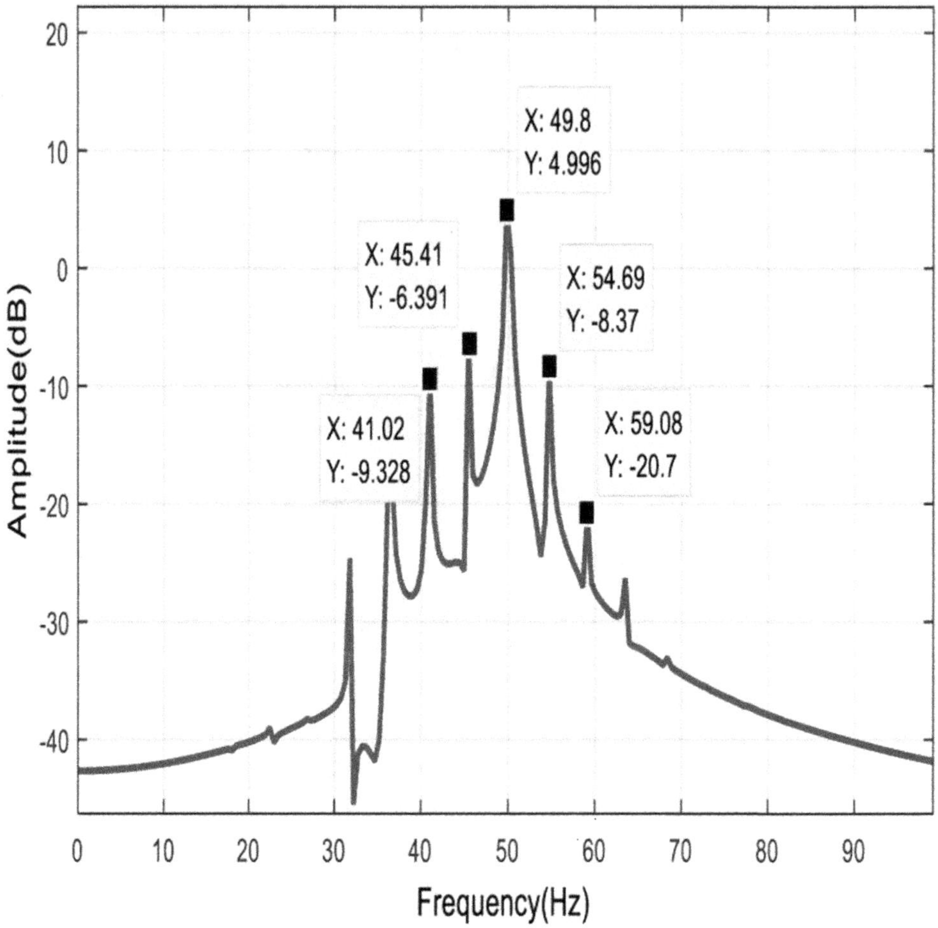

Fault Identification and Severity Estimation

The goal is to develop a cage motor monitoring system which is able from a given stator current vector, to identify the fault studied (the bars fault in this case) and to determine the severity. It is seen previously that this type of fault is manifested by the creation of characteristic harmonics in the spectrum of the stator current. Consequently, the location of these harmonics makes it possible to identify the nature of this fault. While the estimation of the severity is done by the ANFIS networks previously formed on the basis of the ratios of the amplitudes of the harmonics characteristic of the fault by the amplitude of the fundamental one.

Figure 7. Spectrum of the motor current with two broken bars fault under a load torque 7N.m

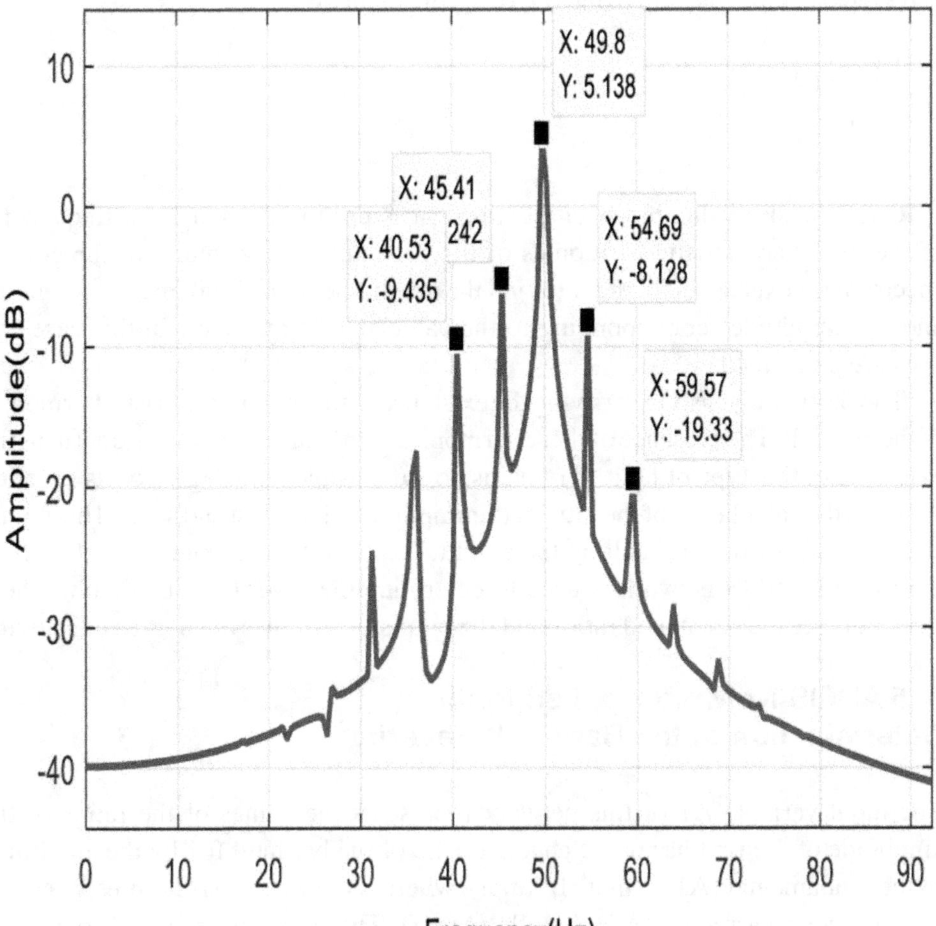

The envisaged system, in this case, is based on the spectral analysis of the stator current by FFT and the two ANFIS networks formed for the case of broken bars fault. The developed program under Matlab environment is described by the following steps as shown:

- The first step is to read the data file of the given stator current.
- The second step consists in applying the spectral analysis of this current by the FFT.

Since the characteristic harmonics of the faults are related to the slip, in other words to the speed, the next step is to calculate the rotational speed of the motor and then calculate the slip using the following equation:

$$g = \frac{n_s - n}{n_s}$$

Once the slip value has been obtained, it is used to calculate the frequencies of the two characteristic harmonics of the bar fault. The scanning of the current spectrum is next carried in order to find the amplitude of the fundamental harmonic and the amplitudes corresponding to the calculated frequencies. In this case two tests are performed.

The first test consists in verifying the existence of the two characteristic harmonics of the bar fault. The presence of these harmonics in the current spectrum confirms the existence of this type of fault. It remains now to evaluate the degree of its severity.

First, the calculation of the ratio of the amplitude of the first and second harmonic characteristic of the bar fault by the amplitude of the fundamental is made. Then, the trained ANFIS network is used to determine the severity of this fault, where the inputs are the calculated ratios and the output determines the degree of severity.

IV.5 ANFIS Network Applied to the Determination of the Bar Fault Severity

The input variable A1 of this network represents the values of the ratios of the amplitude of the first harmonic characteristic of the bar fault fb1 by the amplitude of the fundamental (A1=Ampfb1/Ampf), whereas the variable A2 represents those of the second harmonic (A2=Ampfb2/Ampf). These values are equal to 155 for each variable.

The output variable takes five values describing the degree of severity of the bar fault:

- **0:** Healthy motor
- **1:** One broken bar fault
- **2:** Two broken bars fault

After several tests, the number of membership functions of the following Gaussian type, for each entry, is adopted:

- **The Variable A1:** 9 membership functions (Figure 8.a).
- **The Variable A2:** 9 membership functions (Figure 8.b).

As a result, the number of fuzzy rules equals 81 rules. The learning takes place after 1200 iterations. The obtained results related to the learning values (inputs / outputs) and the error between the desired output and the estimated output are illustrated by Figure 9, where it can be noticed the convergence of the error

Figure 8. Membership functions for the bar fault case

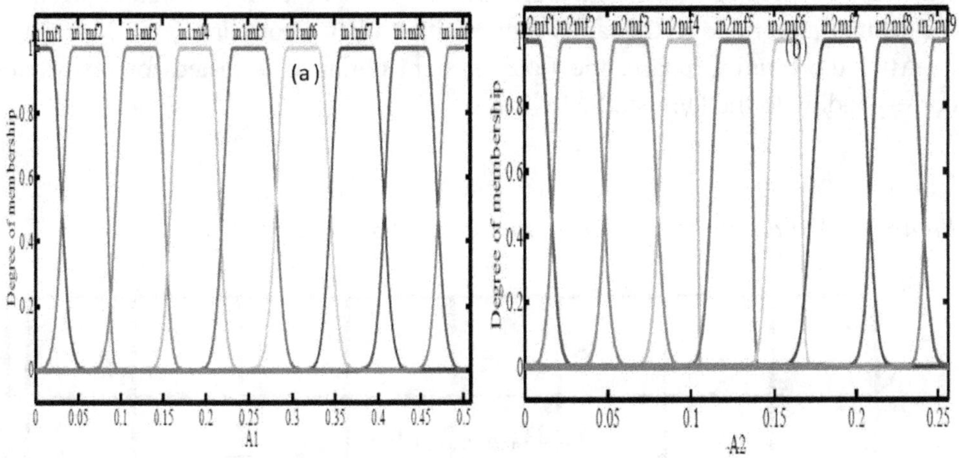

Figure 9. Area representing the network output according to these two inputs

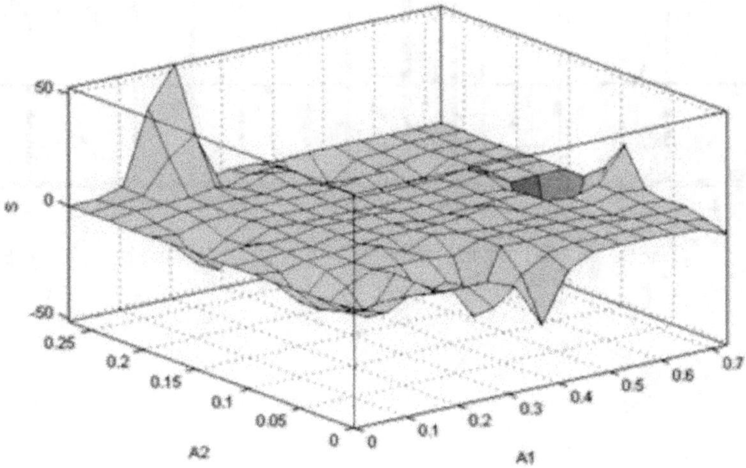

Validation Test of the Elaborated Monitoring System

Once the monitoring system has been implemented, a validation test is performed in order to assess its performance. In this respect, the stator current data derived from the motor simulation results which have not been used to develop the training database for the recognition test of the formed NFN are presented to this monitoring system in order to identify the nature of the fault if it exists and to determine its severity. Otherwise, the monitoring system must designate that the case is of a healthy motor. It is about several stator current vectors of different operating cases conducted on the cage motor and for all severities of the faults previously taken, which is known in advance what represents each one of them (the stator current of the healthy motor and of the faulty motor with broken rotor bars).

After the training phase, the fuzzy system obtained is tested for the values corresponding to the fault studied:

Figure 10. Validation test of the elaborated monitoring system

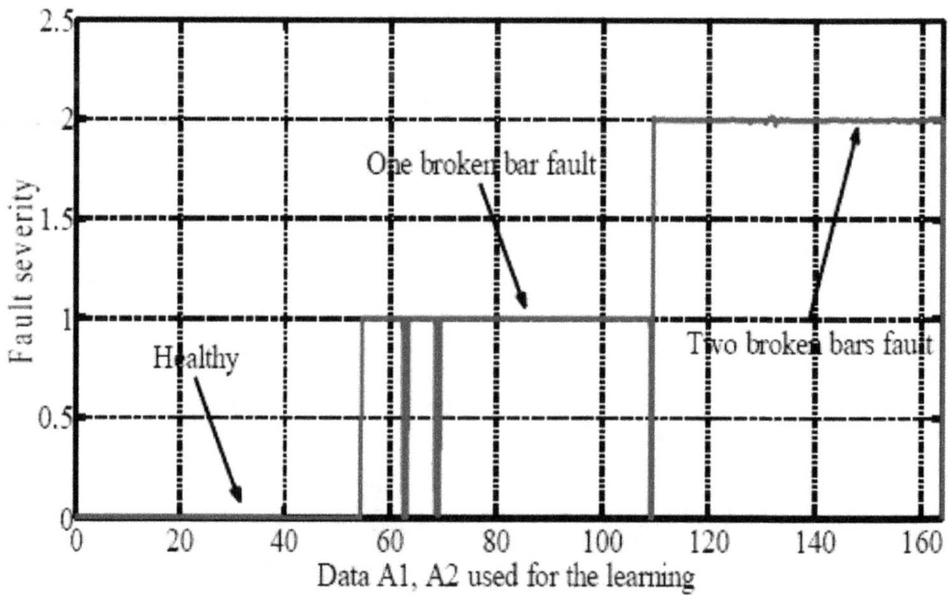

Table 1. Machine Parameters

Rs=7.828 [Ω]	Resistance of stator phase
P=1	Number of pole
J=0.006093 [Kg.m^2]	Moment of inertia
Nr=16	Number of rotor bars
Ns=160	Number of turns
R=0.03575 [m]	Gap average radius
L=0.065 [m]	Length of the machine
Lsl=0.018 [H]	Leakage inductance of stator phase
Rb=150e-6[Ω]	Resistance of rotor bar
Re=72e-6/Nr[Ω]	Resistance of short-circuit ring segment
Lb=1e-7 [H]	Leakage inductance of rotor bar
Le=1e-7 [H]	Leakage inductance of short-circuit ring segment
U_n: 220/380 [V]	Nominal voltage
I_n: 4,5/2,6 [A].	Rated current
N_r=2850 [tr/mn].	Rated speed
P_n=1,1 [kW].	Nominal power
e = 0,00025 [m]	Air gap
C_n= 7 [N.m]	Rated torque
$f_{s=}$ 50 [Hz] n n_s	Rated frequency Synchronous rotor Synchronous rotor speeds

CONCLUSION

In this chapter, the development of a cage motor monitoring system is proposed. This system is based on both the stator current spectrum technique and the ANFIS-NFN. The proposed system is not only able to identify the state of the motor operation (the healthy motor or the faulty motor with broken rotor bars faults) but also to determine the severity of each fault.

The implementation of the monitoring system consists first of all in forming one neuro-fuzzy of the ANFIS type used to determine the severities of the fault studied. To do this, a database is built using the motor simulation results with and without faults. This database is served as input / output for the learning of these networks. Once the learning phase is done, a validation test is performed in order to extract the performance of each network.

REFERENCES

Asfani, D. A., Muhammadc, A. K., Purnomo, M. H., & Hivama, T. (2012). Temporary short circuit detection in induction motor winding using combination of wavelet transform and neural network. *Expert Systems with Applications*, *39*(5), 5367–5375. doi:10.1016/j.eswa.2011.11.048

Benouzza, N. (2006). *Approche des Vecteurs de Park Appliquée à la Détection des Défauts dans les Moteurs Asynchrones à Cage d'Ecureuil* (Thèse de Doctorat). Université des Sciences et de la Technologie d'Oran.

Benouzza, N., Benyettou, A., & Bendiabdallah, A. (2004). An Advanced PARK's Vectors Approach for Rotor Cage Diagnosis. In *Proceeding of CNGE 2004 01*. Université Ibn Khaldoun de Tiaret.

Benouzza, N., Benyettou, A., Bendiabdellah, A., Kraloua, B., & Toumi, D. (2001). Rotor cage faults diagnosis in 3-phase induction motors, by Park's vector approach. *4th Jordanian International Electrical Electronics Engineering Conference 2001*.

Bessam, B., Menacer, A., Boumehraz, M., & Cherif, H. (2016). Detection of broken rotor bar faults in induction motor at low load using neural network. *ISA Transactions*, *64*, 241–246. doi:10.1016/j.isatra.2016.06.004 PMID:27329853

Bessam, B., Menacer, A., Boumehraz, M., & Cherif, H. (2017). Wavelet transform and neural network techniques for inter-turn short circuit diagnosis and location in induction motor. *International Journal of System Assurance Engineering and Management*, *8*(S1), 478–488. doi:10.100713198-015-0400-4

Chen, Z., Meng, J., Liang, B., & Guo, D. (2007) Fault Diagnosis of AC Squirrel cage Asynchronous Motors based on Wavelet Packet-Neural Network. *EUROCON 2007, The International Conference on "Computer as a Tool"*. 10.1109/EURCON.2007.4400425

Costa, C. D., Kashiwagi, M., & Mathias, M. H. (2015). *Rotor failure detection of induction motors by wavelet transform and fourier transform in non stationary condition*. Case Studies in Mechanical Systems and Signal Processing.

Laribi, S. (2005). *Synthèse des méthodes de diagnostic appliquées à la détection des défauts dans les machines asynchrones, Mémoire de Magister*. Algérie: USTO Oran.

Laribi, S. (2016). Contribution à la Commande et Diagnostic des Défauts Statoriques et Rotoriques de la Machine Asynchrone à Cage d'Ecureuil. Université des Sciences et de la Technologie d'Oran Mohamed Boudiaf.

Laribi, S., Bendiabdallah, A., Cherif, B., Boukezata, B., Meradi, S., & Mihoub, Y. (2017). *Use of Neuro-Fuzzy Technique in Diagnosis of Rotor Faults of Cage Induction Motor.* The 5th International Conference on Electrical Engineering, Boumerdes (ICEE-B), Boumerdes, Algeria.

Laribi, S., Bendiabdallah, A., Cherif, B., Boukezata, B., Meradi, S., & Mihoub, Y. (2018) Induction Machine Rotor and Stator Faults Detection by Applying the DTW and N-F Network. IEEE ICIT 2018, Lyon, France.

Laribi, S. & Bendiabdellah, A. (2010). Stator short circuit and broken bar faults diagnosis of an indirect vector control squirrel cage induction motor. *International Review of Electrical Engineering, 5*(5), 2088-2094.

Laribi, S., Benouzza, N., Bendiabdellah, A., & Toumi, D. (2006). Détection des Défauts de Barres Rotoriques dans les Moteurs Asynchrones à Cage d'Ecureuil par l'Emploi de Université des Sciences et de la Technologie d'Oran Mohamed Boudiaf, Analyse Spectrale des Grandeurs Electriques. ICEEA'06 Université de Sidi Belabbes, 136-141.

Meo, S., Gentile, G., & Ometto, A. (2003). *Induction motor current signature analysis to diagnostics of stator short circuits.* IEEE-IAS SDEMPED 2003-Symposium on Diagnostics for Electric Machines, Power Electronics and Drives, Atlanta, GA.

Petrovic, T. K., & Terzic, B. (2009). Sensorless speed detection of squirrel-cage induction machines using stator neutral point voltage harmonics. *Mechanical Systems and Signal Processing, 23*(3), 931–939. doi:10.1016/j.ymssp.2008.06.005

Chapter 11
A State-of-the-Art Critique Review of the Role of Decision-Making Techniques in NDT for Condition Monitoring

Ashish Khaira
Samrat Ashok Technological Institute, India

Ravi K. Dwivedi
Maulana Azad National Institute of Technology, India

ABSTRACT

Nondestructive testing (NDT) is a vital tool in maintenance. Each NDT technique has some benefits and hindrances; therefore, the selection is crucial. Generally, the selection of a technique relies on operating personnel experience, and very few research papers shows uses of the decision-making (DM) approach. It was highlighted by various researchers that if a proper DM approach is used, it will save time and increase fault detection reliability. By keeping this fact in mind, this chapter is an attempt to provide a detailed review of research work from the year 2000-2018 that covered the role of DM techniques while making combinations of NDT for effective condition monitoring. It observed from the literature that very few researchers effectively utilized the power of DM tool. The researcher can use the outcome of this work as a beacon and improve it further.

DOI: 10.4018/978-1-5225-6989-3.ch011

INTRODUCTION

As per Escaler et al. (2006) and Garnier et al. (2011), the proactiveness in fault detection is necessary for increasing overall useful life as components become more trouble-free to ensure timely availability to plan corrective measures which are essential to ensure optimum availability of machine/components. Non-destructive testing (NDT) as the name suggests, is a technique in which component tested for flaw without disintegrating or altering its present structure. It was highlighted by Willcox et al. (2003), D'Orazioa et al. (2008) and Krautkrämer et al. (2013) that NDT is an efficient fault identification methodology that helps in maintaining key performance indicators like Mean Time to Repair (MTTR), Mean Time Between Failure (MTBF) etc. in optimum level. As per Khaira et al. (2015), these techniques have their application in composite testing, defect detection, thickness measurement, corrosion detection etc.

Decision making is present since human existence, as a human they makes decision in there day-by-day activities like "what to do?, what to wear?, what to eat?" etc. In the routine, the decision making is based on persons own experience or expert suggestion but in the industrial environment, there is a strong urge to use some mathematical models. As per Fântână et al. (2013), Toloie et al. (2011) and Ghosh et al. (2009), Multiple-Criteria Decision Making (MCDM) is useful when alternative with best values is not available as each alternative score has multiple conflicting criteria's.

The organization of this research paper is as follows: In section 2, review of the role of decision-making techniques in NDTs In section 3, the result and discussion part covered and then, the conclusions drawn in the last section. The contribution and significance of this research study are of twofold; firstly, the research work from the year 2000-2018 by different researchers reviewed, that includes the combinations of NDT with decision making; secondly, the areas remain untouched were also explored to facilitate other researchers about research gaps in this area.

REVIEW OF ROLE OF DECISION MAKING TECHNIQUES IN NDTS

This section covered the key extracts for the reviewed literature related to decision-making in NDTs. It covered the type of decision making techniques, domain & area of work etc.

D. Horn et al. (2000) investigated statistical methods for combining the results of two complementary inspection techniques, eddy-current and ultrasonic testing for Zr-Nb (2.5%) pressure-tube billets. The reliability of rejection/acceptance decisions based on combined information compared with that based on each inspection technique individually. Combination methods, ranging from a simple OR of individual decisions to more sophisticated data fusion algorithms, are assessed. This paper showed a comparison of the reliability of combined ET and UT measurements to that of each technique taken individually from a database of 108 artificial flaws. This paper also showed the use of probability density function for the likelihood that particular signal amplitude produced by a given phenomenon, such as a flaw of a specific depth, by a PDF which may be obtained from an analysis of the rates at which various signal amplitudes are produced from a narrow range of flaw sizes. The mean amplitude then corresponds to the average of the first moments of the flaw size distributions. An average, weighted by the widths, or second moments, of the flaw size distributions, brings still more a priori information into the decision process. The weighted average and variance is commonly taken as –

$$\mu = \frac{\sum_i x_i / \sigma_i^2}{\sum_i 1 / \sigma_i^2}, \sigma^2 = \frac{1}{\sum_i 1 / \sigma_i^2} \quad (1)$$

Zoubir-Mehdi Sbartaï et al. (2012) defined several indices for the evaluation of the quality and the degree of sensitivity of each NDT measurement, including electrical resistivity methods (Wenner and four-probe square array), GPR (Ground Penetrating Radar), direct wave analysis(arrival time and amplitude, Capacitive Method (3 frequencies of measurement), Ultrasonic surface waves method, with and without contact (velocity, attenuation, quality factor), Ultrasonic waves in direct transmission (UPV), Impact echo (first peak frequency) regarding each concrete property. The indices defined using two statistical approaches (variance analysis and multiple. regression). Based on these indices, a ranking between the NDT parameters became promising for identifying the mainly relevant. The second part used a restricted set of NDT parameters, the correlation and the complementarily between NDT parameters analyzed by a data mining technique (principle component analysis). The two steps (an assortment of individual parameters and the use of best combination) described with numerous practical examples, which demonstrate how the concrete properties can be better assessed in practice. It is also revealed how the uncertainty on the estimated properties can be quantified. The degree of sensitivity for other NDT techniques can also be evaluated by the same approach by concentrating on particular machine component. Although the outcome presented here restricted to porosity and saturation evaluation, the same methodology can be used, following exactly the same steps, for additional properties such as Young's modulus, compressive strength, etc. Moreover, additional NDT techniques can be analyzed following the same methodology. Multiple linear regressions were statistically recognized and chosen for modeling the associations between NDT parameters and the concrete porosity and saturation pair.

$$NDT_{(i)}=\{(A_i(\underset{-}{+}E_a).S)+(B_i(\underset{-}{+}E_b).P)+(C_i(\underset{-}{+}E_c))\} \tag{2}$$

where $NDT_{(i)}$ is the response of the NDT parameter i (i = 1–52); S is the degree of saturation and P is the porosity (both expressed as percentages); A_i, B_i and, C_i are the regression coefficients of the empirical model concerning the parameter $NDT_{(i)}$; E is the standard error on the coefficients, calculated using the following equation:

$$E=\sqrt{\frac{1}{n-2}\sum(Y-y)^2-\frac{\sum(X-x)(Y-y)^2}{\sum(X-x)^2}} \tag{3}$$

in which n is the number of samples; X and Y are mean of x and y values, respectively, on the series of n samples; and x and y are the values of the sample.

M.C. Carnero (2005) proposed a model that carries out the decision-making in relation to the diagnostic techniques and instrumentation selection in the predictive maintenance programs. The model uses a tool combination belonging to operational research, such as analytic hierarchy process (AHP) and factor analysis (FA). The model has been tested in screw compressors when lubricant and vibration analyses are integrated. Also covers pair wise comparison matrix and Eigen-vectors in a PMP based on the lubricant and vibration analyses integration in tabulated form. After explaining the basic concept a case study of a screw compressor with the integration of lubricant and vibration analyses is also explained in this paper. This makes this paper good for further research work. The reason for selecting screw compressor for this study is well explained. Factor analysis (FA) is a statistical procedure used to determine the basic essential variables underlying many interrelated variables; a method of processing data comprising too many variables to allow direct analysis. Steps in an FA include:

1. Selecting and measuring a group of variables.
2. Preparing the correlation matrix.
3. Determining the number of components or factors to be considered.
4. Extracting a set of components or factors from the correlation matrix.
5. Rotating the components or factors to increase interpretability.
6. Interpreting the results.

The correlation matrix R can be diagonalized accomplishing and pre-multiplying it with the matrix V and its transpose V'. The matrix of eigenvectors V pre-multiplied by its transpose produces the identity matrix:

$$V'V = I \tag{4}$$

$$VLV' = R \tag{5}$$

The correlation matrix can be decomposed,

$$R = \left(V\sqrt{L}\right)\left(\sqrt{L}V'\right) \tag{6}$$

If $V\sqrt{L}$ is called A (factor loading matrix), then

$$R=AA' \tag{7}$$

MaCarmen Carnero (2006) proposed an evaluation that carries out the decision-making to the feasibility of the setting up. The evaluation system uses a tool combination belonging to operational research, such as analytic hierarchy process, decision rules, and Bayesian tools. The test of this Evaluation System was conducted in a petrochemical plant and in a food industry. This analysis would enable a relationship between the results obtained by the Model, and each activity sector to be established. A hierarchy must be constructed to apply AHP. In this hierarchy the relationship between the goal, criteria, sub- criteria and alternatives established. In a decision-making problem M alternatives A_i (i=1,2, 3,.,M) and N criteria C_j (j=1,2,3,.,N) considered. In order to determine the relative importance of the alternatives with regard to each of the criteria or between two criteria, linguistic terms used that include the judgments of the decision maker. The quantified judgment on pairs of criteria C_i and C_j are represented by an n-by-n matrix A:

$$A=\begin{bmatrix} a_{11} & a_{12} & \cdots & a_{1n} \\ a_{21} & a_{22} & \cdots & a_{2n} \end{bmatrix} \tag{8}$$

where the a_{ij} is the relative importance of C_i to C_j. The entries a_{ij} defined by the following rules:

1. If $a_{ij} = \alpha$, then $a_{ji} = 1/\alpha$, $\alpha \neq 0$.
2. If C_i judged to be of equal relative importance as C_j, then $a_{ij} = a_{ji} = 1$; $a_{ii} = 1$ for all i.

R. Teti et al. (2010) followed a recent update of the literature on tool condition monitoring and documents the work of the cutting scientific technical committee in CIRP (CIRP is Spanish abbreviation and English full form is International Academy of Production Engg). This paper covered the past contributions of CIRP in these areas and provides an up-to-date comprehensive survey of sensor technologies, signal processing, and decision- making strategies, including Neural Networks, Fuzzy Logic, Genetic Algorithms, and Hybrid Systems for process monitoring. Application examples of industrial processes including reconfigurable sensor systems reported. The capabilities of the various cognitive methods are briefly reviewed before presenting their applications relevant to the scopes of this paper. This paper also covers about MT-Connect, an open software standard for data exchange and communication between manufacturing equipment launched by The Association for Manufacturing Technology. This part is not found in most of the literature. This paper also compares MT-Connect with Cyber Monitor and Cyber Tool Management. Furthermore, the stand-alone sensors transformation used primarily as diagnostic devices in a manufacturing process, with sensors that are a part of an intelligent system for process, tool, and machine monitoring and control.

Andrew K.S. Jardine et al. (2006) summarized recent research and development in machinery diagnostics and prognostics implementing CBM. Various techniques, models, and algorithms reviewed by following the three main steps of a CBM program, namely data acquisition, data processing and maintenance decision-making, with emphasis on the last two steps. Different techniques for multiple sensor data fusion have also been discussed. This paper also summarizes here a short list of references to provide an introductory familiarity with the ideas and methodologies in this area. Various techniques, models, and algorithms discussed in a detailed manner with more emphasis on data processing and maintenance decision-making. Different techniques for multiple sensor data fusion have also been covered in an effective way. Furthermore, next-generation diagnostic and prognostic systems will likely focus more on various aspects of continuous monitoring and automatic diagnostics and prognostics. This information would be invaluable for model building and model validation as well.

Edwin Vijay Kumar et al. (2009) ascertained the whole system health and maintenance need with a degree of certainty by means of condition-monitoring data through hierarchical fuzzy inference system. The projected approach uses a hierarchical system health assessment by means of fuzzy inference on condition-monitoring data collected over a period. The planned approach applied to a large electric motor

(500kW), which utilized four subsystems i.e. power transmission system, an electromagnetic system, ventilation system and support system. Fuzzy set of condition-monitoring information of every condition indicator on every subsystem used to determine the degree of match with the expert opinion fuzzy set, thus inferring the requirement for a periodical overhaul. Subjective specialist opinion and quantitative condition-monitoring data evaluated using hierarchical fuzzy inference system with a rule base. The paper provides a chance to assess quantitative condition-monitoring data and subjective/qualitative domain expertise by means of fuzzy sets. Furthermore, the approach is effective and can be applied to a variety of process equipment to ensure reliability through effective predictive maintenance. It was shown in this paper that the certainty of every subsystem's health is not the identical at the end of 600 days of monitoring and can be classified as very good, good, marginal and sick. The data collected over a phase of time for a given health indicator can be utilized to build up a probability distribution for the variable, which is utilized to attain a normalized fuzzy set. If indicator variable, x_i, (temperature or vibration, etc. . . .) is some point on the parametric domain for which, possibility, $p(x_j)$ is greatest, then we can describe a membership grade function for a set A, μ_A (x) as:

$$\mu_A(x) = \frac{p(x_i)}{p(x_j)} \tag{9}$$

This result in the change of condition signifying variable into a convex normalized fuzzy number "A" with a membership grade function mA(x), thereby characterizing the dynamic performance of the Motor Health Indicating parameters. Once these grades have been ascertained, the degree of match, e.g. for arbitrary fuzzy sets, say, A and B, defined as:

$$DM(A,B) = \frac{\sum \mu_{AIB}(x)_x}{\sum \mu_A(x)_x} \tag{10}$$

Where "x" is the value of a parameter, "A" can be a field data fuzzy possibility set and "B" can be a fuzzy set for expert opinion for a known parameter indicator level. The Degree of Match (DM) determines can be effectively used to allocate certainty of the field data with expert opinion.

Iraklis Lazakis et al. (2016) presented the INCASS (Inspection Capabilities for Enhanced Ship Safety) project that integrated with robotic-automated platforms for online or on-demand ship inspection activities to reach an innovative solution to the ship inspection regime and selecting the software and hardware tools that can implement or facilitate specific inspection tasks, to provide input to the Decision Support System (DSS).

Maria Del Pilar et al. (2017) proposed an analytic hierarchy process based structured and generic model for assisting in the decision-making process for acquiring a combination of condition monitoring (CM) system for an oil-immersed power transformer in order to improve its maintainability. The assessment was performed by pairwise comparisons, and a sensitivity analysis (what-if analysis) was used to identify the implications of changing the criteria weights.

RESULT OF REVIEW

Table 1 shows details about domains and Decision Making Techniques which already applied in different-2 fields. This table is very useful for upcoming researchers to identify research gaps in this field. Table 2 (3, 5, 7, 19, 20) summarizes the description about various key NDT techniques along with their limitations.

Table 1. Details of domains and decision making techniques applied

Author	Year	Domain	Component	Optimization/Decision Making Technique
D. Horn et al.	2000	Eddy-Current and Ultrasonic Testing	Zr-Nb(2.5%) pressure-tube billets	Averaging Method, Data Fusion Method (classical inference, Bayesian inference, and the Dempster–Shafer combination rules)
Zoubir-Mehdi Sbartaï et al.	2012	Electrical resistivity methods (Wenner and four-probe square array),	Concrete Properties	Variance analysis and multiple. Regression.
M.C. Carnero	2005	Lubricant and Vibration Analyses	Screw Compressors	Analytic hierarchy process (AHP) and factor analysis (FA)
MaCarmen Carnero	2006	Generalized	Generalized	Analytic Hierarchy Process (AHP)
R. Teti et al.	2010	Advanced monitoring	Generalized	Neural networks, Fuzzy logic, genetic algorithms, hybrid systems
Andrew K.S. Jardine et al.	2005	Generalized	Generalized	Statistical approaches, AI approaches, Other approaches
Edwin Vijay Kumar et al.	2008	Machine condition monitoring and fault diagnostics	Electric Motor	Fuzzy Logic

Table 2. Comparison of NDT techniques

NDT Techniques	Limitations
Magnetic Particle Testing In this the inspector inspects the surface the principle of magnetic flux induction in sample. It is faster process and as number of stages is less, therefore, easily automated.	• It is difficult to detect diagonal defects. • Probe size affects sensitivity
Radiography In this the inspector inspects the surface using short wavelength radiation passing through the testing specimen. This method is suitable for large areas inspection and images are compatible for computer analysis.	• Both side access is needed • Highly skilled personnel needed
Thermography In this the inspector inspects using infrared cameras that detects radiation and produce images of that radiation, which are known as thermograms. Here testing can be carried out from distances.	• Weather affects the results • Equipments are expensive
Vibration Analysis In this the inspector inspects using vibration tester that converts mechanical vibrations into an electronic signal and then signals analysis by means of computer.	Data interpretation is difficult.

It is necessary to know the distribution of year-wise paper in order to know the actual position of usage of decision-making techniques. Figure 1 shows a crescent tendency in the articles time distribution. The graph clearly indicates that most of the work done in 2010-14.

For a researcher, it is essential to know the areas in which these techniques used and areas those are remains untouched. Figure 2 shows graph between a number of research papers and domain. This graph shown in figure 2 clearly indicates that most of the work done in thermal/infrared monitoring of machinery and least work is on lubrication, eddy current, and vibration testing etc.

Figure 3 shows plots between decision-making techniques v/s number of research paper. This plot is very useful to identify the areas which are remaining almost untouched. From this pie chart, it is evident that neural networks are mostly used while fuzzy logic, genetic algorithms are less frequently used.

Figure 4 shows plots between areas of research v/s number of research papers. It clearly indicates that most of the time decision-making techniques used mostly for review as compared to real life application. Figure 5 shows the name of journal/ conference v/s number of research papers. It is evident from the graph that most of the time researcher preferred NDT & E International journal & decision support system for publishing their NDT decision making based papers.

Figure 1. Year of publication v/s number of papers

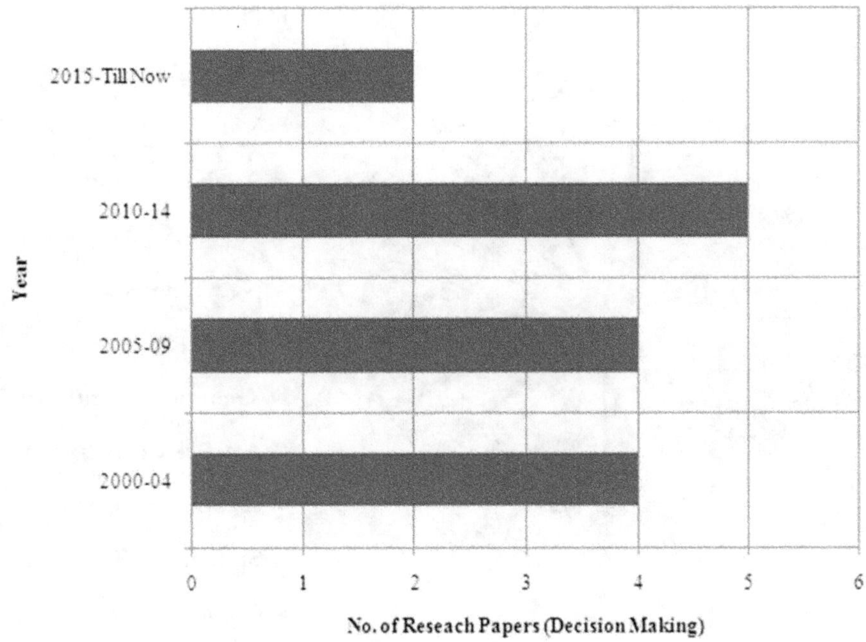

Figure 2. Domain v/s number of research papers

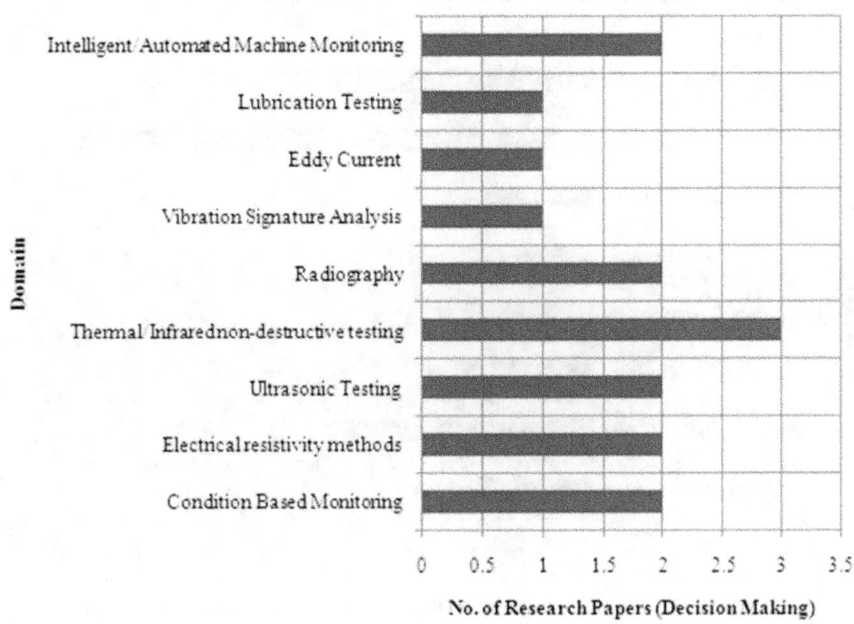

Figure 3. Decision making techniques v/s number of research papers

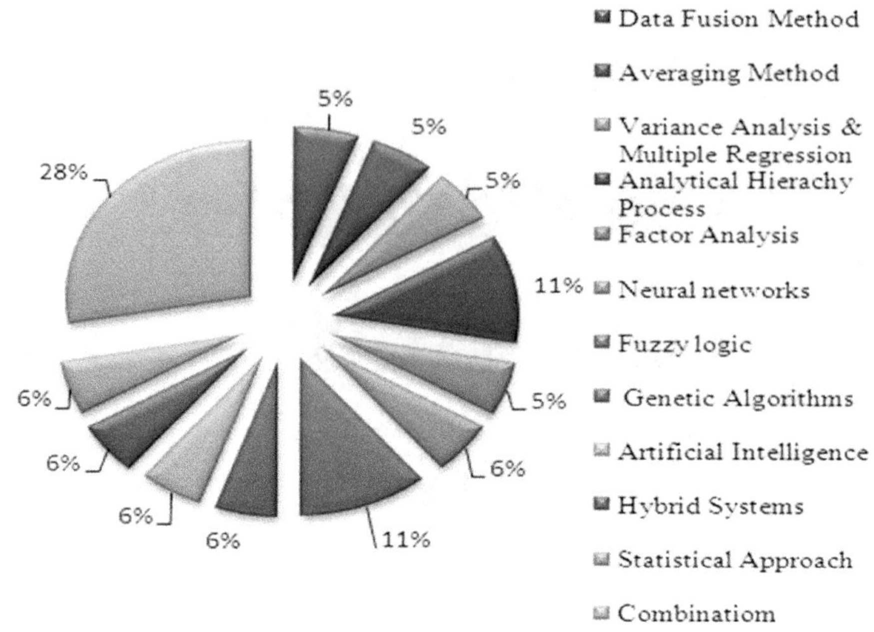

**For a more accurate representation see the electronic version.*

Figure 4. Area of research v/s number of research papers

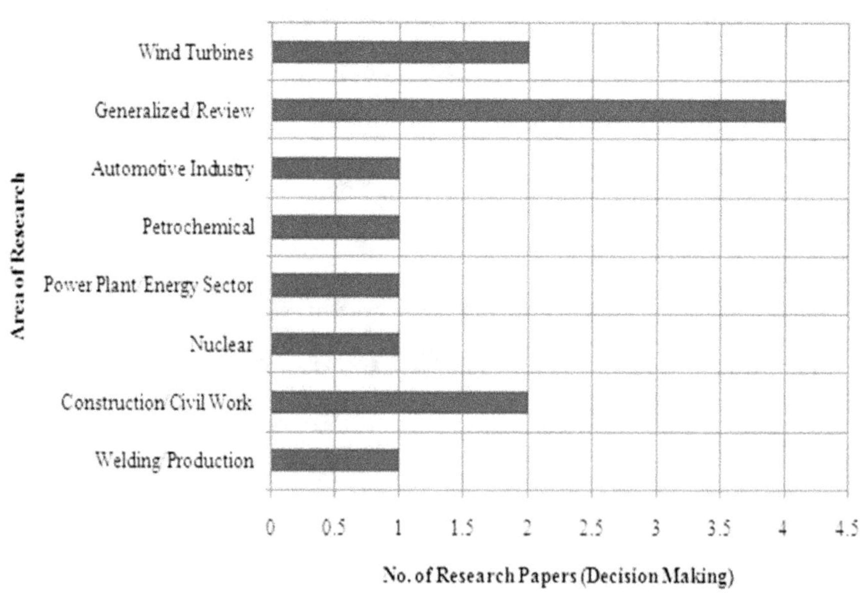

Figure 5. Name of journal/conference v/s number of research papers

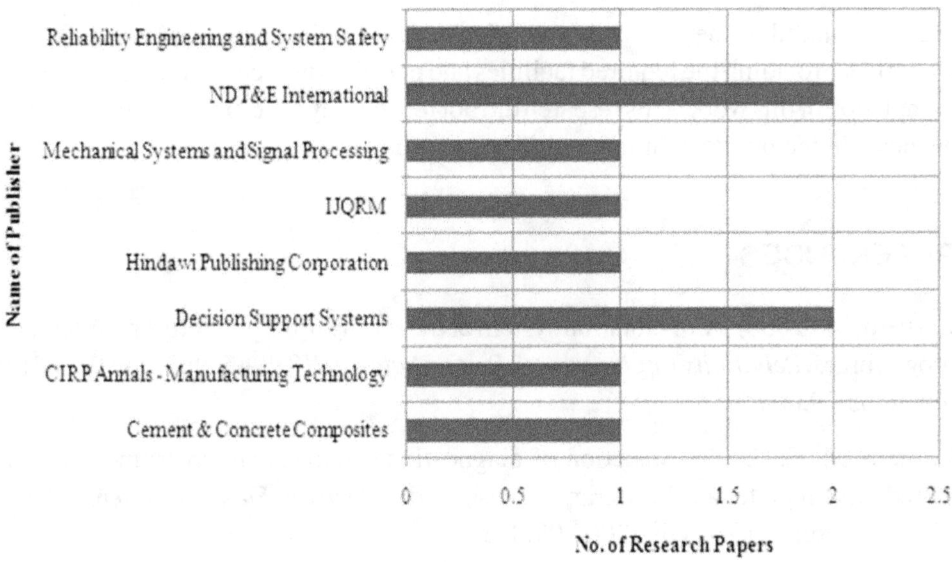

CONCLUSION

As per the objective discussed in the abstract, this research work covered research papers, dealing with decision making in NDT (Non-Destructive testing) techniques along with some practical examples. This paper outlines that mathematically based approaches are necessary for making a combination of NDTs, to ensure higher reliability in fault detection/assessment of defects in industrial equipment/components. The following list shows more outcomes:

1. Very few of the research papers use the decision-making approach. If they use a proper decision-making approach then it will save time and increase the reliability of fault detection.
2. Automotive, energy sector etc. were less covered.

The future work includes:

1. Decision making can apply with lubrication, magnetic testing too.
2. The AHP, PROMETHEE etc. can further explore to use with NDTs.

ACKNOWLEDGMENT

We are thankful to the department of Mechanical Engineering of MANIT-Bhopal (India) for providing the required facilities and permissions needed for the successful completion of this work. This research did not receive any specific grant from funding agencies in the public, commercial, or not-for-profit sectors.

REFERENCES

Carnero, M. (2006). An evaluation system of the setting up of predictive maintenance programmes. *Reliability Engineering & System Safety*, *91*(8), 945–963. doi:10.1016/j.ress.2005.09.003

Carnero, M. C. (2005). Selection of diagnostic techniques and instrumentation in a predictive maintenance program. A case study. *Decision Support Systems*, *38*(4), 539–555. doi:10.1016/j.dss.2003.09.003

D'Orazioa, T., & Leoa, M. (2008). Ultrasonic Testing for Flaw Detection in Mild Steel Sample. *NDT & E International*, *41*, 145–154.

Escaler, X., Egusquiza, E., Farhat, M., Avellan, F., & Coussirat, M. (2006). Detection of cavitation in hydraulic turbines. *Mechanical Systems and Signal Processing*, *20*(4), 983–1007. doi:10.1016/j.ymssp.2004.08.006

Fântână, G. I., Oae, S. A., & Gurau, A. M. (2013). Decision making using the analytic hierarchy process. *International Conferice on MEQAPS*, *13*, 119–124.

Garnier, C., Pastor, M. L., Eyma, F., & Lorrain, B. (2011). The detection of aeronautical defects in situ on composite structures using Non Destructive Testing. *Composite Structures*, *93*(5), 1328–1336. doi:10.1016/j.compstruct.2010.10.017

Ghosh, D., & Roy, S. (2009). A decision-making framework for process plant maintenance. *European Journal of Industrial Engineering*, *4*(1), 78–98. doi:10.1504/EJIE.2010.029571

Hernandez, M. D. P. C., & Labib, A. (2017). Selecting a condition monitoring system for enhancing effectiveness of power transformer maintenance. *Journal of Quality in Maintenance Engineering*, *23*(4), 400–414. doi:10.1108/JQME-07-2015-0027

Horn, D., & Mayo, W. R. (2000). NDE reliability gains from combining eddy-current and ultrasonic testing. *NDT & E International*, *33*(6), 351–362. doi:10.1016/S0963-8695(99)00058-4

Jardine, A. K., Lin, D., & Banjevic, D. (2006). A review on machinery diagnostics and prognostics implementing condition-based maintenance. *Mechanical Systems and Signal Processing*, *20*(7), 1483–1510. doi:10.1016/j.ymssp.2005.09.012

Khaira, A., Srivastava, S., & Suhane, A. (2015). Analysis of relation between ultrasonic testing and microstructure: A step towards highly reliable fault detection. *Engineering Review*, *35*(2), 87–96.

Kim, J. W., & Park, S. (2018). Magnetic Flux Leakage Sensing and Artificial Neural Network Pattern Recognition-Based Automated Damage Detection and Quantification for Wire Rope Non-Destructive Evaluation. *Sensors (Basel)*, *18*(1), 109. PMID:29301294

Krautkrämer, J., & Krautkrämer, H. (2013). *Ultrasonic testing of materials*. Springer Science & Business Media.

Lazakis, I., Dikis, K., Michala, A. L., & Theotokatos, G. (2016). Advanced ship systems condition monitoring for enhanced inspection, maintenance and decision making in ship operations. *Transportation Research Procedia*, *14*, 1679–1688. doi:10.1016/j.trpro.2016.05.133

Marani, R., Palumbo, D., Renò, V., Galietti, U., Stella, E., & D'Orazio, T. (2018). Modeling and classification of defects in CFRP laminates by thermal non-destructive testing. *Composites. Part B, Engineering*, *135*, 129–141. doi:10.1016/j.compositesb.2017.10.010

Sbartaï, Z. M., Breysse, D., Larget, M., & Balayssac, J. P. (2012). Combining NDT techniques for improved evaluation of concrete properties. *Cement and Concrete Composites*, *34*(6), 725–733. doi:10.1016/j.cemconcomp.2012.03.005

Teti, R., Jemielniak, K., O'Donnell, G., & Dornfeld, D. (2010). Advanced monitoring of machining operations. *CIRP Annals-Manufacturing Technology*, *59*(2), 717–739. doi:10.1016/j.cirp.2010.05.010

Toloie-Eshlaghy, A., & Homayonfar, M. (2011). MCDM methodologies and applications: A literature review from 1999 to 2009. *Research Journal of International Studies*, *21*, 86–137.

Vijay Kumar, E., Chaturvedi, S. K., & Deshpandé, A. W. (2009). Maintenance of industrial equipment: Degree of certainty with fuzzy modelling using predictive maintenance. *International Journal of Quality & Reliability Management*, *26*(2), 196–211. doi:10.1108/02656710910928824

Willcox, M., & Downes, G. (2003). *A brief description of NDT techniques*. Toronto: NDT Equipment Limited.

Chapter 12
Identifying Parameters for Nonlinear Degradation-Based Prognostics:
A Case Study With Exponential Degradation on Induction Motors

Feng Yang
*Institute of High Performance Computing (A*STAR), Singapore*

Mohamed Salahuddin
*Institute of High Performance Computing (A*STAR), Singapore*

ABSTRACT

Prognostics and health management (PHM) methodologies are increasingly playing active roles in improving the availability, reliability, efficiency, productivity, and safety of systems in many industries. In predicting the remaining useful life (RUL), this chapter introduces a prognostics framework with health index (HI) formulation, with specific emphasis on incorporating and validating nonlinear HI degradations. The key issue to the success of this framework is how to identify appropriate parameters in describing the behavior of the nonlinear HI degradations. Using exponential HI degradation as an example in predicting the RULs of induction motors, this chapter discusses three different explorations in verifying the existence of good parameter values as well as identifying the appropriate parameters automatically. Comprehensive experiments were carried out with degradation process (DP) data from eight induction motors, and it was discovered that good parameters can be automatically determined with the proposed parameter identification method.

DOI: 10.4018/978-1-5225-6989-3.ch012

INTRODUCTION

Condition-based maintenance (or predictive maintenance) has received increased attention in many industries, due to its promise of cost savings over preventive maintenance (i.e. maintenance conducted routinely in a time-specific manner) and corrective maintenance (i.e. maintenance only conducted upon failure). By predicting the future trends of health using process information from past, current and even future (such as the planned operating condition), predictive maintenance allows for convenient and flexible scheduling of maintenance works. Undoubtedly, this will reduce costs, with other potential advantages including increased equipment lifetime and availability, increased plant safety and productivity, reduced disruption and downtime to equipment, reduced human work-load and optimized inventory handling. In achieving predictive maintenance, prognostics modelling is a requisite, which generally aims to predict the Remaining Useful Life (RUL) of equipment.

Among existing prognostics techniques, the use of data-driven prognostics is gaining much attention in both academia and industry. This is primarily due to its non-requirement of any prior knowledge related to the equipment's degradation as well as the increasing availability of sensory data in practice. In essence, data-driven prognostics approaches can be divided into either direct or Health Index (HI)-based, depending on whether a HI is formulated and used as an intermediate expression of the health condition for predicting the RULs (Yang et al., 2016). Direct prognostics approaches attempt to predict the RULs directly from input sensory data, using modelling methods such as Support Vector Regression (SVR) (Khelif et al., 2017; Loutas, Roulias, & Georgoulas, 2013), Neural Network (NN) (Liu et al., 2010), Gaussian process regression (Goebel, Saha, & Saxena, 2008), and ensemble of multiple models (Hu et al., 2012; Baraldi, Mangili, & Zio, 2012; Patil et al., 2015). Although these approaches have been developed and applied to several practical problems, showing their simplicity and efficiency in implementation, the prediction accuracies generated from these direct approaches are generally not as effective as those obtained from HI-based approaches (see for comparison examples in Yang et al. (2016) and Khelif et al. (2017)).

Alternatively, HI-based prognostics approaches predict RUL in a two-stage process, by firstly predicting a HI from the input sensory data for representing the intrinsic health status and then followed by mapping the HI to the RUL. Compared to direct prognostics approaches, the merit of HI-based prognostics lies in its ability to 'manipulate' and improve the HI, thus, resulting in improved accuracy in predicting the final RUL. The prominence and effectiveness of HI-based prognostics for RUL prediction has been argued and shown in several existing work (Bechhoefer, 2012; He et al., 2012; Liu, Wang, & Tian, 2015; Tian et al., 2016; Liu et al., 2017). One specific case of 'manipulating' HI is in Yang et al. (2016), where the HI was assumed to be degrading linearly and was dynamically smoothed, using current and past information as well as domain knowledge of the HI. The success was illustrated in predicting RULs of eight induction motors, with substantial improvement in the prediction accuracy.

Based on the works of Yang et al. (2016), this chapter introduces a generic formulation of HI-based prognostics, with the ability to consider both linear and nonlinear HI degradations. In this formulation, modelling the degradation behaviour of HI is a pre-requisite, implying that the HI degradation curve needs to be determined a prior. In consideration of simplicity and ease of implementation, linear degradation of health has been studied and gained success in many literatures (Wang et al., 2011; Si et al., 2013; Yang et al., 2016). However, in most real and practical systems, nonlinearity is ubiquitous as most processes degrade in a nonlinear manner, which has driven the efforts in utilizing nonlinear degradation for prognostics (Si et al., 2012; Feng et al., 2013; Si, 2015; Yin & Zhu, 2015).

The major stumbling issue in employing nonlinear HI degradation lies in its complexity in choosing the nonlinear mode (such as exponential) as well as the corresponding mode parameters. Dealing with this issue is the main focus of this chapter, with specific focus on identifying the parameters within a nonlinear degradation mode. Upon introducing the HI-based RUL prediction framework with implementation of nonlinear HI degradation (specifically exponential HI for implementation), the rest of this chapter describes the three different explorations that were carried out in determining the parameters, and the appropriate way of identifying the parameters.

RUL PREDICTION BASED ON NONLINEAR HI DEGRADATION

The adopted HI-based RUL prediction process, incorporated with nonlinear degradation, is depicted in Figure 1, with more details explained thereafter. The RUL prediction process begins with the Degradation Process (DP) data, which are generally time series signals collected from sensors. DP data are then processed segment by segment (e.g. every 0.5 second), and each segment of the DP data are then pushed into Feature Engineering to generate a set of features representing the data segment in a simpler way. When necessary, signal pre-processing, such as denoising, missing value correction, etc, can be applied before transforming the data segments into features. All of the generated features are then put into HI Prediction Modelling to have the predicted HI, which is then dynamically smoothed based on the past HIs and domain knowledge. Upon this, the smoothed HI is then mapped back to obtain the corresponding RUL prediction. Here, when multiple complete DP data are available for modelling the RUL prediction, multiple RULs will then be

Figure 1. Process and ensemble implementation of RUL prediction with HI formulation

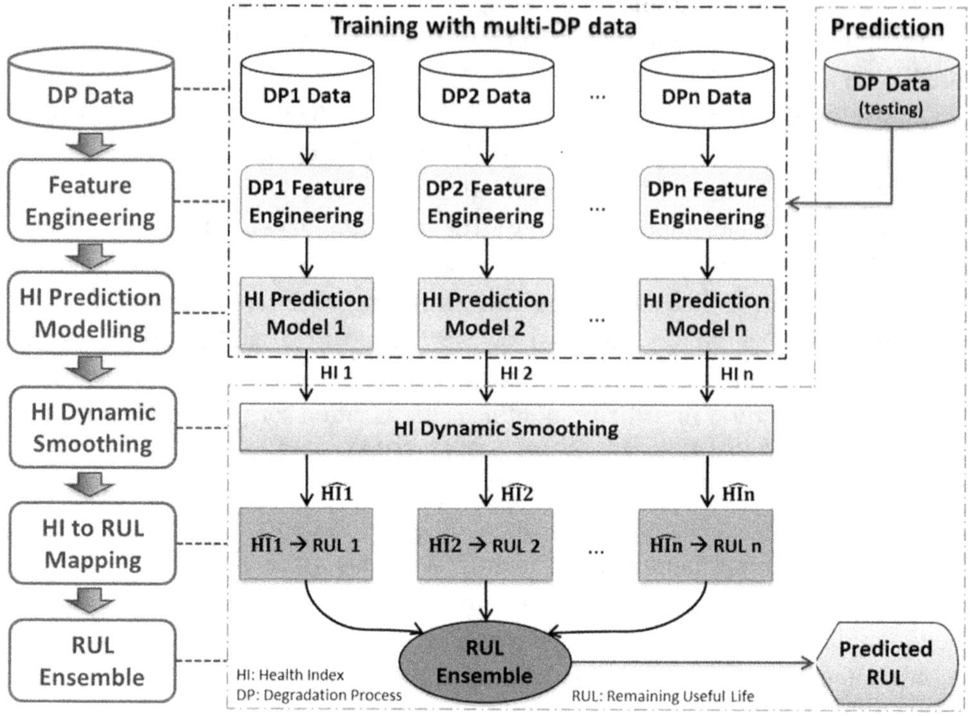

predicted, with the finally predicted RUL obtained from the ensemble (e.g. average) of the multiple RULs.

DP Data

DP is a process whereby a system's health degrades over time (see illustration in Figure 2). Without loss of generality, the life time of a typical system is cycle based and at each cycle, a short period of time series data is available for analysis. For example, in some real systems, a cycle can be 1 day, where at each cycle, 1 minute of time series data are collected. It can then be generally assumed that the system's health remains unchanged during that short period of data collection. A complete DP including all of the data collected from the beginning of the system to the time when the system's health degrades failure level (i.e. falls into failure). Figure 2 illustrates a complete DP where the health level degrades from 100% healthy into failure, and multiple channels of signals were recorded at each collection of the time t.

In the experimental studies conducted in this chapter, the DP data were obtained from University of Tennessee USA, produced by Professor J. Wesley Hines's team

Figure 2. Illustration of the degradation process (DP) and data collection

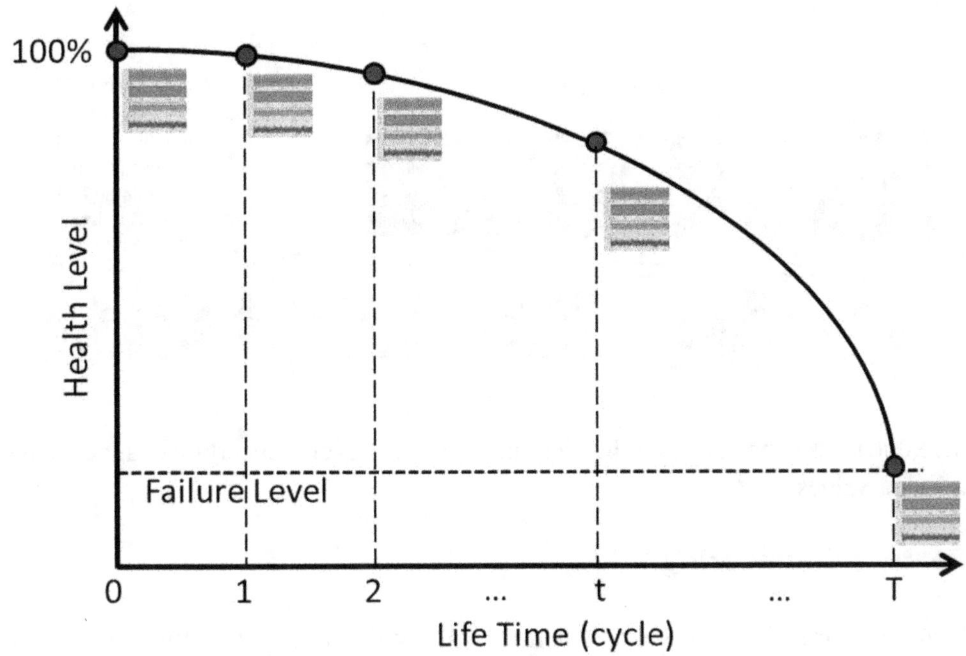

with the Emerson general-purpose 3-phase induction motors undergoing the designed accelerated thermal aging process. Each motor underwent cycles of thermal aging until it completely failed, with raw data collected at the end of each thermal aging cycle. Figure 3 depicts the timeline of the accelerated thermal aging process as well as the data collected from the steady-state condition. In total, 8 seconds (2 seconds x 4) of time series data were recorded at each cycle, with sampling frequency of 10 kHz. The first cycle of data were collected before each new motor went for first thermal aging, such to get the 100% healthy data In this chapter, 11 channels of signals, from each of the 8 motors, were used, including 3-phase voltage (Voltage 1, 2, 3), 3-phase current (Current 1, 2, 3), two-direction vibration (Vibration 1, 2), temperature and acoustic.

From the data collected, it was found that missing values were observed in a few cycles. The details of these missing values and the corresponding lifetime of each motor are given in Table 1. In dealing with the missing value problem, the simplest replacement method was used as the pre-processing, i.e. the missing values in the cycle from a particular channel are replaced by values from its previous cycle. Taking the missing values in Current 2@Cycle 5 for example, the missing values will be replaced by Current 2@Cycle 4. The replacement method was employed

Figure 3. Timeline of the accelerated thermal aging and data collection (adapted from Sharp (2012))

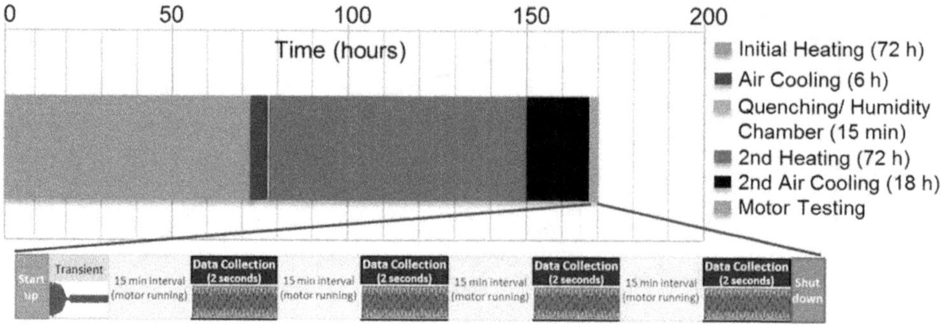

here due to its simplicity and the fact that only a few cycles at certain channels have missing values.

Feature Engineering

Feature engineering transforms a segment of raw signals into a feature vector with reduced data size, whilst preserving the necessary information in the raw signals. This is achieved in two steps: 1) feature extraction to generate a set of features and 2) feature selection to refine the features by removing the irrelevant/noisy features, whilst only keeping those informative ones. In reference to the studies by Yang et al. (2016), time-domain and time-frequency domain feature extraction methods were found to achieve comparable performances, with both of them outperforming the frequency-domain feature extraction method. In this chapter, for the sake of

Table 1. Details of missing values in the eight motors' data

Motor	Lifetime(cycle)	Missing Values
Motor 1	18	Current 2@Cycle 5,6; Voltage 3@Cycle 5, 6, 7
Motor 2	28	Current 2@Cycle 5, 6; Voltage 3@Cycle 5, 6, 7
Motor 3	26	Current1@Cycle 7; Current2@Cycle 5,6,7; Current3@Cycle 7; Voltage 3@Cycle 5,6,7
Motor 4	29	Current2@Cycle 5,6; Voltage 3@Cycle 5,6,7
Motor 5	28	Current2@Cycle 2; Voltage 3@Cycle 2,3
Motor 6	27	Current2@Cycle 5; Voltage 3@Cycle 5,6,7
Motor 7	27	Current2@Cycle 5,6; Voltage 3 @Cycle 5,6,7
Motor 8	25	Current2@Cycle 5; Voltage 3@Cycle 5,6

simplicity, only time-domain feature extraction method is used, with details of the corresponding features given in Table 2.

Upon feature extraction, feature selection is then conducted, in selecting the most useful features. Although there are many different types of feature selection methods, in this chapter, for simplicity and consistency, the feature ranking method of Fisher's ratio is employed. In essence, Fisher's ratio uses the following equation to calculate the importance of each single feature, X_j and then ranks all of the features based on their individual importance (Yang & Mao, 2011).

Table 2. Details of features by time-domain feature extraction (Wu, Yang, & Zhou, 2012)

Feature	Equation	Feature	Equation
Average amplitude of vibration	$p_1 = \frac{1}{n}\sum_{i=1}^{n} x(i)$	**kurtosis coefficient**	$p_7 = \frac{\sum_{i=1}^{n}\left(x(i) - p_1\right)^4}{(n-1)p_2^4}$
standard deviation	$p_2 = \left(\frac{\sum_{i=1}^{n}\left(x(i) - p_1\right)^2}{n-1}\right)^{1/2}$	**peak factor**	$p_8 = \frac{p_5}{p_3}$
Root-mean-square amplitude	$p_3 = \left(\frac{1}{n}\sum_{i=1}^{n} x(i)^2\right)^{1/2}$	**margin factor**	$p_9 = \frac{p_5}{p_4}$
Square of mean of rooted absolute amplitude	$p_4 = \left(\frac{1}{n}\sum_{i=1}^{n}\sqrt{\lvert x(i)\rvert}\right)^2$	**waveform factor**	$p_{10} = \frac{p_3}{\frac{1}{n}\sum_{i=1}^{n}\lvert x(i)\rvert}$
peak value	$p_5 = max\lvert x(i)\rvert$	**impulse factor**	$p_{11} = \frac{p_5}{\frac{1}{n}\sum_{i=1}^{n}\lvert x(i)\rvert}$
Skewness coefficient	$p_6 = \frac{\sum_{i=1}^{n}\left(x(i) - p_1\right)^3}{(n-1)p_2^3}$		

$$FR\left(X_j\right) = \frac{\left(m_{j(1)} - m_{j(2)}\right)^2}{\sigma^2_{j(1)} + \sigma^2_{j(2)}} \tag{1}$$

where $m_{j(c)}$ is the mean and $\sigma^2_{j(c)}$ the variance of feature X_j in the class *c*, for *c*=1, 2 (e.g. healthy vs. faulty). The larger the $FR\left(X_j\right)$, the higher discriminative capability the feature X_j has. Thus, feature selection using Fisher's ratio will provide a list of the top features which will then be used in the modelling step for predicting the HI.

HI Prediction Modelling

In the developed HI-based RUL prediction framework, HI ($\in[0,1]$) is an intermediate variable representing the health state of a system within the degradation process, where HI=100% implies completely healthy whilst HI=0% implies total failure of the system. In training a model for predicting the HI, there are two issues that needs to be addressed: 1) how the HI degrades over time and 2) which algorithm to build the relationship from instances (an instance is a feature vector from feature engineering) to the HIs. Working out the first issue (e.g. exponential degradation with identified parameters) makes it able to determine the training data's HI values that are not known from the raw measures but required as the input for training HI prediction model. As discussed in the following, this chapter used the nonlinear (specifically exponential) HI degradation and neural network for the HI prediction modelling.

HI Degradation Mode

The HI degradation mode refers to the form in which the system's health degrades over time. In contrast to linear degradation that was generally used as approximations in some simplified scenarios, nonlinear degradation mode appears to be more popular in real world systems. The two types of nonlinear degradation curves, namely, convex and concave, are illustrated together with linear degradation in Figure 4. In convex degradation, the health level decreases over time in an accelerated speed, with example like the growth of fatigue cracks of a surface (Lu & Meeker, 1993). On the contrary, in concave degradation, the health level decreases over time in a decreased speed, with examples in chemical processes with a limited amount of material to react, such as in the growth of failure-causing conducting filaments of chlorine-copper compounds in printed-circuit boards (Meeker & LuValle, 1995).

Figure 4. Possible modes for univariate health degradation curves (adapted from Meeker & Escobar (1998))

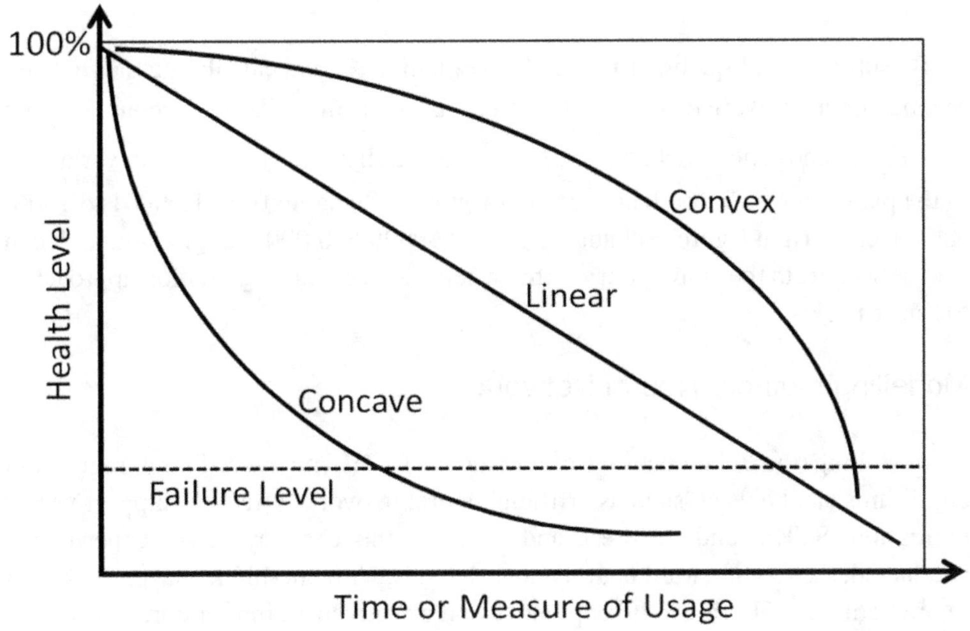

The focus of this chapter is on utilizing nonlinear HI degradation for predicting the RUL. Among the various nonlinear degradation forms, exponential degradation is one of the most popular forms used in describing the degradation of a system (Kim et al., 2016; Nguyen et al., 2017; Gebraeel, 2006; Fan et al., 2013; Wang et al., 2016; Wang et al., 2018; Peng et al., 2017). Here, the exponential degradation of the HI is chosen, with the mathematical formulation expressed in Equation (2).

$$HI\left(t\right) = a \times exp\left(b \times t\right) + c \tag{2}$$

where t is the time index and a, b, c are the parameters. Without generality, a degradation process can be assumed to have 100% of health at the beginning (i.e. $HI\left(0\right) = HI_{begin} = 100\%$) and drops to very low health, like 10%, at the last time T for data collection before falling into failure (i.e. $HI\left(T\right) = HI_{end} = 10\%$). This leads to:

$$a = HI_{begin} - c \tag{3}$$

$$b = \frac{1}{T}\log\left(\frac{HI_{end} - c}{HI_{begin} - c}\right) \tag{4}$$

By substituting Equation (3) and (4) to Equation (2) and calculating the first and second order derivatives of $HI\left(t\right)$, it can be seen that $HI\left(t\right)$ is convex when $c > HI_{begin}$ and concave when $c < HI_{end}$. Figure 5 shows the HI curves for variations in the parameter c, for both convex (left figure, c changing from 1.00001 to 1000) and concave (right figure, c changing from 0.99999 to 0.0001) degradations. It can also be seen from the convex curves that when $c \to \infty$, the degradation approaches to linear HI.

Modelling Method: Neural Network

In modelling the relationship from Instance to HI, many different regression algorithms can be used, such as artificial neural networks (ANNs), support vector regression (SVR), random forest, and so on. In this chapter, for the experiments conducted, the feed-forward neural network (NN), with one-hidden layer, is chosen in predicting the HI, due to its popularity and effectiveness in literatures as well as its theoretical capability of modeling arbitrarily complex functions (Haykin, 1999). Figure 6 illustrates the one hidden layer feed-forward neural network with I input nodes, K hidden nodes and J output nodes. In training the NN, the sigmoid activation function in hidden neurons and the backpropagation (abbreviation for "backward propagation of errors") method will be used in conjunction with an optimization method (e.g. Levenberg-Marquardt (Matlab, 2017) in updating the weights.

Figure 5. Exponential degradation HI curves (left: convex and right: concave)

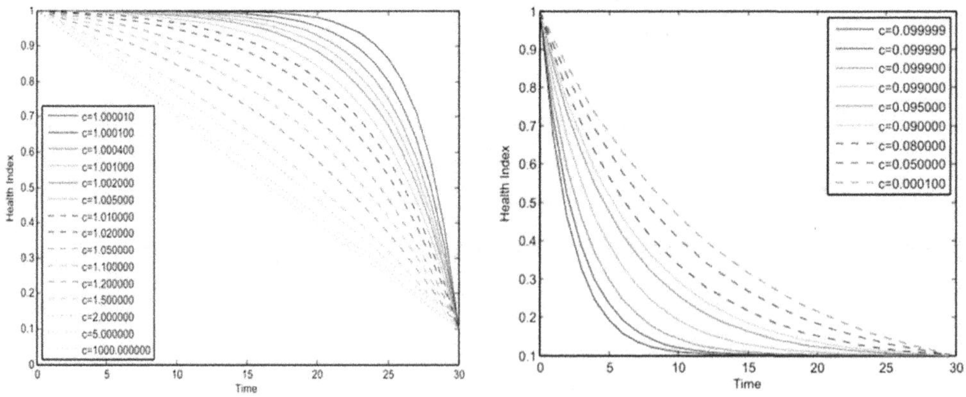

**For a more accurate representation see the electronic version.*

Figure 6. Feedforward neural network with one hidden layer

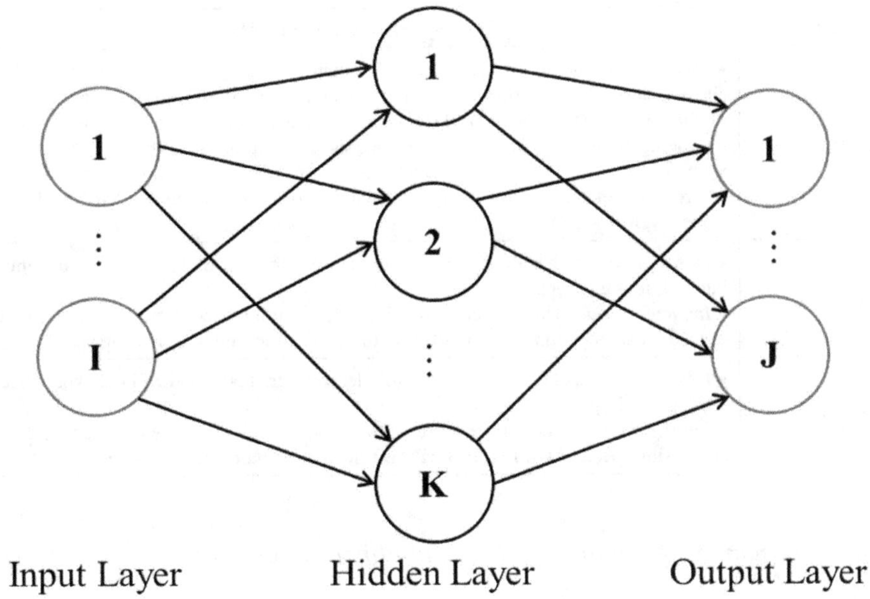

HI Dynamic Smoothing

In the RUL prediction process shown in Figure 1, the predicted HIs from each individual HI prediction model are dynamically smoothed, by making use of both the current and past HI values. This particular step is with the idea of incorporating the past information in enhancing current predictions, mimicking the natural decision-making procedure of human beings where correct decisions are generally made by considering current and past information. As used in linear HI degradation, the 3 assumed rules (domain knowledge), namely, consistency, monotonicity and gradualness, are also applied in the dynamic smoothing under nonlinear HI degradation. Table 3 gives the details of description as well as implementation of these rules.

Figure 7 describes the HI dynamic smoothing process that has been implemented for exponential degradation of HI. In STEP 1, upon defining the correction method, the outliers are first corrected. STEP 2 is then to have adjustments reducing inconsistencies according to the first rule of "consistency". In STEP 3, further adjustments are conducted based on the other two rules of "monotonicity" and "gradualness", by restricting the current HI value between minimum and maximum drops. In STEP 4, the current predicted together with the past adjusted HI values are used to fit a piece-wise weighted linear regression model which is then used to predict the smoothed HI value at current time. STEP 5 is to update the intermediate

Table 3. Rules for HI dynamic smoothing

Rule	Description and Implementation
Consistency	• *Description*: The health of a motor will remain constant or only having very slight changes during short periods of running time, e.g. 20 seconds. Therefore, multiple predictions from the same model (in the experiments, HI is predicted every 0.5 seconds, producing 40 predictions in the 20 seconds signals) will be consistent. • *Implementation*: Outliers are corrected and restricted to [*mean-std, mean+std*] where *mean* and *std* are the average and standard deviation of the corrected predictions respectively.
Monotonicity	• *Description*: The health of a motor will not improve over time, assuming no manual interventions e.g. repairs. • *Implementation*: The predicted HI value at time *t* is adjusted to ensure that it does not exceed HI value at time *t-1*. In fulfilling this rule, a minimum decreasing speed is set.
Gradualness	• *Description*: The health of a motor will decrease gradually and will not experience a sudden big drop. • *Implementation*: For nonlinear HI degradation, this is realized by restricting the decreasing speed of HI. In fulfilling this rule, a maximum decreasing speed is set.

Figure 7. Dynamic smoothing of HI with nonlinear degradation

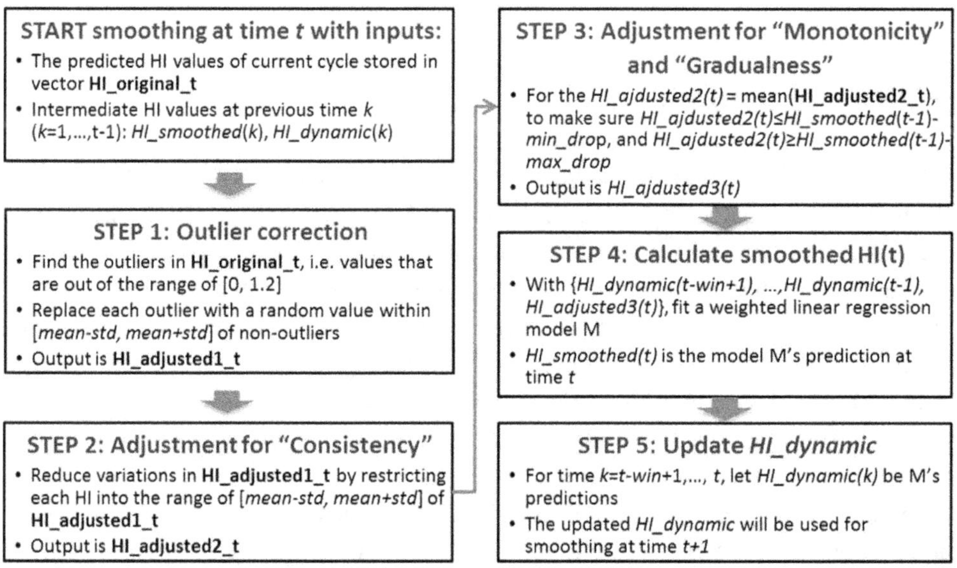

values of *HI_dynamic* for subsequent smoothing of HIs. In comparison with linear HI degradation, the primary difference here is in STEP 3, where the maximum and minimum drops of HI at current time are defined to limit the range of HI drop. These two values, namely, *max_drop* and *min_drop* at time *t*, are calculated by restricting the drop speed from the previous time *t-1*, using the expected drop speed from the

HI degradation curve of the training motor. In the implementation, the maximum and minimum drop speeds were set to 1.3 and 0.8 times of the expected speed respectively. The window size, *win*, in fitting the piece-wise linear regression model was set to 5 (the same value as in the implementation with linear HI degradation).

Figure 8 illustrates with an example the advantage of HI dynamic smoothing, by plotting and compare the prediction curves of the originally predicted (see 'Original'), the dynamically smoothed (see 'Smoothed'), and the true HI values (see 'True'). HI values from the intermediate variable, *HI_dynamic*, are also included (see 'Dynamic'). The comparison between the smoothed and original HI curves clearly indicates the advantage of the adoption of HI dynamic smoothing strategy, based on the observation from the width of the CI and their closeness to the true HIs.

HI to RUL Mapping

With a determined exponential HI degradation form, mapping from HI to RUL can be carried out in a direct calculation. By using the HI degradation curve of a training

Figure 8. Advantage of HI dynamic smoothing

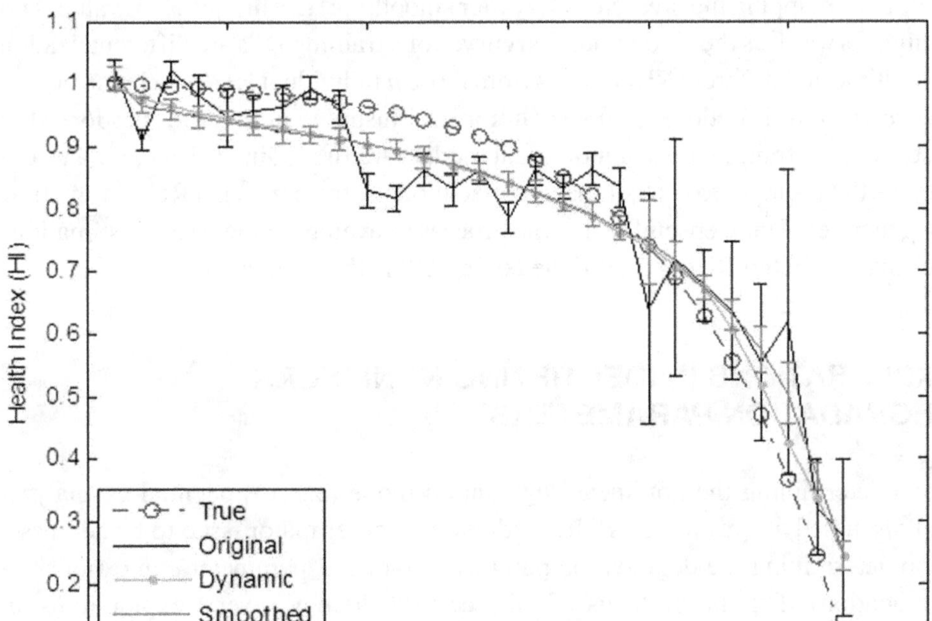

DP as well as the predicted HI value of a testing DP at time t, the corresponding predicted RUL value is analytically calculated from Equation (2).

$$\hat{t} = \frac{1}{b}\ln\left(\frac{\widehat{HI}\left(t\right) - c}{a}\right) \tag{5}$$

$$\widehat{RUL}\left(t\right) = T - \hat{t} \tag{6}$$

where $\widehat{HI}\left(t\right)$ is the predicted HI of the testing DP at time t, *T* is the total lifetime of the training DP and $\widehat{RUL}\left(t\right)$ is the estimated RUL of the testing DP at time t.

RUL Ensemble

Equation (5) and (6) give the RUL estimation based on one training DP data. However, when there are multiple training DPs, say *n* DPs, where each DP has a different lifetime *T*, mapping the predicted HIs (after smoothing) to a unique RUL value is not straightforward, as the HI degradation curves of *n* training DPs are different, leading to multiple and different RUL values from those *n* individual HI degradation curves. Hence, the multi-model ensemble solution, as illustrated in Figure 1, is adopted. In essence, for *n* training DPs, *n* individual models are first trained. Then, for a testing DP, the data is input to each of the *n* trained models, from which *n* RUL estimations are generated. Upon ensemble, i.e. the process of averaging the *n* RUL estimations, the final predicted RUL value of the testing DP is then obtained.

EXPLORATIONS IN IDENTIFYING NONLINEAR DEGRADATION PARAMETERS

Upon determining the nonlinear degradation mode (e.g. exponential as analyzed and discussed above) for the RUL prediction, another major issue to be addressed is on determining the degradation parameters (such as parameter *c* in exponential degradation). This is the focus of this section whereby several explorations are conducted in finding the most appropriate parameter values. Experiments were carried out on the 8 induction motors' data in the case of exponential degradation, where a set of parameter c values were predefined (convex: c changes from 1.00001 to 1000; concave: c changes from 0.099999 to 0.0001, see the corresponding degradation curves in Figure 5). A summary of the explorations is shown in Table 4, with more details given thereafter.

In each experiment, the 8 induction motors' data were separated into training and testing data i.e. set aside one motor's data as testing data (e.g. motor 1) and the rest of the 7 motors' data as training data (e.g. motor 2 to 8). In evaluating the performance of the RUL prediction, the Root Mean Square Error (RMSE) is used. In addition, the Confidence Interval (CI) for the predicted RUL at each time index t is also calculated. Given the predicted RUL (r_t) and true RUL (r_t^*) at the time index t, the calculations for the RMSE and CI are given below.

- Root Mean Square Error (RMSE)

$$RMSE = \sqrt{\frac{1}{n}\sum_{t=1}^{n}\left(r_t - r_t^*\right)^2} \tag{7}$$

- Confidence Interval (CI) is defined as an interval that contains 95% of the predicted RULs at the same time index

Table 4. Summary of the explorations in identifying the parameter c for the RUL prediction

Exploration	Description & Observation
Exploration 1: Manually assigned parameter c value for all of the training DPs	• *Description*: All of the n training DPs share the same degradation parameter c value. • *Observation*: Very good performance was achieved when parameter c was assigned some particular values.
Exploration 2: Grid search to determine parameter c value for each of the individual training DP	• *Description*: To find the best parameter c value for each individual training DP, by grid search from a predefined set of c values. • *Observation*: Each training DP had the best fitted degradation curve; however, the prediction performance on a testing DP was not as good as expected, possibly due to 'overfitting' problem, i.e. the training DPs were overfitted.
Exploration 3: Grid search to determine parameter c value for all of the training DPs	• *Description*: To find the best parameter c value for all of the n training DPs, by grid search from a predefined set of c values. • *Observation*: Good performance on testing DPs was achieved as compared to Exploration 2, but not as good as the best performance achieved in Exploration 1.

$$\Pr\left(r_t - \frac{CI_t}{2} \le r_t^l \le r_t + \frac{CI_t}{2}\right) = 0.95 \tag{8}$$

where r_t^l is the predicted RUL of the instance l at the time index t.

Exploration 1: Manually Assigning Parameter *c* Value for All Training DPs

The purpose of this exploration is to find out whether good RUL prediction performances can be achieved by choosing some appropriate parameter *c* value. For this purpose, each parameter *c* value in the pre-defined set of c values is manually assigned for all of the training DPs (i.e. the 8 induction motors' degradation data). With the assigned parameter *c* values, the degradation curve of each training DP and the corresponding HI value at each time index *t* can then be calculated (refer to Equation (2) to (4)). The calculated HI values of all of the training DPs are then used to train the HI prediction models, which will be used for predicting the HIs of a testing DP and the corresponding RULs (refer to Figure 1 for the process).

Thus, with each given value of the parameter *c*, each motor's data was used once as the testing data and the corresponding RMSE of the RUL predictions were then evaluated. The weighted average (weights determined by the corresponding lifetimes as given in Table 1) of the RMSEs of the testing 8 motors were calculated and the results shown in Figure 9 for each of the predefined parameter *c* values. In addition to nonlinear HI degradation (for both convex and concave), the performance with linear HI degradation (weighted RMSE = 2.74, see the yellow line) was also plotted in the figure acting as a benchmark.

From Figure 9, it can be seen that very good performances (with the weighted average of RMSE lower than 2.0) were achieved as compared to the performance with linear HI degradation (weighted average of RMSE=2.74), indicating the possibility of choosing some good parameter *c* values for the nonlinear (exponential) HI degradation in predicting the RUL. In addition, it was also discovered that all of the good performances were obtained with the convex HI degradations, which implies that the health degradation of the induction motors undergoing accelerated thermal aging should follow convex shape rather than concave shape of degradation.

Exploration 2: Grid Search to Determine a *c* Value for Each Individual Training DP

Rather than manually assigning parameter *c* values for the training DPs, this as well as the next explorations attempt to automatically search for a good parameter *c* value,

Figure 9. Weighted RMSE by varying parameter c in exponential degradation

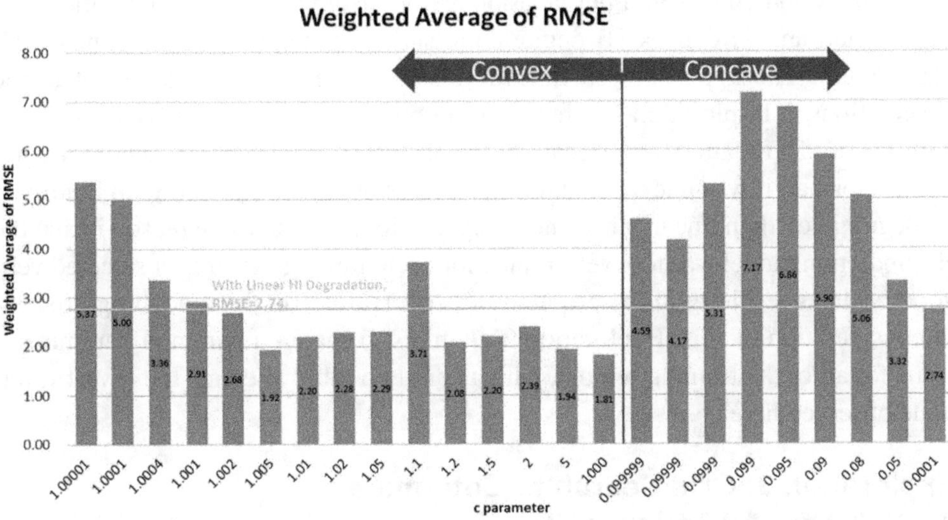

based on the training DPs. The difference between these two explorations is that Exploration 2 tries to search for a particular *c* value for each training DP (i.e. each training DP has its own parameter *c* value), whilst Exploration 3 tries to search for all of the training DPs (i.e. all training DPs share the same *c* value).

Thus, within the predefined set of parameter *c* values for both convex and concave degradations, grid search was employed for the parameter search. The procedure is described in the following steps.

Step 1: For each candidate *c* value, the corresponding HI values are calculated (using Equation (2) to (4)).

Step 2: The HI prediction model is built, i.e. trained with the training motor's data (input: instances at different time indexes; output: the corresponding HI values).

Step 3: The trained HI prediction model is then applied to the training motor's data to get the predicted HI values.

Step 4: The RMSE performance is evaluated with the calculated HI values and the predicted HI values.

Step 5: Repeat Step 1 to Step 4 to get the RMSE performance for each candidate *c* value, and the *c* value with the lowest RMSE is chosen as the best *c* value.

The RMSE performances, for both convex and concave degradations, for each of the testing motor is shown in Table 5. For comparison purpose, the benchmark performance from linear HI degradation is also provided. From the performance results, it was discovered that 1) in most of the cases, the performances with both convex and concave HI degradation are not as good as the benchmark linear degradation and 2) convex HI degradation seems to outperform the concave HI degradation. The latter showed that the results obtained were consistent to the observations in Exploration 1, where very good performances were only achieved with convex degradation (i.e. concave didn't perform as well as convex degradation). For the former, it was unexpected to have even worse performance with nonlinear HI degradation than linear using linear degradation. One possible reason is that the HI prediction model was too well trained for each training motor as it selected very good c values resulting in the problem like 'overfitting' in the modelling process. In consideration of this, Exploration 3 attempts to search a common parameter c value for all of the training motors, which could possibly alleviate the 'overfitting' issue observed here.

Exploration 3: Grid Search to Determine a *c* Value for All Training DPs

Rather than searching a particular c value for each of the training motor in Exploration 2, Exploration 3 tries to choose a parameter c value for all of the training motors using grid search. Here, all data from the training motors were used in the grid search to determine the common parameter c value shared by all of the training motors. The grid search used in Exploration 3 is similar to that in Exploration 2, with the only difference being all of the training motors' data (not just an individual training motor's data) were used in Step 2 to Step 4.

The performance results, together with results for Exploration 2, are shown in Table 5. From the results, the following were observed: 1) In Exploration 3, in most cases (specifically 6 out of 8 cases), the RUL predictions with convex degradation achieved better performance as compared to the performance with linear HI degradation. This can be more clearly seen from the weighted average of RMSE across the 8 motors, i.e. 2.04 vs. 2.74, a 25.5% improvement in weighted average of RMSE. 2) Similar to the observations in Exploration 1 and 2, RUL predictions with concave degradation were not good as convex degradation. In fact, using concave degradation performed worse than using linear degradation, further implying the unsuitability of using concave degradation in describing the induction motors' degradation process.

Results Discussion

By manually assigning a parameter c in Exploration 1, the experimental results showed the existence of good parameter c values in exponential HI degradation for predicting the RUL. This led to Exploration 2 and 3, further attempts to find

Table 5. RMSE of RUL predictions in Exploration 2 and Exploration 3

Motor	Linear Degradation	Exploration 2		Exploration 3	
		Convex	Concave	Convex	Concave
1	6.76	3.26	6.56	5.91	5.35
2	2.38	2.70	5.36	1.73	4.88
3	0.92	1.68	3.59	1.03	4.92
4	3.46	4.69	4.97	2.78	5.26
5	4.31	3.35	5.24	2.66	4.25
6	1.49	2.99	4.52	0.72	4.79
7	2.48	2.56	4.24	1.26	3.70
8	1.15	3.60	3.47	1.35	4.75
Weighted Average	**2.74**	**3.11**	**4.69**	***2.04***	**4.71**

the appropriate parameter c values in an automatic manner based on the available training DPs' data. These two explorations were similarly conducted using grid search, with the difference being that Exploration 2 tried to find a unique c value for each of the individual DP, whilst Exploration 3 attempted to find a common c value shared by all of the training DPs. The RMSE results obtained in Exploration 2 were not good (in fact, even worse than those with linear HI degradation), which is unexpected. One possible reason is that the degradation of each of the individual DP was too well trained by finding the best matched c value, resulting in the problem of 'overfitting' as faced in many training-testing processes. By searching a common parameter c value for all of the training DPs in Exploration 3, a best match to the whole training DPs but not to each individual DP was obtained. This is like the measure of 'regularization' which was generally applied in alleviating/solving the overfitting problem. The very good performance was achieved in Exploration 3, proving the effectiveness of the method used in that exploration.

As to convex and concave degradations, the common observation from the three explorations was that RUL prediction with convex HI degradation performed better than that with concave HI degradation. This does not imply that concave is not as good as convex, but it only indicates that for the studied problem of motors' accelerated degradation, convex degradation is more suitable than concave degradation in describing the health degradation of motors.

In considering the performance from all of the explorations, it is suggested that the method described in Exploration 3 is used to identify the parameter c in employing the nonlinear HI based RUL prediction, for predicting the RUL of induction motors undergoing thermal aging. For better observing and comparing the results, the

RUL predictions by the different methods on one motor are illustrated below in Figure 10. The RUL with the corresponding CI at each time (in cycle) was plotted. It can be seen that with the parameter identification method in Exploration 3, the predicted RULs are much closer to the true RULs, as compared to method used in Exploration 2 as well as the RUL prediction with linear HI degradation. Another benefit obtained by using nonlinear HI is that narrower CI (i.e. higher confidence) was achieved as compared to RUL prediction using linear HI.

CONLCUSION

Identifying the appropriate parameter values is a key component in applying the nonlinear Health Index (HI)-based prognostics for predicting Remaining Useful Life (RUL). Focusing on the parameter identification issue, this chapter used the exponential HI degradation as an example and had three explorations for finding the

Figure 10. Illustration of RUL predictions

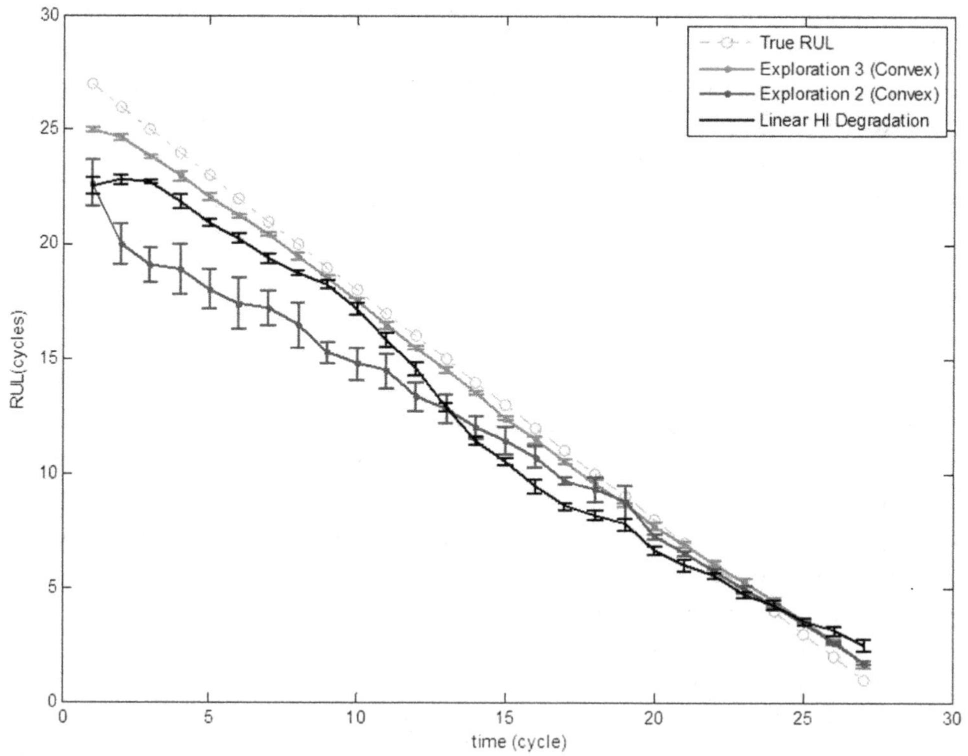

best parameter value. By predefining a set of candidate parameter values, Exploration 1 tested the performance of each candidate and verified that good parameter values exist for achieving high accuracy of the RUL predictions. In order to automatically obtain the best parameter value based on available training data, Exploration 2 and 3 tried two different ways using grid search. In the scenario with multiple training Degradation Processes (DPs), Exploration 2 tried to find the best parameter value for each training DP (i.e. each training DP has its own mode of nonlinear degradation curve), whilst Exploration 3 had a regularized way to find a common parameter value for all training DPs (i.e. all training DPs share the same mode of degradation curve). Experimental results showed that parameter identification as in Exploration 2 didn't achieve good performance, implying the problem of 'overfitting' (i.e. training DPs were too well fitted by finding each DP's own parameter value). In contrast, parameter identification as in Exploration 3 achieved very good performance, and hence it is suggested be the way of determining the parameter value within the introduced prognostics framework.

ACKNOWLEDGMENT

The authors would like to acknowledge that the datasets used in this chapter were obtained from the University of Tennessee.

REFERENCES

Baraldi, P., Mangili, F., & Zio, E. (2012). A kalman filter-based ensemble approach with application to turbine creep prognostics. *IEEE Transactions on Reliability*, *61*(4), 966–977. doi:10.1109/TR.2012.2221037

Bechhoefer, E. (2012). Data driven prognostics for rotating machinery. In S. Kadry (Ed.), *Diagnostics and Prognostics of Engineering Systems: Methods and Techniques* (pp. 120–133). IGI Global.

Fan, Z., Liu, G., Si, X., Zhang, Q., & Zhang, Q. (2013). Degradation data-driven approach for remaining useful life estimation. *Journal of Systems Engineering and Electronics*, *24*(1), 173–182. doi:10.1109/JSEE.2013.00022

Feng, L., Wang, H., Si, X., & Zou, H. (2013). A State-Space-Based Prognostic Model for Hidden and Age-Dependent Nonlinear Degradation Process. *IEEE Transactions on Automation Science and Engineering*, *10*(4), 1072–1086. doi:10.1109/TASE.2012.2227960

Gebraeel, N. (2006). Sensory-Updated Residual Life Distributions for Components With Exponential Degradation Patterns. *IEEE Transactions on Automation Science and Engineering*, *3*(4), 382–393. doi:10.1109/TASE.2006.876609

Goebel, K., Saha, B., & Saxena, A. (2008). A comparison of three data-driven techniques for prognostics. In *Proceedings of 62nd Meeting Society for Machinery Failure Prevention Technology (MFPT)* (pp. 119–131). Virginia Beach, VA: Academic Press.

Haykin, S. (1999). *Neural Networks: A Comprehensive Foundation*. Englewood Cliffs, NJ: Prentice-Hall.

He, D., Bechhoefer, E., Ma, J., & Zhu, J. (2012). A Particle Filtering Based Approach for Gear Prognostics. In S. Kadry (Ed.), *Diagnostics and Prognostics of Engineering Systems: Methods and Techniques* (pp. 120–133). IGI Global.

Hu, C., Youn, B. D., Wang, P., & Yoon, J. T. (2012). Ensemble of data-driven prognostic algorithms for robust prediction of remaining useful life. *Reliability Engineering & System Safety*, *103*, 120–135. doi:10.1016/j.ress.2012.03.008

Khelif, R., Chebel-Morello, B., Malinowski, S., Laajili, E., Fnaiech, F., & Zerhouni, N. (2017). Direct Remaining Useful Life Estimation Based on Support Vector Regression. *IEEE Transactions on Industrial Electronics*, *64*(3), 2276–2285. doi:10.1109/TIE.2016.2623260

Kim, T., Oh, H., Kim, H., & Youn, B. D. (2016). An Online-Applicable Model for Predicting Health Degradation of PEM Fuel Cells With Root Cause Analysis. *IEEE Transactions on Industrial Electronics*, *63*(11), 7094–7103. doi:10.1109/TIE.2016.2586022

Liu, J., Saxena, A., Goebel, K., Saha, B., & Wang, W. (2010). An adaptive recurrent neural network for remaining useful life prediction of lithium-ion batteries. *Proceedings of Annual Conference of the Prognostics and Health Management Society*.

Liu, S., Wang, Y., & Tian, F. (2015). Prognosis of underground cable via online data-driven method with field data. *IEEE Transactions on Industrial Electronics*, *62*(12), 7786–7794. doi:10.1109/TIE.2015.2458300

Liu, Z., Liu, T., Han, J., Bu, S., Tang, X., & Pecht, M. (2017). Signal Model-Based Fault Coding for Diagnostics and Prognostics of Analog Electronic Circuits. *IEEE Transactions on Industrial Electronics*, *64*(1), 605–614. doi:10.1109/TIE.2016.2599142

Loutas, T., Roulias, D., & Georgoulas, G. (2013). Remaining useful life estimation in rolling bearings utilizing data-driven probabilistic e-support vectors regression. *IEEE Transactions on Reliability*, *62*(4), 821–832. doi:10.1109/TR.2013.2285318

Lu, C. J., & Meeker, W. Q. (1993). Using degradation measures to estimate a time-to-failure distribution. *Technometrics*, *35*(2), 543–559. doi:10.1080/00401706.1993.10485038

Matlab. (2017). *Matlab R2017b document*. Retrieved on 11-Nov-2017 from: https://www.mathworks.com/help/nnet/ref/trainlm.html

Meeker, W. Q., & Escobar, L. A. (1998). *Statistical Methods for Reliability Data*. John Wiley & Sons Inc.

Meeker, W. Q., & LuValle, M. J. (1995). An Accelerated Life Test Model Based on Reliability Kinetics. *Technometrics*, *37*(2), 133–146. doi:10.1080/00401706.1995.10484298

Nguyen, V., Seshadrinath, J., Wang, D., Nadarajan, S., & Vaiyapuri, V. (2017). Model-Based Diagnosis and RUL Estimation of Induction Machines under Inter-Turn Fault. *IEEE Transactions on Industry Applications*, *53*(3), 2690–2701. doi:10.1109/TIA.2017.2669195

Patil, M. A., Tagade, P., Hariharan, K. S., Kolake, S. M., Song, T., Yeo, T., & Doo, S. (2015). A novel multistage support vector machine based approach for li ion battery remaining useful life estimation. *Applied Energy*, *159*, 285–297. doi:10.1016/j.apenergy.2015.08.119

Peng, W., Li, Y. F., Yang, Y. J., Mi, J., & Huang, H. Z. (2017). Bayesian Degradation Analysis With Inverse Gaussian Process Models Under Time-Varying Degradation Rates. *IEEE Transactions on Reliability*, *66*(1), 84–96. doi:10.1109/TR.2016.2635149

Sharp, M. E. (2012). *Prognostic approaches using transient monitoring methods* (Ph.D. dissertation). University of Tennessee, Knoxville, TN.

Si, X. S. (2015). An Adaptive Prognostic Approach via Nonlinear Degradation Modeling: Application to Battery Data. *IEEE Transactions on Industrial Electronics*, *62*(8), 5082–5096. doi:10.1109/TIE.2015.2393840

Si, X. S., Wang, W., Hu, C. H., Chen, M. Y., & Zhou, D. H. (2013). A Wiener-process-based degradation model with a recursive filter algorithm for remaining useful life estimation. *Mechanical Systems and Signal Processing*, *35*(1-2), 219–237. doi:10.1016/j.ymssp.2012.08.016

Si, X. S., Wang, W., Hu, C. H., Zhou, D. H., & Pecht, M. G. (2012). Remaining useful life estimation based on a nonlinear diffusion degradation process. *IEEE Transactions on Reliability*, *61*(1), 50–67. doi:10.1109/TR.2011.2182221

Tian, J., Morillo, C., Azarian, M. H., & Pecht, M. (2016). Motor Bearing Fault Detection Using Spectral Kurtosis-Based Feature Extraction Coupled With K-Nearest Neighbor Distance Analysis. *IEEE Transactions on Industrial Electronics*, *63*(3), 1793–1803. doi:10.1109/TIE.2015.2509913

Wang, W., Carr, M., Xu, W., & Kobbacy, A. K. H. (2011). A model for residual life prediction based on Brownian motion with an adaptive drift. *Microelectronics and Reliability*, *51*(2), 285–293. doi:10.1016/j.microrel.2010.09.013

Wang, Y., Peng, Y., Zi, Y., Jin, X., & Tsui, K. L. (2016). A Two-Stage Data-Driven-Based Prognostic Approach for Bearing Degradation Problem. *IEEE Transactions on Industrial Informatics*, *12*(3), 924–932. doi:10.1109/TII.2016.2535368

Wang, Z. Q., Hu, C. H., & Fan, H. D. (2018). Real-Time Remaining Useful Life Prediction for a Nonlinear Degrading System in Service: Application to Bearing Data. *IEEE/ASME Transactions on Mechatronics*, *23*(1), 211–222. doi:10.1109/TMECH.2017.2666199

Wu, Q., Yang, X., & Zhou, Q. (2012). Pattern recognition and its application in fault diagnosis of electromechanical system. *Journal of Information and Computational Science*, *9*(8), 2221–2228.

Yang, F., Habibullah, M. S., Zhang, T., Xu, Z., Lim, P., & Nadarajan, S. (2016). Health Index-Based Prognostics for Remaining Useful Life Predictions in Electrical Machines. *IEEE Transactions on Industrial Electronics*, *63*(4), 2633–2644. doi:10.1109/TIE.2016.2515054

Yang, F., & Mao, K. Z. (2011). Robust feature selection for microarray data based on multi-criterion fusion. *IEEE/ACM Transactions on Computational Biology and Bioinformatics*, *8*(4), 1080–1092. doi:10.1109/TCBB.2010.103 PMID:21566255

Yin, S., & Zhu, X. (2015). Intelligent particle filter and its application to fault detection of nonlinear system. *IEEE Transactions on Industrial Electronics*, *62*(6), 3852–3861. doi:10.1109/TIE.2015.2399396

Chapter 13
Sensor and Its Application

Saifur Rahman
Najran University, Saudi Arabia

Abdullah S. Alwadie
Najran University, Saudi Arabia

S. Hasan Saeed
Integral University, India

Faizan A. Khan
Integral University, India

ABSTRACT

Electronic nose systems are used to deliver a pattern response to a listed odor, and pattern recognition software is used to perform odor recognition and discrimination by using a series of sensors. The method of electronic noses generally includes time taking measurements in a non-standard test and error process. The sensory panel problem can be solved by electronic nose. For this purpose, a sensor model is used to design sensor array. The generated signal of these sensor array is used further to classify a mixture of two gases using principle component analysis (PCA)-based classification analysis. During classification, the efficiency of PCA classification has been checked over the different signal preprocessing technique. Continuous real monitoring of odor is done at specific sites in the field over hours, days, weeks, or even months. An electronic machine can also avoid many other troubles linked with the employ of human panels. Each and every variability, adaptation (becoming minimum sensitive during extended exposure), and revelation to hazardous compounds all come to mind.

DOI: 10.4018/978-1-5225-6989-3.ch013

INTRODUCTION

Some twenty years ago, "electronic noses" went for a great deal of development, since the inception of the first applications of solid state gas sensors. Around a thousand articles on this subject have been published over the last 4 years, mainly in relation to the food and beverage industry (Hudon, Guy & Hermia, 2000), but also concerning environmental, agricultural, and medical topics, in the automotive industry, etc. Electronic noses automatically detect and recognize odors, vapors and gases. Electronic noses (e-nose) comprise of a set of gas sensors with dissimilar selectivity shapes, a signal collecting unit and pattern recognition software applied to computer (Rahman, Usmani & Saeed, 2013). The identification of gases and smells are of main interest because it has a significant number of possible applications, few of which are linked to quality control in foods and cosmetics industries and to checking air quality. In the past 15 years, electronic noses mimic the human olfactory method and have developed rapidly. Many number of companies have now developed commercial instruments, which consist of an array of non-selective gas/odour sensor with a pattern recognition engine (Gardner & Bartlett, 1999; Gardner & Hines, 1997). The design of model is generally a test and error process through which the number and type of smell sensors and the pattern recognition tools are selected under the constraints determined by each application.

The e-nose attempts to emulate the mammalian nose by using an array of sensors that can simulate mammalian olfactory responses to aromas (Hudon, Guy & Hermia, 2000). The odour molecules are drawn into the e-nose using sampling techniques such as headspace sampling, diffusion methods, bubblers or pre-concentrators (Pearce et al., 2003). The odor sample is drawn across the sensor array and induces a reversible physical and/or chemical change in the sensing material, which causes an associated change in electrical properties, such as conductivity (G) (Singh, Hines & Garner, 1996).

DESCRIPTION OF ERROR MODELS

Figure 1 shows sensor A, B, C and D are designed at different parameters. Figure 2 shows the block diagram of the designed sensor and figure 3 shows the Simulink model for mixture of two gases for each sensor in which the sensor parameter used for feature extraction is sensor element conductance Gs. Feature evaluation Gs is evaluated using following equation:

$$G_S = G_{0T}e^{-EA0/KT} + K_{1T}e^{-EA1/KT}C_1^{n1KT} + K_{2T}e^{-EA2/KT}C_2^{n2KT} + \ldots K_{mixT}e^{-EAmix/KT}C_1^{n1KT}C_2^{n2KT} \quad (1)$$

Figure 1. Sensor Array Simulink Model for four different sensors

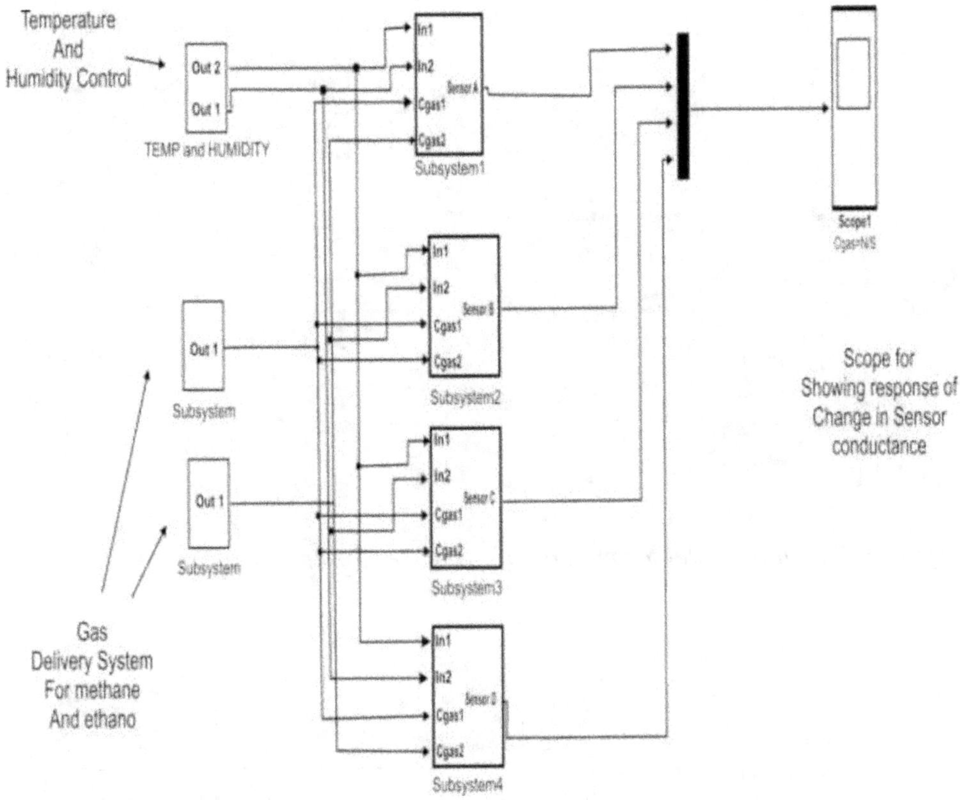

(Llobet et al., 2001)

Where K = Boltzmann's constant, T = Absolute Temperature, C1 & C2 =Concentration of gas1 & gas2, K_{1T} &K_{2T}= pre exponential factors of sensitivities to ethanol and methane, n1 & n2= pre factors of power law exponent for oxides, Kmix= pre exponential factor of the change in conductance caused by the interaction of the two species, EA= activation energies (Llobet et al., 2001). The dependence on the baseline and power law factor on the temperature has been studied elsewhere (Clifford & Tuma, 1983; Vilanova et al., 1998).

These facts were used to regulate the values of the parameters of the MATLAB sensor models, by applying an average fitting process in Matlab. The values of the fixed (static) model parameters for the four tin oxide sensors is shown in the table below. Formerly the sensor parameters of models were set and the simulation process was as follows.

Figure 2. Block diagram of designed sensor

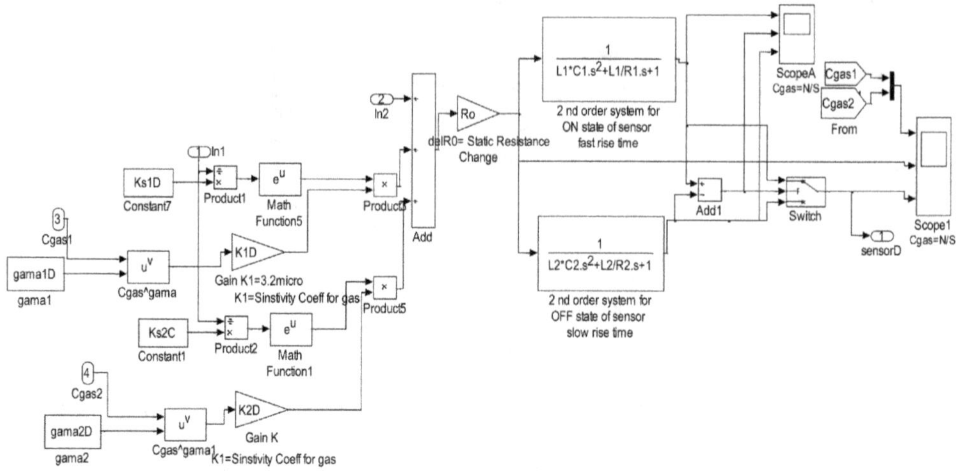

Figure 3. Simulink model for mixture of two gases for each sensor

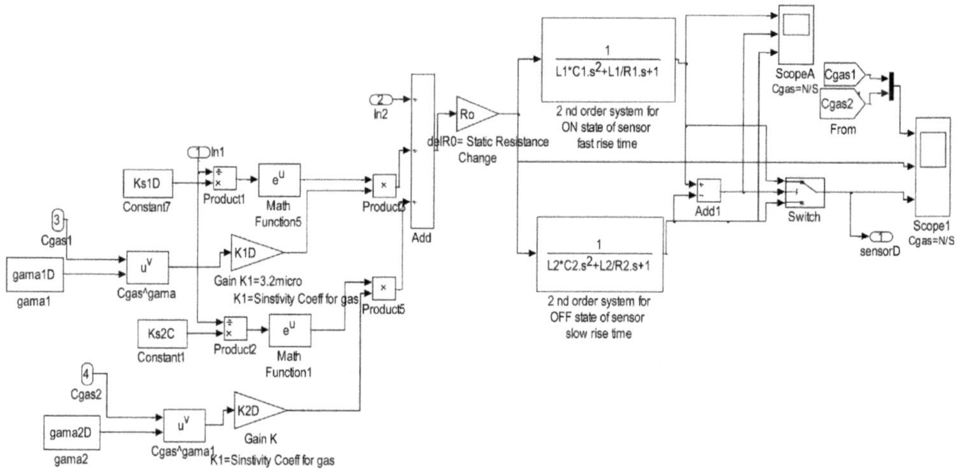

1. Random and systematic errors were applied to change the values of the parameters in Matlab models. Random errors were supposed to have a Gaussian distribution.
2. By iteratively executing MATLAB simulations by changing the model parameters, a huge response data set was achieved. Characteristic were refined from the responsive sensor.

Table 1. Values of Simulink static model parameter for the four Tin oxide gas sensors in the presence of ethanol, methane and their binary mixtures (Llobet et al., 2001)

Parameter	Sensor 1	Sensor 2	Sensor 3	Sensor 4
G_0	0.16	0.32	0.12	0.23
E_{a0}	$5*10^{-3}$	$4.3*10^{-3}$	$3.9*10^{-3}$	$5.6*10^{-3}$
K_1	$6.14*10^{-4}$	$3.14*10^{-3}$	$8.3*10^{-3}$	$1.7*10^{-3}$
EA_1	$5.31*10^{-3}$	$2.31*10^{-2}$	$2.1*10^{-2}$	$2.23*10^{-2}$
n_1	13.95	11.03	5.97	10.26
K_2	$9.8*10^{-5}$	$2.8*10^{-4}$	$1.6*10^{-4}$	$1.72*10^{-3}$
EA_2	$6.2*10^{-3}$	$9.21*10^{-3}$	$3.11*10^{-3}$	$1.02*10^{-3}$
n_2	6.83	7.55	6.54	4.05
K_{mix}	$-5.2*10^{-6}$	$-1.2*10^{-6}$	$-2.5*10^{-5}$	$-1*10^{-5}$
EA_{mix}	$3.31*10^{-3}$	$2.87*10^{-3}$	$4.35*10^{-2}$	$2.19*10^{-3}$

3. These properties were key in to a PARC identification engine (Principal Component Analysis), which was simulated by MATLAB as to discuss system performance with relation to the electronic nose discrimination skill.

For Cgas1 (Ethanol) & Cgas2 (Methane) two different model delivery system has been designed. From all four sensors both samples are passed through. The duration of sample contact is taken as 500 sec to 1500sec.

RESULTS AND DESCRIPTION

The aforementioned models were applied to perform a Monte Carlo simulation, coupled with principle component analysis (PCA), for a four element tin oxide sensor array to detect two gas methane and ethanol. In Figure 4 shown below one can see the change in the static and dynamic response of sensor A in presence of both gases.

Temperature and humidity is controlled as shown in Figure 5 below. Temperature is kept at 50^0 C and randomly changed as:

$$V_h' = V_h (1 + K_h \alpha_h) \quad (2)$$

Where V_h indicates heating voltage, K deviation And alpha random value. Similar to variation in T we also randomly input the humidity.

Figure 4. Response of sensor A in presence of both gases

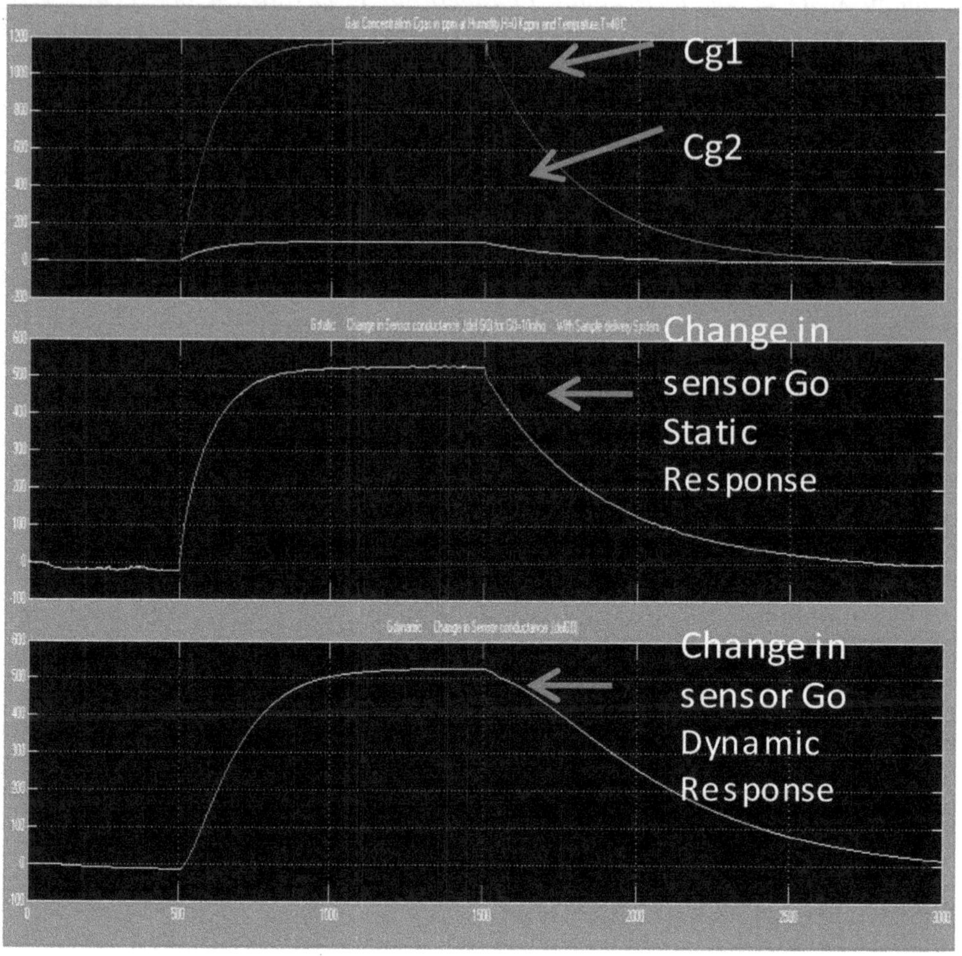

**For a more accurate representation see the electronic version.*

The different variations in delG can be recorded from the response of sensors. The response recorded are for three different cases.

Case 1: In presence of ethanol ie Cg1=10ppm and Cg2=0 ppm.
Case 2: In presence of methane ie Cg1=0ppm and Cg2=100 ppm.
Case 3: In presence of both ethanol and methane ie Cg1=10ppm and Cg2=100 ppm.

The concentration of gases varied by following relation:

$$C'_{eth} = C_{eth}\,(1 + K_{eth}\alpha_{eth}) \tag{3}$$

Figure 5. Block of T & H control

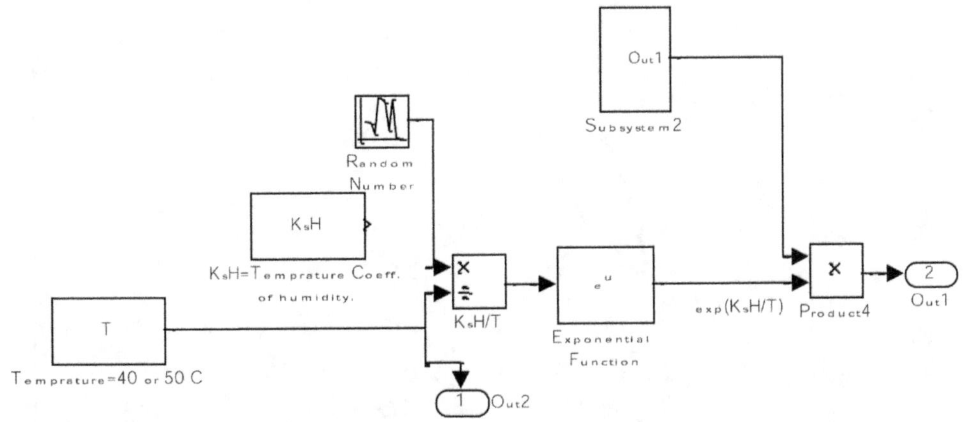

$$C'_{met} = C_{met}\,(1 + K_{met}\alpha_{met}) \tag{4}$$

Where C'_{eth} is the concentration of ethanol gas and C'_{met} is the concentration of methane gas.

K=0.1 and alpha random value with deviation 1.

For all 3 cases all the 8 parameters of temperature and humidity are varied. These parameters are:Go,Eao,K_1,EA_1,n_1, K_2,EA_2,n_2. Hence we generated 10x3 response for training the classifier model using value of del GoA (conductance of sensor A), del GoB(conductance of sensor B), del GoC(conductance of sensor C) & del GoD(conductance of sensor D).

In Figure 6 the response of delG is shown for case one for error model when there is variation in temperature there is small change can be recorded in the response of Del GB.

From Figure 7 and 8 we can see the response of Del G for case one error model having variation in humidity and KsH, there is no change in the response of the conductance of all the four sensors.

From Figure 9 it can be seen that for cluster plot obtained for 3 cases for 10 error models even when random error of 10% is added all the three cases can easily be discriminated.

From Figure 10 it can be seen that for cluster plot obtained for 3 cases for 10 error models when random error of 20% is added all the three cases cannot be easily discriminated.

Figure 6. Response of delG for case one for error model with variation in temp

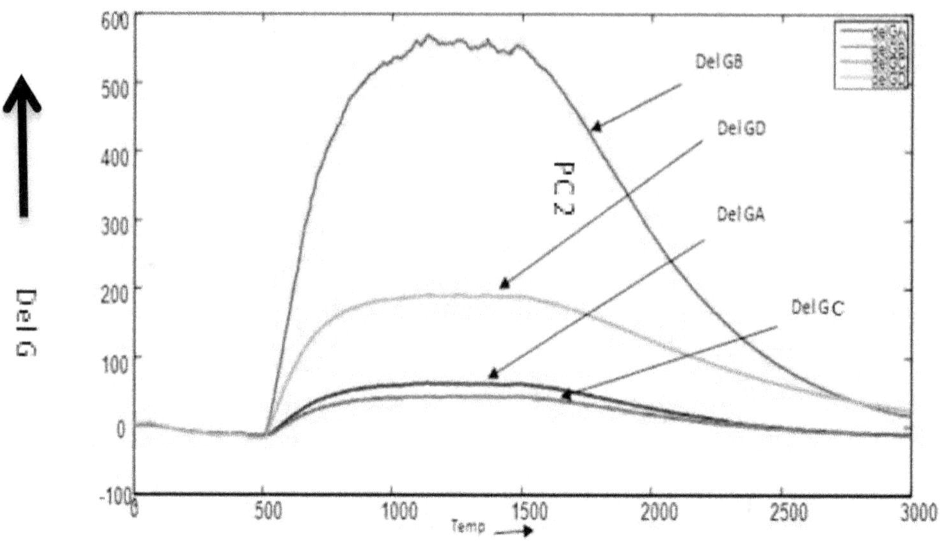

**For a more accurate representation see the electronic version.*

Figure 7. Response of delG for case one for error model with variation in humidity

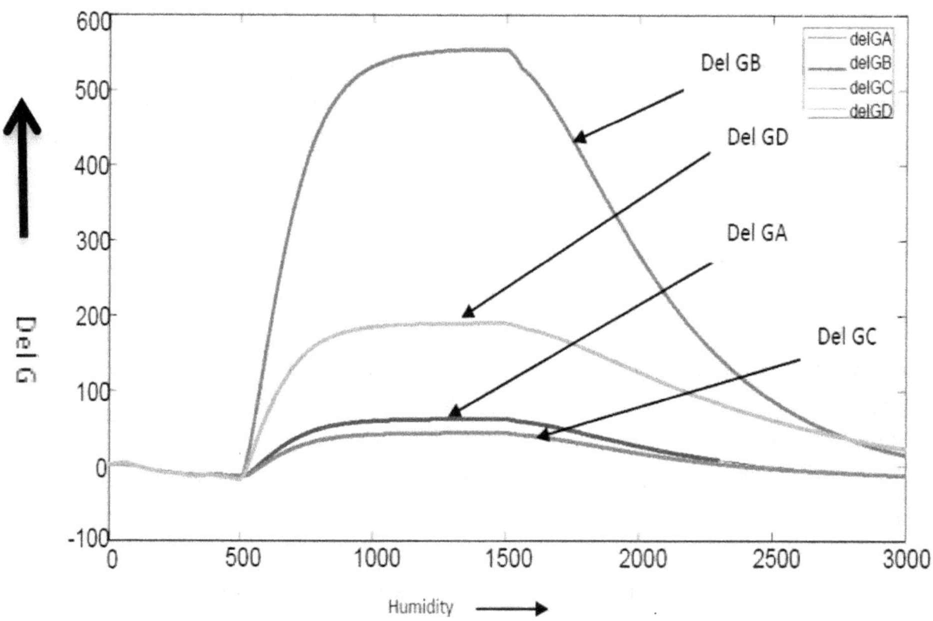

**For a more accurate representation see the electronic version.*

Figure 8. Response of delG for case one for error model with variation in KsH

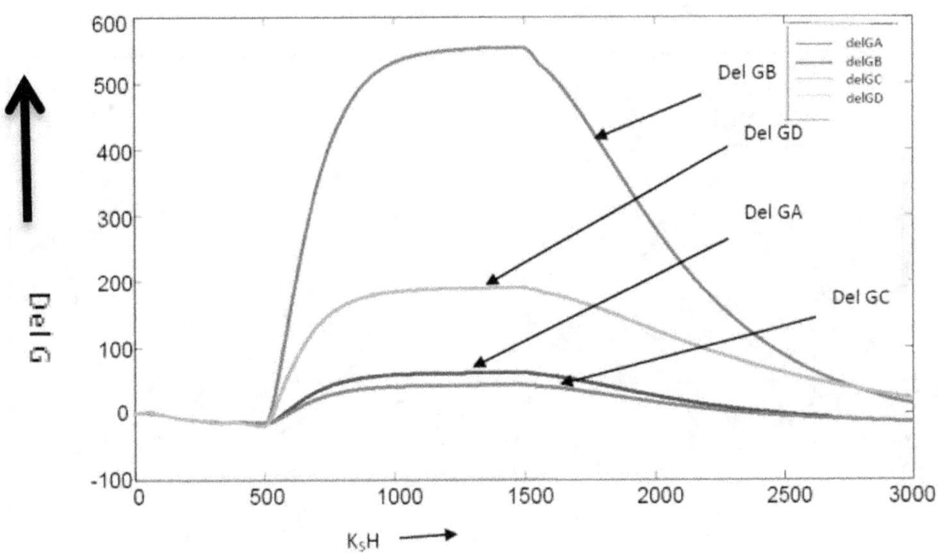

**For a more accurate representation see the electronic version.*

Figure 9. Cluster Plot obtained for 3 cases for 10 error models at random error of 10%

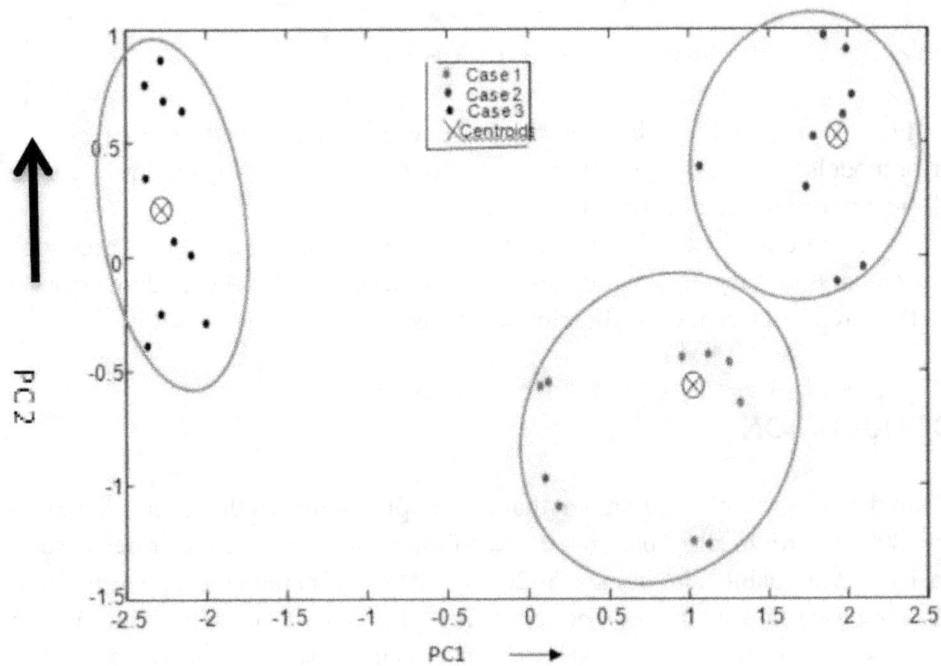

**For a more accurate representation see the electronic version.*

Figure 10. Cluster Plot obtained for 3 cases for 10 error models at random error of 20%

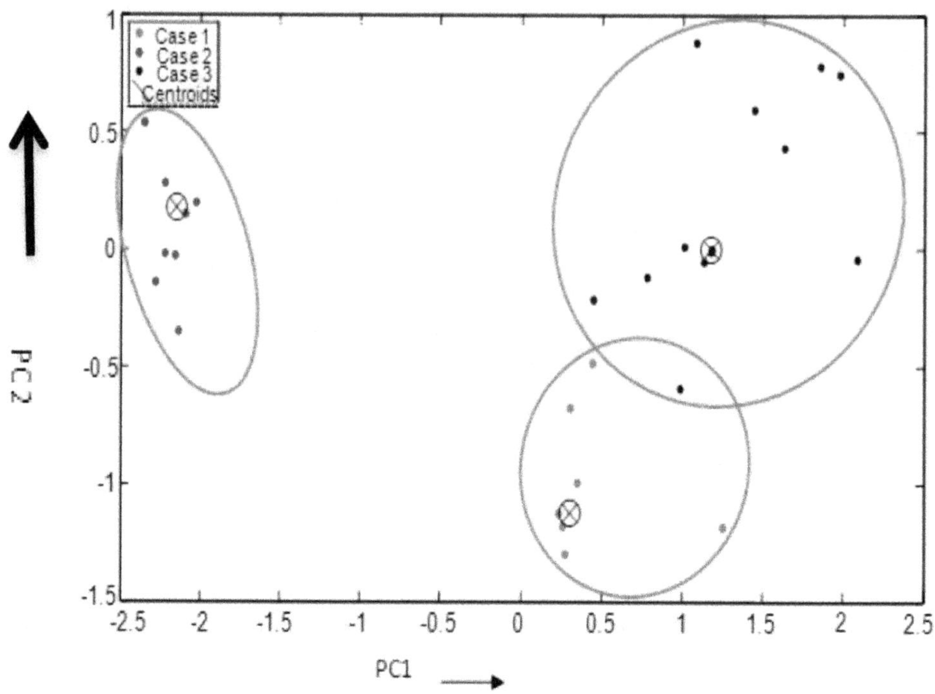

**For a more accurate representation see the electronic version.*

From Figure 11 it can be seen that for cluster plot obtained for 3 cases for 10 error models on varying dynamic response coeff. (L;C and R) by random error 10% all the three cases can be discriminated.

From Figure 12 it can be seen that for cluster plot obtained for 3 cases for 10 error models on varying dynamic response coeff. (L;C and R) by random error 20% all the three cases cannot be discriminated easily

CONCLUSION

From the above results it is shown that cluster plot indicates that even we add 10% and 20% of error in all 10 parameters even though the clusters for all three cases are easily discriminable. But in case of 20% deviation of parameter model dispersion has increased in principal component values. In the case of variation in dynamic response coefficient, the dispersion of cluster is increased in both 10% and 20% of deviation in parameters but still the clusters were discriminable.

Figure 11. Cluster Plot obtained for 3 cases for 10 error models on varying dynamic response coeff. (L;C and R)by 10%

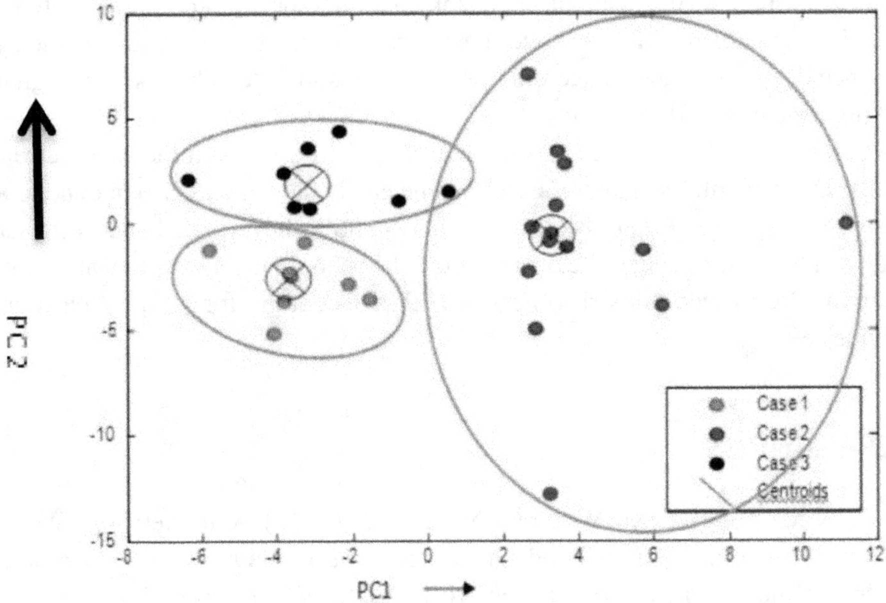

**For a more accurate representation see the electronic version.*

Figure 12. Cluster Plot obtained for 3 cases for 10 error models on varying dynamic response coeff. (L;C and R) by 20%

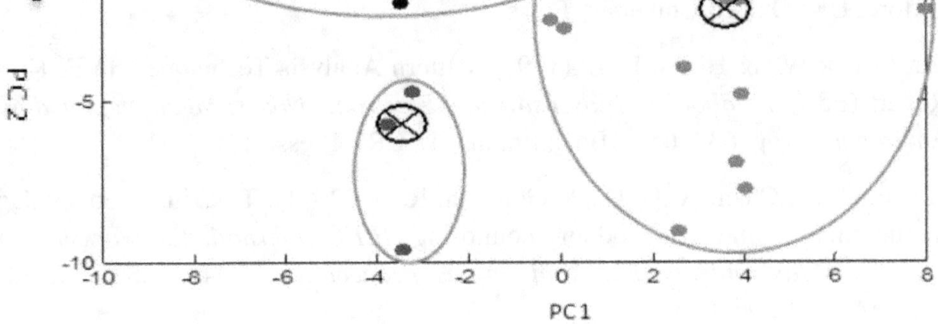

**For a more accurate representation see the electronic version.*

In the meanwhile, mandarin fruit ripeness procedure on the tree, the respiration lessens, meaning the generated vapours gets lower down, which fumes reach in a fewer quantity, the mean signal of sensor series decrease. Vapour associated with fruit also contains methane. Above sensor array is sensitive to methane among the presence of different gas hence the lowering in response can be taken as signal for picking time of mandarin fruit.

Presence of methane indicates the natural gas in the coal mines and petroleum industries accidental leakage of gases can be easily detected as rise in conductance of sensor response. Ethanol is present in alcoholic beverages hence this sensor array can be used to check the drivers and other people using respiration or during breathing. Beverages industries check the Ethanol contents for quality measurement of wines etc.

REFERENCES

Bi, Y., Lv, M., Song, C., Xu, W., Guan, N., & Yi, W. (2015). AutoDietary: A Wearable Acoustic Sensor System for Food Intake Recognition in Daily Life. *IEEE Sensors Journal, Volume*, *16*(3), 806–816. doi:10.1109/JSEN.2015.2469095

Blackburn, C. (2006). *Food spoilage microorganisms*. Boca Raton, FL: CRC Press LLC. doi:10.1201/9781439824573

Clifford, P. K., & Tuma, D. T. (1983). Characteristics of Semiconductors Gas Sensors II. Transient response to temperature change. *Sensors and Actuators*, *3*, 255–281. doi:10.1016/0250-6874(82)80027-9

Gardner, J. W., & Bartlett, P. N. (1996). Performance definition and standerdisation of electronic noses. *Sensors and Actuators. B, Chemical*, *33*(1-3), 60–67. doi:10.1016/0925-4005(96)01819-9

Gardner, J. W., & Bartlett, P. N. (1999). *Electronic Noses: Principles and applications*. Oxford, UK: Oxford University Press.

Gardner, J. W., & Hines, E. L. (1997). Pattern Analysis Techniques. In E. Kress Rogers (Ed.), *Handbook of Biosensors and Electronic Noses: Medicine, Food and Environment* (pp. 633–652). Boca Raton, FL: CRC Press.

Green, G. C., Chan, A. D. C., & Goubran, R. A. (2011). Tracking food spoilage in the smart home using odour monitoring. *IEEE International Workshop on Medical Measurements and Applications Proceedings*, 284 – 287. 10.1109/MeMeA.2011.5966685

Harun, F.K.C., Covington, J.A., & Gardner, J.W. (2012). Mimicking the biological olfactory system: A Portable electronic Mucosa. *Nanobiotechnology, 6*(2), 45 – 51.

Hudon, G., Guy, C., & Hermia, J. (2000). Measurement of odor intensity by an electronic nose. *Journal of the Air & Waste Management Association, 50*(10), 1750–1758. doi:10.1080/10473289.2000.10464202 PMID:11288303

Llobet, E., Rubio, J., Vilanova, X., Brezmes, J., Correig, X., Gardner, J. W., & Hines, E. L. (2001). Electronic Nose Simulation Tool Centered On PSpice. *Elsevier, Sensors and Actuators, 76*(1-3), 419–429. doi:10.1016/S0925-4005(01)00650-5

McWilliams, A., Beigi, P., Srinidhi, A., Lam, S., & MacAulay, C. E. (2015). Sex and Smoking Status Effects on the Early Detection of Early Lung Cancer in High-Risk Smokers Using an Electronic Nose. *IEEE Transactions on Biomedical Engineering, Volume, 62*(8), 2044–2054. doi:10.1109/TBME.2015.2409092 PMID:25775482

Pearce, T. C., Schiffman, S. S., Nagle, H. T., & Gardner, J. W. (2003). *Handbook of Machine olfaction*. Weinheim: WileyVCH.

Rahman, S., Saeed, S. H., Siddiqui, M. J., & Rahman, S. (2013). Polymeric chemoresistor model use in design and simulation of sensor. *Multimedia, Signal Processing and Communication Technologies (IMPACT), 2013 International Conference on*, 273 – 277. 10.1109/MSPCT.2013.6782134

Rahman, S., Usmani, T., & Saeed, S.H. (2013). Review Of Electronic Nose and application. *International Journal of Computing and Corporate Research, 3*(2).

Singh, S., Hines, E. L., & Garner, J. W. (1996). Fuzzy neural computing of coffee and tainted-water data from an electronic nose. *Sensors and Actuators. B, Chemical, 6*(3), 185–190. doi:10.1016/0925-4005(96)80047-5

Torri, L., Sinelli, N., & Limbo, S. (2010). Shelf life evaluation of fresh-cut pineapple by using an electronic nose. *Postharvest Biol. Technol., 56*, 239-245.

Vilanova, X., Llobat, E., Brezmes, J., & Correig, X. (1998). *Influence of moisture and working temperature on the steady state and transient response of thick film tin oxide gas sensors, Internal Report GSM 98/01*. Universitat Rovira i Virgili.

Wang, Y., Wang, J., Zhou, B., & Lu, Q. (2009). Monitoring storage time and quality attribute of egg based on electronic nose. *Analytica Chimica Acta, 650*(2), 183–188. doi:10.1016/j.aca.2009.07.049 PMID:19720190

ADDITIONAL READING

Orts, J., Llobet, E., Vilanova, X., Brezems, J., & Correig, X. (1999). Selective methane detection under varying moisture conditions using static and dynamic sensor signals. *Sensors and Actuators. B, Chemical*, *60*(2-3), 106–117. doi:10.1016/S0925-4005(99)00257-9

Saeed, S. H., Abbas, Z., & Gopal, B. (2009) Experimental Use of Electronic Nose For Analysis of Volatile Organic Compound (VOC). in IEEE digital library organized by International Conference on Multimedia, Signal Processing and Communication Technologies (IMPACT 2009), 14-16 March.

VanGeloven, P., Honore, M., Roggen, J., Leppavuori, S., & Rantala, T. (1991). The influence of relative humidity on the response of tin oxide gas sensors to carbon monoxide. *Sensors and Actuators. B, Chemical*, *4*(1-2), 185–188. doi:10.1016/0925-4005(91)80196-Q

Compilation of References

Abellan-Nebot, J. V., & Subirón, F. R. (2010). A review of machining monitoring systems based on artificial intelligence process models. *International Journal of Advanced Manufacturing Technology*, *47*(1-4), 237–257. doi:10.100700170-009-2191-8

Acosta, G. G., Verucchi, C. J., & Gelso, E. R. (2004). A current monitoring system for diagnosing electrical failures in induction motors. *Mechanical Systems and Signal Processing*, *20*(1), 953–965.

Agrawal, D., Yadav, N., & Saini, S. (2015). Condition Monitoring of Slip-ring Induction Motor. *International Journal of Innovative Research in Advanced Engineering*, *2*(3), 78–84.

Aldeen, M., & Crusca, F. (2006). Observer-based fault detection and identification scheme for power systems. *IEE Proceedings. Generation, Transmission and Distribution*, *153*(1), 71–79. doi:10.1049/ip-gtd:20045101

Alhelou, H. H., & Golshan, M. E. H. (2016, May). Hierarchical plug-in EV control based on primary frequency response in interconnected smart grid. In Electrical Engineering (ICEE), 2016 24th Iranian Conference on (pp. 561-566). IEEE. doi:10.1109/IranianCEE.2016.7585585

Alhelou, H. H., Golshan, M. H., & Askari-Marnani, J. (2018). Robust sensor fault detection and isolation scheme for interconnected smart power systems in presence of RER and EVs using unknown input observer. *International Journal of Electrical Power & Energy Systems*, *99*, 682–694. doi:10.1016/j.ijepes.2018.02.013

Alhelou, H. S. H., Golshan, M. E. H., & Fini, M. H. (2015, December). Multi agent electric vehicle control based primary frequency support for future smart micro-grid. In *Smart Grid Conference (SGC)*, 2015 (pp. 22-27). IEEE.

Ali, J. B., Fnaiech, N., Saidi, L., Chebel-Morello, B., & Fnaiech, F. (2015). Application of empirical mode decomposition and artificial neural network for automatic bearing fault diagnosis based on vibration signals. *Applied Acoustics*, *89*, 16–27. doi:10.1016/j.apacoust.2014.08.016

Alwan, H. O., & Farhan, N. M. (2017). Design of Data Aquistion interface circuit used in Detection Inter-turn Fault in Motor based on Motor Current Signature Analysis (MCSA) Technique. Int. *Journal of Engineering Research and Application*, *7*(6), 79–89. doi:10.9790/9622-0706017989

Andersen, P., Bendtsen, J. D., Pedersen, T. S., Mortensen, J. H., & Nielsen, R. J. (2006). Observer-based coal mill control using oxygen measurements. IFAC Proceedings Volumes, 39(7), 171-176.

Antonino-Daviu, J. A., Riera-Guasp, M., Folch, J. R., & Palomares, M. P. M. (2006). Validation of a new method for the diagnosis of rotor bar failures via wavelet transform in industrial induction machines. *IEEE Transactions on Industry Applications*, *42*(4), 990–996. doi:10.1109/TIA.2006.876082

Aroui, T., Koubaa, Y., & Toumi, A. (2007). Application of feedforward neural network for induction machine rotor faults diagnostics using stator current. *J. Electr. Syst.*, *3*, 213–226.

Asfani, D. A., Muhammadc, A. K., Purnomo, M. H., & Hivama, T. (2012). Temporary short circuit detection in induction motor winding using combination of wavelet transform and neural network. *Expert Systems with Applications*, *39*(5), 5367–5375. doi:10.1016/j.eswa.2011.11.048

Bachir, S., Tnani, S., Trigeassou, J. C., & Champenois, G. (2006). Diagnosis by parameter estimation of stator and rotor faults occurring in induction machines. *IEEE Transactions on Industrial Electronics*, *53*(3), 963–973. doi:10.1109/TIE.2006.874258

Baraldi, P., Mangili, F., & Zio, E. (2012). A kalman filter-based ensemble approach with application to turbine creep prognostics. *IEEE Transactions on Reliability*, *61*(4), 966–977. doi:10.1109/TR.2012.2221037

Bayar, H., & Terz, U. K. (2016). Grid-connected induction generator interturn fault analysis using a PCA-ANN-based algorithm. *Turkish Journal of Electrical Engineering and Computer Sciences*, *24*, 1340–1350. doi:10.3906/elk-1312-145

Bechhoefer, E. (2012). Data driven prognostics for rotating machinery. In S. Kadry (Ed.), *Diagnostics and Prognostics of Engineering Systems: Methods and Techniques* (pp. 120–133). IGI Global.

Benbouzid, M. E. H. (2000). A Review of induction motors signature analysis as a medium for faults detection. *IEEE Transactions on Industrial Electronics*, *47*(5), 984–993. doi:10.1109/41.873206

Benbouzid, M. E. H., & Kliman, G. B. (2003). What stator current processing-based technique to use for induction motor rotor faults diagnosis? *IEEE Transactions on Energy Conversion*, *18*(2), 238–244. doi:10.1109/TEC.2003.811741

Benítez, H. D., Loaiza, H., Caicedo, E., Ibarra-Castanedo, C., Bendada, A., & Maldague, X. (2009). Defect characterization in infrared non-destructive testing with learning machines. *NDT & E International*, *42*(7), 630–643. doi:10.1016/j.ndteint.2009.05.004

Benouzza, N. (2006). *Approche des Vecteurs de Park Appliquée à la Détection des Défauts dans les Moteurs Asynchrones à Cage d'Ecureuil* (Thèse de Doctorat). Université des Sciences et de la Technologie d'Oran.

Benouzza, N., Benyettou, A., & Bendiabdallah, A. (2004). An Advanced PARK's Vectors Approach for Rotor Cage Diagnosis. In *Proceeding of CNGE 2004 01*. Université Ibn Khaldoun de Tiaret.

Benouzza, N., Benyettou, A., Bendiabdellah, A., Kraloua, B., & Toumi, D. (2001). Rotor cage faults diagnosis in 3-phase induction motors, by Park's vector approach. *4th Jordanian International Electrical Electronics Engineering Conference 2001*.

Bessam, B., Menacer, A., Boumehraz, M., & Cherif, H. (2016). Detection of broken rotor bar faults in induction motor at low load using neural network. *ISA Transactions*, *64*, 241–246. doi:10.1016/j.isatra.2016.06.004 PMID:27329853

Bessam, B., Menacer, A., Boumehraz, M., & Cherif, H. (2017). Wavelet transform and neural network techniques for inter-turn short circuit diagnosis and location in induction motor. *International Journal of System Assurance Engineering and Management*, *8*(S1), 478–488. doi:10.100713198-015-0400-4

Bhardwaj, K., & Agarwal, A. (2012). Fault Diagnosis of Three Phase Induction Motor Using Fuzzy Logic Controller and Sequence Analyzer. *MIT International Journal of Electrical and Instrumentation Engineering*, *2*(2), 112–118.

Bhowmik, P. S., Pradhan, S., & Prakash, M. (2013). Fault Diagnostic and Monitoring Methods of Induction Motor: A Review. *International Journal of Applied Control Electrical and Electronics Engineering*, *1*, 1–18.

Bimber, O., & Raskar, R. (2005). *Spatial Augmented Reality: Merging Real and Virtual Worlds*. A. K. Peters, Ltd. doi:10.1201/b10624

Binns, K. J., & Dye, M. (1973). Identification of principal factors affecting unbalanced magnetic pull in cage induction motors. *Proceedings of IEE – Pt. B, 120*, 349 – 354.

Birkfellner, W., Figl, M., Huber, K., Watzinger, F., Wanschitz, F., Hummel, J., ... Ewers, R. (2002). A head-mounted operating binocular for augmented reality visualization in medicine-design and initial evaluation. *IEEE Transactions on Medical Imaging*, *21*(8), 991–997. doi:10.1109/TMI.2002.803099 PMID:12472271

Bi, Y., Lv, M., Song, C., Xu, W., Guan, N., & Yi, W. (2015). AutoDietary: A Wearable Acoustic Sensor System for Food Intake Recognition in Daily Life. *IEEE Sensors Journal, Volume*, *16*(3), 806–816. doi:10.1109/JSEN.2015.2469095

Blackburn, C. (2006). *Food spoilage microorganisms*. Boca Raton, FL: CRC Press LLC. doi:10.1201/9781439824573

Blackburn, J. L., & Domin, T. J. (2015). *Protective relaying: principles and applications*. CRC press.

Blodt, M., Chabert, M., Regnier, J., & Faucher, J. (2006). Mechanical Load Fault Detection in Induction Motors by Stator Current Time-Frequency Analysis. *IEEE Transactions on Industry Applications*, *42*(6), 1454–1463. doi:10.1109/TIA.2006.882631

Bonaldi, E. L., Lambert-Torresm, G., da Silva, J. G. B., de Oliveira, L. E. L., & da Silva, L. E. B. (2012). *Predictive Maintenance by Electrical Signature Analysis to Induction Motors*. INTECH Open Access Publisher.

Boudinar, A. H., Benouzza, N., Bendiabdellah, A., & Khodja, M. E. A. (2016). Induction motor bearing fault analysis using Root-MUSIC method. IEEE Trans. on Industry Applications, 52(5), 3851-3860.

Boudinar, A. H., Benouzza, N., & Bendiabdellah, A. (2015). *Induction motor cracked rotor bars fault analysis using an improved Root-MUSIC method. 3rd International Conference on Control, Engineering and Information Technology (CEIT)*, Tlemcen, Algeria. 10.1109/CEIT.2015.7233035

Boyer, S. A. (2009). *SCADA: supervisory control and data acquisition*. International Society of Automation.

Burrus, C. S., Gopinath, R. A., & Guo, H. (1998). *Introduction to wavelets and wavelet transforms*. Prentice Hall.

Calderita, L. V., Bustos, P., Mejías, C. S., Fernández, F., Viciana, R., & Bandera, A. (2015). Asistente Robótico Socialmente Interactivo para Terapias de Rehabilitación Motriz con Pacientes de Pediatría. *Revista Iberoamericana de Automática e Informática Industrial RIAI*, *12*(1), 99–110. doi:10.1016/j.riai.2014.09.007

Çalis, H., & Çakir, A. (2007). Rotor bar fault diagnosis in three phase induction motors by monitoring fluctuations of motor current zero crossing instants. *Electric Power Systems Research*, *77*(5), 385–392. doi:10.1016/j.epsr.2006.03.017

Caliskan, F., & Genc, I. (2008). A robust fault detection and isolation method in load frequency control loops. *IEEE Transactions on Power Systems*, *23*(4), 1756–1767. doi:10.1109/TPWRS.2008.2004831

Cameron, J., Thomson, W., & Dow, A. (1986). Vibration and current monitoring for detecting air gap eccentricity in large induction motors. *Proceedings of IEE – Pt. B, 133(3)*, 155 – 163.

Campos-Delgado, D. U., & Espinoza-Trejo, D. R. (2011). An observer-based diagnosis scheme for single and simultaneous open-switch faults in induction motor drives. *IEEE Transactions on Industrial Electronics*, *58*(2), 671–679. doi:10.1109/TIE.2010.2047829

Carnero, M. (2006). An evaluation system of the setting up of predictive maintenance programmes. *Reliability Engineering & System Safety*, *91*(8), 945–963. doi:10.1016/j.ress.2005.09.003

Carnero, M. C. (2005). Selection of diagnostic techniques and instrumentation in a predictive maintenance program. A case study. *Decision Support Systems*, *38*(4), 539–555. doi:10.1016/j.dss.2003.09.003

Carvalho, A. A., Silva, R. R., Rebello, J. M. A., & Sagrilo, L. V. S. (2010). Pattern recognition techniques applied to the detection and classification of welding defects by magnetic testing. *Research in Nondestructive Evaluation*, *21*(2), 91–111. doi:10.1080/09349840903381655

Casimir, R., Boutleux, E., Clerc, G., & Yahoui, A. (2006). The use of features selection and nearest neighbours rule for faults diagnostic in induction motors. *Engineering Applications of Artificial Intelligence*, *19*(2), 169–177. doi:10.1016/j.engappai.2005.07.004

Caudell, T. P., & Mizell, D. W. (1992). Augmented reality: An application of heads-up display technology to manual manufacturing processes. *Proceedings of the Twenty-Fifth Hawaii International Conference on*. 10.1109/HICSS.1992.183317

Chaturvedi, D. K., Iqbal, S. M., & Singh, M. P. (2013). Health Monitoring Techniques of Induction Motor: An Overview. *Proceedings of Int. Conf. on Emerging Trends in Engineering and Technology*, 469-477.

Chaturvedi, G. K., & Thomas, D. W. (1982). Bearing fault detection using adaptive noise cancelling. *ASME Journal of Mechanical Design, 104*(2), 280–289. doi:10.1115/1.3256337

Chen, Z., Meng, J., Liang, B., & Guo, D. (2007) Fault Diagnosis of AC Squirrel cage Asynchronous Motors based on Wavelet Packet-Neural Network. *EUROCON 2007, The International Conference on "Computer as a Tool"*. 10.1109/EURCON.2007.4400425

Chen, J., & Patton, R. J. (2012). *Robust model-based fault diagnosis for dynamic systems* (Vol. 3). Springer Science & Business Media.

Chi, H.-L., Kang, S.-C., & Wang, X. (2013). Research trends and opportunities of augmented reality applications in architecture, engineering, and construction. *Automation in Construction, 33*, 116–122. doi:10.1016/j.autcon.2012.12.017

Chudasama, K. J., & Shah, V. (2012). Induction Motor Noninvasive Fault Diagnostic techniques: A Review. *International Journal of Engineering Research and Technology, 1*(5), 1–7.

Clifford, P. K., & Tuma, D. T. (1983). Characteristics of Semiconductors Gas Sensors II. Transient response to temperature change. *Sensors and Actuators, 3*, 255–281. doi:10.1016/0250-6874(82)80027-9

Corbo, M. A., & Malanoski, S. B. (1996). Practical design against torsional vibration. In *Proceedings of the Twenty fifth Turbomachinery Symposium* (pp. 189 – 222). Texas A&M University.

Corne, B., Vervisch, B., Debruyne, C., Knockaert, J. & Desmet, J. (2015). Comparing MCSA with Vibration Analysis in order to detect Bearing Faults - A Case Study. *IEEE Explore*, 1366-1372.

Costa, C. D., Kashiwagi, M., & Mathias, M. H. (2015). *Rotor failure detection of induction motors by wavelet transform and fourier transform in non stationary condition.* Case Studies in Mechanical Systems and Signal Processing.

Cudworth, C. J., Smith, J. R., & Mykura, F. (1984). Mechanical damping of torsional vibrations in turbogenerators due to network disturbances. In *Proceedings of the Third International Conference on Vibration in Rotating Machinery* (pp. 139 – 145). Institution of Mechanical Engineers.

Cupertino, F., Martorana, G., Salvatore, L., & Stasi, S. (2003, September). Diagnostic start-up test to detect induction motor broken bars via Short-Time MUSIC algorithm applied to current Space-Vector. *EPE, 03*, 2–4.

D'Orazioa, T., & Leoa, M. (2008). Ultrasonic Testing for Flaw Detection in Mild Steel Sample. *NDT & E International, 41*, 145–154.

Das, S., Purkait, P., Koley, C., & Chakravorti, S. (2014). Performance of a load-immune classifier for robust identification of minor faults in induction motor stator winding. *Dielectrics and Electrical Insulation. IEEE Transactions on*, *21*, 33–44.

De Angelo, C., Bossio, G., Giaccone, S., Garcia, G. O., Solsona, J., & Valla, M. I. (2006, November). Model based stator fault detection in induction motors. In *IEEE Industrial Electronics, IECON 2006-32nd Annual Conference on* (pp. 1095-1100). IEEE. 10.1109/IECON.2006.347895

De Angelo, C. H., Bossio, G. R., Giaccone, S. J., Valla, M. I., Solsona, J. A., & Garcia, G. O. (2009). Online model-based stator-fault detection and identification in induction motors. *IEEE Transactions on Industrial Electronics*, *56*(11), 4671–4680. doi:10.1109/TIE.2009.2012468

Delgado-Arredondo, P. A., Garcia-Perez, A., Morinigo-Sotelo, D., Osornio-Rios, R. A., Avina-Cervantes, J. G., Rostro-Gonzalez, H., & Romero-Troncoso, R. J. (2015). Comparative study of time-frequency decomposition techniques for fault detection in induction motors using vibration analysis during startup transient. *Shock and Vibration*, *2015*, 1–14. doi:10.1155/2015/708034

Díaz, D. J., Boj, C., & Portalés, C. (2016). Hybridplay: A new technology to foster outdoors physical activity, verbal communication and teamwork. *Sensors (Basel)*, *16*(4), 586. doi:10.339016040586 PMID:27120601

Didier, G., Ternisien, E., Caspary, O., & Razik, H. (2007). A new approach to detect broken rotor bars in induction machines by current spectrum analysis. *Mechanical Systems and Signal Processing*, *21*(2), 1127–1142. doi:10.1016/j.ymssp.2006.03.002

Drif, M. H., & Cardoso, A. J. M. (2014). Stator fault diagnostics in squirrel cage three-phase induction motor drives using the instantaneous active and reactive power signature analyses. *IEEE Transactions on Industrial Informatics*, *10*(2), 1348–1360. doi:10.1109/TII.2014.2307013

Drozdowski, P., & Duda, A. (2014). A. Influence of magnetic saturation effects on the fault detection of induction motors. *Arch. Electr. Eng*, *63*, 489–506. doi:10.2478/aee-2014-0035

Dunleavy, M., & Dede, C. (2014). *Augmented reality teaching and learning. In Handbook of research on educational communications and technology* (pp. 735–745). Springer. doi:10.1007/978-1-4614-3185-5_59

Dzwonkowski, A., & Swędrowski, L. (2012, September). Uncertainty analysis of measuring system for instantaneous power research. *Metrology and Measurement Systems*, *19*(3), 573–582. doi:10.2478/v10178-012-0050-7

Eftekhari, M., Moallem, M., Sadri, S., & Hsieh, M. F. (2013). A novel indicator of stator winding inter-turn fault in induction motor using infrared thermal imaging. *Infrared Physics & Technology*, *61*, 330–336. doi:10.1016/j.infrared.2013.10.001

El Bouchikhi, E., Choqueuse, V., & Benbouzid, M. E. H. (2014). Induction machine faults detection using stator current parametric spectral estimation. Mechanical Systems and Signal Processing, 52-53, 447-464.

El Bouchikhi, E., Choqueuse, V., & Benbouzid, M. E. H. (2012, December). Current Frequency Spectral Subtraction and Its Contribution to Induction Machines' Bearings Condition Monitoring. *IEEE Transactions on Energy Conversion*, *28*(1), 135–144. doi:10.1109/TEC.2012.2227746

Escaler, X., Egusquiza, E., Farhat, M., Avellan, F., & Coussirat, M. (2006). Detection of cavitation in hydraulic turbines. *Mechanical Systems and Signal Processing*, *20*(4), 983–1007. doi:10.1016/j.ymssp.2004.08.006

Espinoza-Trejo, D. R., Campos-Delgado, D. U., Barcenas, E., & Martınez-Lopez, F. J. (2012). Robust fault diagnosis scheme for open-circuit faults in voltage source inverters feeding induction motors by using non-linear proportional-integral observers. *IET Power Electronics*, *5*(7), 1204–1216. doi:10.1049/iet-pel.2011.0309

Fântână, G. I., Oae, S. A., & Gurau, A. M. (2013). Decision making using the analytic hierarchy process. *International Conferice on MEQAPS*, *13*, 119–124.

Fan, Z., Liu, G., Si, X., Zhang, Q., & Zhang, Q. (2013). Degradation data-driven approach for remaining useful life estimation. *Journal of Systems Engineering and Electronics*, *24*(1), 173–182. doi:10.1109/JSEE.2013.00022

Fenger, M., Lloyd, B.A., & Thomson, W.T. (2003). Development of a tool to detect faults in induction motors via current signature analysis. *Cement Industry Technical Conference IEEE Explore*, 37-46.

Feng, L., Wang, H., Si, X., & Zou, H. (2013). A State-Space-Based Prognostic Model for Hidden and Age-Dependent Nonlinear Degradation Process. *IEEE Transactions on Automation Science and Engineering*, *10*(4), 1072–1086. doi:10.1109/TASE.2012.2227960

Filho, P. L., Pederiva, R., & Brito, J. (2014). Detection of stator winding faults in induction machines using flux and vibration analysis. *Mechanical Systems and Signal Processing*, *42*(1-2), 377–387. doi:10.1016/j.ymssp.2013.08.033

Fini, M. H., Yousefi, G. R., & Alhelou, H. H. (2016). Comparative study on the performance of many-objective and single-objective optimisation algorithms in tuning load frequency controllers of multi-area power systems. *IET Generation, Transmission & Distribution*, *10*(12), 2915–2923. doi:10.1049/iet-gtd.2015.1334

Freire, N. M., Estima, J. O., & Cardoso, A. J. (2014). A voltage-based approach without extra hardware for open-circuit fault diagnosis in closed-loop PWM AC regenerative drives. *IEEE Transactions on Industrial Electronics*, *61*(9), 4960–4970. doi:10.1109/TIE.2013.2279383

Galatis, P., Gavalas, D., Kasapakis, V., Pantziou, G., & Zaroliagis, C. (2016). Mobile Augmented Reality Guides in Cultural Heritage. *Proceedings of the 8th EAI International Conference on Mobile Computing, Applications and Services*. 10.4108/eai.30-11-2016.2266954

Gama, R. S., Silva, C. C. J., Nascimento, C. E., Silva, A. M., & Cesar, D. C. (2017). Comparing MCSA with vibration analysis on rotating electrical machines to detect broken bars – a case study. *Journal of Multidisciplinary Engineering Science and Technology*, *4*(8), 7918–7913.

Gao, Z., Colby, R. S., Habetler, T. G., & Harley, R. G. (2008). A Model Reduction Perspective on Thermal Models for Induction Machine Overload Relays. *IEEE Transactions on Industrial Electronics*, *55*(10), 3525–3534. doi:10.1109/TIE.2008.926772

Garcia-Ramirez, A. G., Morales-Hernandez, L. A., Osornio-Rios, R. A., Benitez-Rangel, J. P., Garcia-Perez, A., & de Jesus Romero-Troncoso, R. (2014). Fault detection in induction motors and the impact on the kinematic chain through thermographic analysis. *Electric Power Systems Research*, *114*, 1–9. doi:10.1016/j.epsr.2014.03.031

Gardner, J. W., & Bartlett, P. N. (1996). Performance definition and standerdisation of electronic noses. *Sensors and Actuators. B, Chemical*, *33*(1-3), 60–67. doi:10.1016/0925-4005(96)01819-9

Gardner, J. W., & Bartlett, P. N. (1999). *Electronic Noses: Principles and applications*. Oxford, UK: Oxford University Press.

Gardner, J. W., & Hines, E. L. (1997). Pattern Analysis Techniques. In E. Kress Rogers (Ed.), *Handbook of Biosensors and Electronic Noses: Medicine, Food and Environment* (pp. 633–652). Boca Raton, FL: CRC Press.

Garnier, C., Pastor, M. L., Eyma, F., & Lorrain, B. (2011). The detection of aeronautical defects in situ on composite structures using Non Destructive Testing. *Composite Structures*, *93*(5), 1328–1336. doi:10.1016/j.compstruct.2010.10.017

Gebraeel, N. (2006). Sensory-Updated Residual Life Distributions for Components With Exponential Degradation Patterns. *IEEE Transactions on Automation Science and Engineering*, *3*(4), 382–393. doi:10.1109/TASE.2006.876609

Geethanjali, M., & Hemavathi, R. (2010). Fuzzy Logic Based Protection for Three Phase Induction Motor Rotor faults Using Motor Current Signature Analysis based DSP Algorithm. *Proceedings of the 34th National Systems Conference*, 85.

Genc, I., & Caliskan, F. (2011). Robust Actuator Fault Detection and Isolation in a Multi-Area Interconnected Power System. *Proceedings of the World Congress on Engineering*, 2.

Germen, E., Başaran, M., & Fidan, M. (2014). Sound based induction motor fault diagnosis using Kohonen self-organizing map. *Mechanical Systems and Signal Processing*, *46*(1), 45–58. doi:10.1016/j.ymssp.2013.12.002

Ghosh, D., & Roy, S. (2009). A decision-making framework for process plant maintenance. *European Journal of Industrial Engineering*, *4*(1), 78–98. doi:10.1504/EJIE.2010.029571

Gimeno, J., Morillo, P., Casas, S., & Fernández, M. (2011). An augmented reality (AR) CAD system at construction sites. In *Augmented Reality-Some Emerging Application Areas*. InTech.

Gimeno, J., Portalés, C., Coma, I., Fernández, M., & Martínez, B. (2017). Combining Traditional and Indirect Augmented Reality for Indoor Crowded Environments. A Case Study on the Casa Batlló Museum. *Computers & Graphics*, *69*, 92–103. doi:10.1016/j.cag.2017.09.001

Giner Martínez, F., & Portalés Ricart, C. (2005). *The Augmented User: A Wearable Augmented Reality Interface*. Paper presented at the International Conference on Virtual Systems and Multimedia (VSMM'05), Ghent, Belgium.

Goebel, K., Saha, B., & Saxena, A. (2008). A comparison of three data-driven techniques for prognostics. In *Proceedings of 62nd Meeting Society for Machinery Failure Prevention Technology (MFPT)* (pp. 119–131). Virginia Beach, VA: Academic Press.

Goh, J. Y., Pitter, M. C., See, C. W., Somekh, M. G., & Vanderstraeten, D. (2004). Sub-pixel image correlation: An alternative to SAM and dye penetrant for crack detection and mechanical stress localisation in semiconductor packages. *Microelectronics and Reliability*, *44*(2), 259–267. doi:10.1016/S0026-2714(03)00162-8

Gopinath, S. (2010). Study on electric motor mass unbalance based on vibration monitoring analysis technique. In *2nd International Conference on Mechanical and Electrical Technology*. Singapore: IEEE. 10.1109/ICMET.2010.5598417

Govardhan, T., Choudhury, A., & Paliwal, D. (2017). Vibration analysis of a rolling element bearing with localized defect under dynamic radial load. *Journal of Vibration Engineering & Technologies*, *5*(2), 165–175.

Granda, D., Aguilar, W.G., Arcos-Aviles, D. & Sotomayor, D. (2017). Broken Bar Diagnosis for Squirrel Cage Induction Motors Using Frequency Analysis Based on MCSA and Continuous Wavelet Transform. *MDPI Mathematical and Computational Applications, 22*, 30. doi:10.3390/mca22020030

Green, G. C., Chan, A. D. C., & Goubran, R. A. (2011). Tracking food spoilage in the smart home using odour monitoring. *IEEE International Workshop on Medical Measurements and Applications Proceedings*, 284 – 287. 10.1109/MeMeA.2011.5966685

Greety, J., & Jose, V. (2013). Induction Motor Fault Diagnosis Methods: A Comparative Study. *Proceedings of International Conference on Electrical Engineering*, 63-69.

Guedidi, S., Zouzou, S. E., Laala, W., Sahraoui, M., & Yahia, K. (2011). *Broken Bar Fault Diagnosis of Induction Motors Using MCSA and Neural Network*. IEEE Explore. doi:10.1109/DEMPED.2011.6063690

Hameed, Z., Hong, Y. S., Cho, Y. M., Ahn, S. H., & Song, C. K. (2009). Condition monitoring and fault detection of wind turbines and related algorithms: A review. *Renewable & Sustainable Energy Reviews*, *13*(1), 1–39. doi:10.1016/j.rser.2007.05.008

Harmouche, J., Delpha, C., & Diallo, D. (2015) Improved fault diagnosis of ball bearings based on the global spectrum of vibration signals. *IEEE Transactions on Energy Conversion, 30*, 376-383.

Harun, F.K.C., Covington, J.A., & Gardner, J.W. (2012). Mimicking the biological olfactory system: A Portable electronic Mucosa. *Nanobiotechnology, 6*(2), 45 – 51.

Hayes, M. H. (1991). *Statistical digital signal processing and modelling*. John Wiley & Sons.

Haykin, S. (1999). *Neural Networks: A Comprehensive Foundation*. Englewood Cliffs, NJ: Prentice-Hall.

He, D., Bechhoefer, E., Ma, J., & Zhu, J. (2012). A Particle Filtering Based Approach for Gear Prognostics. In S. Kadry (Ed.), *Diagnostics and Prognostics of Engineering Systems: Methods and Techniques* (pp. 120–133). IGI Global.

Hernandez, M. D. P. C., & Labib, A. (2017). Selecting a condition monitoring system for enhancing effectiveness of power transformer maintenance. *Journal of Quality in Maintenance Engineering*, *23*(4), 400–414. doi:10.1108/JQME-07-2015-0027

He, W., Zi, Y., Chen, B., Wu, F., & He, Z. (2015). Automatic fault feature extraction of mechanical anomaly on induction motor bearing using ensemble super-wavelet transform. *Mechanical Systems and Signal Processing*, *54*, 457–480. doi:10.1016/j.ymssp.2014.09.007

Holbert, K. E., Lin, K., & Karady, G. G. (2006). Enhancement of electrical motor reliability through condition monitoring. In *IFAC Symposium on Power Plants and Power Systems Control* (pp. 255 – 260). Kananaskis, Canada: Academic Press.

Holm-Hansen, B. T., Gao, R. X., & Zhang, L. (2004). Customized wavelet for bearing defect detection. *ASME Journal of Dynamic Systems. Measurement and Control*, *126*(4), 740–745. doi:10.1115/1.1850534

Horn, D., & Mayo, W. R. (2000). NDE reliability gains from combining eddy-current and ultrasonic testing. *NDT & E International*, *33*(6), 351–362. doi:10.1016/S0963-8695(99)00058-4

Hu, C., Youn, B. D., Wang, P., & Yoon, J. T. (2012). Ensemble of data-driven prognostic algorithms for robust prediction of remaining useful life. *Reliability Engineering & System Safety*, *103*, 120–135. doi:10.1016/j.ress.2012.03.008

Hudon, G., Guy, C., & Hermia, J. (2000). Measurement of odor intensity by an electronic nose. *Journal of the Air & Waste Management Association*, *50*(10), 1750–1758. doi:10.1080/10473289.2000.10464202 PMID:11288303

Hwas, A., & Katebi, R. (2012, September). Model-based fault detection and isolation for wind turbine. In *Control (CONTROL), 2012 UKACC International Conference on* (pp. 876-881). IEEE. 10.1109/CONTROL.2012.6334747

Ibrahim, A., Badaoui, M. E., Guillet, F., & Bonnardot, F. (2008, December). A New Bearing Fault Detection Method in Induction Machines Based on Instantaneous Power Factor'. *IEEE Transactions on Industrial Electronics*, *55*(12), 4252–4259. doi:10.1109/TIE.2008.2003211

IEEE. (2006). *IEEE 1415TM-2006 IEEE Guide for Induction Machinery Maintenance Testing and Failure Analysis*. IEEE. Retrieved from: https://www.maintenance.org/fileSendAction/fcType/0/fcOid/399590942962795456/filePointer/399590942964800333/fodoid/399590942964800331/Pages_from_1415-2006-2.pdf

Iorgulescu, M., Alexandru, M., & Beloiu, R. (2012). Noise and vibration monitoring for diagnosis of DC motor's faults. In *13th International Conference on Optimization of Electrical and Electronic Equipmen*. Brasov, Romania: IEEE. 10.1109/OPTIM.2012.6231919

Irfan, M., Saad, N., Ibrahim, R., Asirvadam, V. S., Alwadie, A. S., & Sheikh, M. A. (2017). *An Assessment on the Non-Invasive Methods for Condition Monitoring of Induction Motors*. Fault Diagnosis and Detection Mustafa Demetgul, IntechOpen; doi:10.5772/67917

Ishii, H., Kobayashi, M., & Arita, K. (1994). Iterative design of seamless collaboration media. *Communications of the ACM*, *37*(8), 83–97. doi:10.1145/179606.179687

Jacobina, C. B., Filho, J. E. C., & Lima, A. M. (2000). On-line estimation of the stator resistance of induction machines based on zero-sequence model. *IEEE Transactions on Power Electronics*, *15*(2), 346–353. doi:10.1109/63.838107

Jardine, A. K., Lin, D., & Banjevic, D. (2006). A review on machinery diagnostics and prognostics implementing condition-based maintenance. *Mechanical Systems and Signal Processing*, *20*(7), 1483–1510. doi:10.1016/j.ymssp.2005.09.012

Jeon, S. H., Oh, K. K., & Choi, J. Y. (2002). Flux observer with online tuning of stator and rotor resistances for induction motors. *IEEE Transactions on Industrial Electronics*, *49*(3), 653–664. doi:10.1109/TIE.2002.1005393

Jordan, H. (1950). *The low noise electrical motor*. Essen: Springer Verlag.

Kallesoe, C. S., Izadi-Zamanabadi, R., Vadstrup, P., & Rasmussen, H. (2007). Observer-based estimation of stator-winding faults in delta-connected induction motors: A linear matrix inequality approach. *IEEE Transactions on Industry Applications*, *43*(4), 1022–1031. doi:10.1109/TIA.2007.900494

Karanayil, B., Rahman, M. F., & Grantham, C. (2007). Online Stator and Rotor Resistance Estimation Scheme Using Artificial Neural Networks for Vector Controlled Speed Sensorless Induction Motor Drive. *IEEE Transactions on Industrial Electronics*, *54*(1), 167–176. doi:10.1109/TIE.2006.888778

Karmakar, S., Chattopadhyay, S., Mitra, M., & Sengupta, S. (2016). *Induction Motor and Faults. Induction Motor Fault Diagnosis: Approach through Current Signature Analysis*. Springer. doi:10.1007/978-981-10-0624-1_2

Karmakar, S., Chattopadhyay, S., Mitra, M., & Sengupta, S. (2016). *Induction Motor Fault Diagnosis: Approach through Current Signature Analysis*. Springer. doi:10.1007/978-981-10-0624-1

Kasban, H., Zahran, O., Arafa, H., El-Kordy, M., Elaraby, S. M., & El-Samie, F. A. (2011). Welding defect detection from radiography images with a cepstral approach. *NDT & E International*, *44*(2), 226–231. doi:10.1016/j.ndteint.2010.10.005

Kaufmann, H., & Schmalstieg, D. (2003). Mathematics and geometry education with collaborative augmented reality. *Computers & Graphics*, *27*(3), 339–345. doi:10.1016/S0097-8493(03)00028-1

Khaira, A., Srivastava, S., & Suhane, A. (2015). Analysis of relation between ultrasonic testing and microstructure: A step towards highly reliable fault detection. *Engineering Review*, *35*(2), 87–96.

Kharche, P. P., & Kshirsagar, S. V. (2014). Review of Fault Detection in Rolling Element Bearing. *Int. J. of Innovative Research in Advanced Engineering*, *1*, 169–174.

Khelif, R., Chebel-Morello, B., Malinowski, S., Laajili, E., Fnaiech, F., & Zerhouni, N. (2017). Direct Remaining Useful Life Estimation Based on Support Vector Regression. *IEEE Transactions on Industrial Electronics*, *64*(3), 2276–2285. doi:10.1109/TIE.2016.2623260

Kia, S. H., Henao, H. G., & Capolino, A. (2005). Zoom-MUSIC frequency estimation method for three phase induction machine fault detection. *The 32nd Annual Conference of IEEE,* 2603-2608.

Kia, S. H., Henao, H., & Capolino, G. A. (2007, August). A high-resolution frequency estimation method for three-phase induction machine fault detection. *IEEE Transactions on Industrial Electronics*, *54*(4), 2305–2314. doi:10.1109/TIE.2007.899826

Kilgus, T., Heim, E., Haase, S., Prüfer, S., Müller, M., Seitel, A., ... Schlemmer, H.-P. (2015). Mobile markerless augmented reality and its application in forensic medicine. *International Journal of Computer Assisted Radiology and Surgery*, *10*(5), 573–586. doi:10.100711548-014-1106-9 PMID:25149272

Kim, J. W., & Park, S. (2018). Magnetic Flux Leakage Sensing and Artificial Neural Network Pattern Recognition-Based Automated Damage Detection and Quantification for Wire Rope Non-Destructive Evaluation. *Sensors (Basel)*, *18*(1), 109. PMID:29301294

Kim, T., Oh, H., Kim, H., & Youn, B. D. (2016). An Online-Applicable Model for Predicting Health Degradation of PEM Fuel Cells With Root Cause Analysis. *IEEE Transactions on Industrial Electronics*, *63*(11), 7094–7103. doi:10.1109/TIE.2016.2586022

Kim, Y. H., Youn, Y.-W., Hwang, D.-H., Sun, J.-H., & Kang, D.-S. (2013, September). Resolution Parameter Estimation Method to Identify Broken Rotor Bar Faults in Induction Motors. *IEEE Transactions on Industrial Electronics*, *60*(9), 4103–4117. doi:10.1109/TIE.2012.2227912

Kliman, G. B. (1988). Non-invasive detection of broken rotor bars in operating induction motors. *IEEE Trans. on Energy Conversion, 3*(4), 873-879.

Kompella, D. K. C., Gopala Rao, M. V., Rao, R. S., & Sreenivasu, R. N. (2014). Estimation of bearing faults in induction motor by MCSA using Daubechis wavelet analysis. *Proceedings of international conference on smart electric grid (ISEG14)*, 1–6.

Kompella, D. K. C., Gopala Rao, V. M., & Rayapudi, S. R. (2017). *Bearing fault detection in a 3 phase induction motor using stator current frequency spectral subtraction with various wavelet decomposition techniques. Ain Shams Engineering Journal.* doi:10.1016/j.asej.2017.06.002

Korde, A. (2002). On-Line Condition Monitoring of Motors Using Electrical Signature Analysis. *Recent Advances in Condition Based Plant Maintenance*, 17-18.

Kothari, D. P., & Nagrath, I. J. (2010). *Electric Machines*. New Delhi: Tata McGraw Hill Education Pvt. Ltd.

Kounavis, C. D., Kasimati, A. E., & Zamani, E. D. (2012). Enhancing the tourism experience through mobile augmented reality: Challenges and prospects. *International Journal of Engineering Business Management*, *4*, 10. doi:10.5772/51644

Kral, C., Habetler, T. G., Harley, R. G., Pirker, F., Pascoli, G. Berguggenberger, H., & Fenz, C. (2004). Rotor Temperature Estimation of Squirrel-Cage Induction Motors by Means of a Combined Scheme of Parameter Estimation and a Thermal Equivalent Model. IEEE Transactions on Industry Applications, 40(4), 1049-1057.

Kral, C., Kapeller, H., Pirker, F., & Pascoli, G. (2005). Discrimination of Rotor Faults and Low Frequency Load Torque Modulations of Squirrel Cage Induction Machines by Means of the Vienna Monitoring Method. *IEEE 36th Annual Power Electronics Specialists Conference,* 2861- 2866.

Kral, C., Habetler, T. G., & Harley, R. G. (2004). Detection of Mechanical Imbalances of Induction Machines Without Spectral Analysis of Time-Domain Signals. *IEEE Transactions on Industry Applications*, *40*(4), 1101–1105. doi:10.1109/TIA.2004.830762

Kral, C., Pirker, F., & Pascoli, G. (2008). The Impact of Inertia on Rotor Fault Effects – Theoretical Aspects of the Vienna Monitoring Method. *IEEE Transactions on Power Electronics*, *23*(4), 2136–2142. doi:10.1109/TPEL.2008.925183

Krautkrämer, J., & Krautkrämer, H. (2013). *Ultrasonic testing of materials*. Springer Science & Business Media.

Kucuker, A., & Bayrak, M. (2015). Detection of stator winding fault in induction motor using instantaneous power signature analysis. *Turkish Journal of Electrical Engineering and Computer Sciences*, *23*, 1263–1271. doi:10.3906/elk-1304-72

Laribi, S. & Bendiabdellah, A. (2010). Stator short circuit and broken bar faults diagnosis of an indirect vector control squirrel cage induction motor. *International Review of Electrical Engineering,* *5*(5), 2088-2094.

Laribi, S. (2016). Contribution à la Commande et Diagnostic des Défauts Statoriques et Rotoriques de la Machine Asynchrone à Cage d'Ecureuil. Université des Sciences et de la Technologie d'Oran Mohamed Boudiaf.

Laribi, S., Bendiabdallah, A., Cherif, B., Boukezata, B., Meradi, S., & Mihoub, Y. (2017). *Use of Neuro-Fuzzy Technique in Diagnosis of Rotor Faults of Cage Induction Motor.* The 5th International Conference on Electrical Engineering, Boumerdes (ICEE-B), Boumerdes, Algeria.

Laribi, S., Bendiabdallah, A., Cherif, B., Boukezata, B., Meradi, S., & Mihoub, Y. (2018) Induction Machine Rotor and Stator Faults Detection by Applying the DTW and N-F Network. IEEE ICIT 2018, Lyon, France.

Laribi, S., Benouzza, N., Bendiabdellah, A., & Toumi, D. (2006). Détection des Défauts de Barres Rotoriques dans les Moteurs Asynchrones à Cage d'Ecureuil par l'Emploi de Université des Sciences et de la Technologie d'Oran Mohamed Boudiaf, Analyse Spectrale des Grandeurs Electriques. ICEEA'06 Université de Sidi Belabbes, 136-141.

Laribi, S. (2005). *Synthèse des méthodes de diagnostic appliquées à la détection des défauts dans les machines asynchrones, Mémoire de Magister*. Algérie: USTO Oran.

Lazakis, I., Dikis, K., Michala, A. L., & Theotokatos, G. (2016). Advanced ship systems condition monitoring for enhanced inspection, maintenance and decision making in ship operations. *Transportation Research Procedia, 14*, 1679–1688. doi:10.1016/j.trpro.2016.05.133

Lee, S. B., Tallam, R. M., & Habetler, T. G. (2003). A robust, on-line turn- fault detection technique for induction machines based on monitoring the sequence component impedance matrix. *IEEE Transactions on Power Electronics, 18*(3), 865–872. doi:10.1109/TPEL.2003.810848

Leite, V. C., da Silva, J. G. B., Veloso, G. F. C., da Silva, L. E. B., Lambert-Torres, G., & Bonaldi, E. L. (2015). Detection of localized bearing faults in induction machines by spectral kurtosis and envelope analysis of stator current. *IEEE Transactions on Industrial Electronics, 62*(3), 1855–1865. doi:10.1109/TIE.2014.2345330

Linaza, M. T., Gutierrez, A., & García, A. (2013). *Pervasive augmented reality games to experience tourism destinations. In Information and Communication Technologies in Tourism 2014* (pp. 497–509). Springer. doi:10.1007/978-3-319-03973-2_36

Lin, P., & Gu, J. (2011). Research on Transformer Condition-based Maintenance System using the Method of Fuzzy Comprehensive Evaluation. *World Academy of Science, Engineering and Technology, 5*(8), 452–457.

Liu, J., Saxena, A., Goebel, K., Saha, B., & Wang, W. (2010). An adaptive recurrent neural network for remaining useful life prediction of lithium-ion batteries. *Proceedings of Annual Conference of the Prognostics and Health Management Society.*

Liu, Q., & Wirdelius, H. (2007). A 2D model of ultrasonic wave propagation in an anisotropic weld. *NDT & E International, 40*(3), 229–238. doi:10.1016/j.ndteint.2006.10.004

Liu, S., Wang, Y., & Tian, F. (2015). Prognosis of underground cable via online data-driven method with field data. *IEEE Transactions on Industrial Electronics, 62*(12), 7786–7794. doi:10.1109/TIE.2015.2458300

Liu, Z., Liu, T., Han, J., Bu, S., Tang, X., & Pecht, M. (2017). Signal Model-Based Fault Coding for Diagnostics and Prognostics of Analog Electronic Circuits. *IEEE Transactions on Industrial Electronics, 64*(1), 605–614. doi:10.1109/TIE.2016.2599142

Llobet, E., Rubio, J., Vilanova, X., Brezmes, J., Correig, X., Gardner, J. W., & Hines, E. L. (2001). Electronic Nose Simulation Tool Centered On PSpice. *Elsevier, Sensors and Actuators, 76*(1-3), 419–429. doi:10.1016/S0925-4005(01)00650-5

Loutas, T., Roulias, D., & Georgoulas, G. (2013). Remaining useful life estimation in rolling bearings utilizing data-driven probabilistic e-support vectors regression. *IEEE Transactions on Reliability*, *62*(4), 821–832. doi:10.1109/TR.2013.2285318

Lu, C. J., & Meeker, W. Q. (1993). Using degradation measures to estimate a time-to-failure distribution. *Technometrics*, *35*(2), 543–559. doi:10.1080/00401706.1993.10485038

Makdisie, C., Haidar, B., & Alhelou, H. H. (2018). An Optimal Photovoltaic Conversion System for Future Smart Grids. In *Handbook of Research on Power and Energy System Optimization* (pp. 601–657). IGI Global. doi:10.4018/978-1-5225-3935-3.ch018

Marani, R., Palumbo, D., Renò, V., Galietti, U., Stella, E., & D'Orazio, T. (2018). Modeling and classification of defects in CFRP laminates by thermal non-destructive testing. *Composites. Part B, Engineering*, *135*, 129–141. doi:10.1016/j.compositesb.2017.10.010

Marple, S. L. (1987). *Digital Spectral Analysis with Applications*. Prentice-Hall Signal Processing Series.

Márquez, F. P. G., Tobias, A. M., Pérez, J. M. P., & Papaelias, M. (2012). Condition monitoring of wind turbines: Techniques and methods. *Renewable Energy*, *46*, 169–178. doi:10.1016/j.renene.2012.03.003

Matlab. (2017). *Matlab R2017b document*. Retrieved on 11-Nov-2017 from: https://www.mathworks.com/help/nnet/ref/trainlm.html

McFadden, P. D., & Wang, W. (1990). *Time-frequency domain analysis of vibration signals for machinery diagnostics (I) Introduction to Wigner-Ville distribution.* Report no. 1859/90, University of Oxford

McFadden, P. D., & Smith, J. D. (1984). Model for the vibration produced by a single point defect in a rolling element bearing. *Journal of Sound and Vibration*, *96*(1), 69–82. doi:10.1016/0022-460X(84)90595-9

McWilliams, A., Beigi, P., Srinidhi, A., Lam, S., & MacAulay, C. E. (2015). Sex and Smoking Status Effects on the Early Detection of Early Lung Cancer in High-Risk Smokers Using an Electronic Nose. *IEEE Transactions on Biomedical Engineering, Volume*, *62*(8), 2044–2054. doi:10.1109/TBME.2015.2409092 PMID:25775482

Meeker, W. Q., & Escobar, L. A. (1998). *Statistical Methods for Reliability Data*. John Wiley & Sons Inc.

Meeker, W. Q., & LuValle, M. J. (1995). An Accelerated Life Test Model Based on Reliability Kinetics. *Technometrics*, *37*(2), 133–146. doi:10.1080/00401706.1995.10484298

Mehala, N., & Dahia, R. (2007). Motor Current Signature Analysis and its Applications in Induction Motor Fault Diagnosis. *International Journal of Systems Applications, Engineering and Development*, 2(1), 29–35.

Mehla, N., & Dahiya, R. (2007). An Approach of Condition Monitoring of Induction Motor Using MCSA. International Journal of Systems Applications. *Engineering & Development*, *1*(1), 13–17.

Meo, S., Gentile, G., & Ometto, A. (2003). *Induction motor current signature analysis to diagnostics of stator short circuits*. IEEE-IAS SDEMPED 2003-Symposium on Diagnostics for Electric Machines, Power Electronics and Drives, Atlanta, GA.

Merizalde, Y., Hernández-Callejo, L., & Duque-Perez, O. (2017). State of the Art and Trends in the Monitoring, Detection and Diagnosis of Failures in Electric Induction Motors. *MDPI Energies*, *10*(7), 1056. doi:10.3390/en10071056

Messaoudi, M., & Sbita, L. (2010). Multiple faults Diagnosis in Induction Motor Using the MCSA Method. *International Journal of Signal and Image Processing*, *1*(3), 190–195.

Meža, S., Turk, Ž., & Dolenc, M. (2014). Component based engineering of a mobile BIM-based augmented reality system. *Automation in Construction*, *42*, 1–12. doi:10.1016/j.autcon.2014.02.011

Milgram, P., & Kishino, F. (1994). A taxonomy of mixed reality visual displays. *IEICE Transactions on Information and Systems*, *77*(12), 1321–1329.

Miljković, D. & Zagreb, H. (2015). *Brief Review of Motor Current Signature Analysis*. Academic Press.

Mirafzal, B., Povinelli, R. J., & Demerdash, N. A. (2006). Interturn fault diagnosis in induction motors using the pendulous oscillation phenomenon. *IEEE Transactions on Energy Conversion*, *21*(4), 871–882. doi:10.1109/TEC.2005.853767

Mortazavizadeh, S. A., & Mousavi, S. M. G. (2014). A review on condition monitoring and diagnostic techniques of rotating electrical machines. *Physical Science International Journal*, *4*(3), 310–338. doi:10.9734/PSIJ/2014/4837

Najafabadi, T. A., Salmasi, F. R., & Jabehdar-Maralani, P. (2011). Detection and isolation of speed-, DC-link voltage-, and current-sensor faults based on an adaptive observer in induction-motor drives. *IEEE Transactions on Industrial Electronics*, *58*(5), 1662–1672. doi:10.1109/TIE.2010.2055775

Nakra, B. C., Yadava, G. S., & Thuestad, L. (1989). *Vibration measurement and analysis*. New Delhi, India: National Productivity Council.

Nandi, S., Ahmed, S., & Toliyat, H. (2001). Detection of rotor slot and other eccentricity related harmonics in a three phase induction motor with different rotor cages. *IEEE Transactions on Energy Conservation*, *16*(3), 253–260. doi:10.1109/60.937205

Nandi, S., Toliyat, H., & Li, X. (2005). Condition monitoring and fault diagnosis of electrical motors—A review. *IEEE Transactions on Energy Conversion*, *20*(4), 719–729. doi:10.1109/TEC.2005.847955

Narwade, S., Kulkarni, P., & Partil, C. (2013). Fault detection of induction motor using current and vibration monitoring. *Int. J. Adv. Comput*, *3*, 272.

Nasiri, M. R., Mahjoob, M. J., & Vahid-Alizadeh, H. (2011, April). Vibration signature analysis for detecting cavitation in centrifugal pumps using neural networks. In *Mechatronics (ICM), 2011 IEEE International Conference on* (pp. 632-635). IEEE. 10.1109/ICMECH.2011.5971192

Negoita, A., Scutaru, G., & Ionescu, R. M. (2010). A brief review of monitoring techniques for for rotating electrical machines. *Bulletin of the Transalvania University of Brasov*, *3*(2), 277–284.

Nguyen, V., Seshadrinath, J., Wang, D., Nadarajan, S., & Vaiyapuri, V. (2017). Model-Based Diagnosis and RUL Estimation of Induction Machines under Inter-Turn Fault. *IEEE Transactions on Industry Applications*, *53*(3), 2690–2701. doi:10.1109/TIA.2017.2669195

Odgaard, P. F., & Mataji, B. (2006). Fault detection in coal mills used in power plants. *IFAC Proceedings Volumes, 39*(7), 177-182.

Odgaard, P. F., Lin, B., & Jørgensen, S. B. (2007). Observer-based and regression model-based detection of emerging faults in coal mills. Fault Detection, Supervision and Safety of Technical Processes 2006, 687-692. doi:10.1016/B978-008044485-7/50116-0

Odgaard, P. F., & Mataji, B. (2008). Observer-based fault detection and moisture estimating in coal mills. *Control Engineering Practice*, *16*(8), 909–921. doi:10.1016/j.conengprac.2007.10.008

Odgaard, P. F., Stoustrup, J., Nielsen, R., & Damgaard, C. (2009, March). Observer based detection of sensor faults in wind turbines. *Proceedings of European wind energy conference*, 4421-4430.

Ojaghi, M., Sabouri, M., & Faiz, J. (2014). Diagnosis methods for stator winding faults in three-phase squirrel-cage induction motors. *International Transactions on Electrical Energy Systems*, *24*(6), 891–912. doi:10.1002/etep.1750

Orman, M., & Pinto, C. (2013). Vibration, sound and thermal analysis as a condition monitoring methods for electric motors. *Applied Mechanics and Materials*, *239-240*, 462–467. doi:10.4028/www.scientific.net/AMM.239-240.462

Ozelgin, I. (2008). Analysis of magnetic flux density for airgap eccentricity and bearing faults. *International Journal of Systems Applications, Engineering & Development*, *2*, 162–169.

Padovese, L., & Vicente, S. A. (2004). Optimizing probabilistic neural networks by the use of genetic algorithms for rolling bearing fault diagnosis. *CEP*, *5508*, 900.

Paliwal, D. (2015). *Identification of single and multiple defects in rolling element bearing through wavelet transform* (Unpublished doctoral dissertation). Sir Padampat Singhania University, Udaipur, India.

Paliwal, D., Choudhury, A., & Govardhan, T. (2014). Identification of faults through wavelet transform vis-à-vis fast Fourier transform of noisy vibration signals emanated from defective rolling element bearings. *Frontiers of Mechanical Engineering*, *9*(2), 130–141. doi:10.100711465-014-0298-6

Pandey, K., Zope, P., & Suralkar, S. (2012). Review on fault diagnosis in three-phase induction motor. In MEDHA–2012, Proceedings published by International Journal of Computer Applications. IJCA.

Patil, M. A., Tagade, P., Hariharan, K. S., Kolake, S. M., Song, T., Yeo, T., & Doo, S. (2015). A novel multistage support vector machine based approach for li ion battery remaining useful life estimation. *Applied Energy*, *159*, 285–297. doi:10.1016/j.apenergy.2015.08.119

Pearce, T. C., Schiffman, S. S., Nagle, H. T., & Gardner, J. W. (2003). *Handbook of Machine olfaction*. Weinheim: WileyVCH.

Peng, W., Li, Y. F., Yang, Y. J., Mi, J., & Huang, H. Z. (2017). Bayesian Degradation Analysis With Inverse Gaussian Process Models Under Time-Varying Degradation Rates. *IEEE Transactions on Reliability*, *66*(1), 84–96. doi:10.1109/TR.2016.2635149

Petrovic, T. K., & Terzic, B. (2009). Sensorless speed detection of squirrel-cage induction machines using stator neutral point voltage harmonics. *Mechanical Systems and Signal Processing*, *23*(3), 931–939. doi:10.1016/j.ymssp.2008.06.005

Picazo-Ródenas, M. J., Antonino-Daviu, J., Climente-Alarcon, V., Royo-Pastor, R., & Mota-Villar, A. (2015). Combination of noninvasive approaches for general assessment of induction motors. *IEEE Transactions on Industry Applications*, *51*(3), 2172–2180. doi:10.1109/TIA.2014.2382880

Portalés, C., Gimeno, J., Casas, S., Olanda, R., & Giner, F. (2016). Interacting with augmented reality mirrors. In J. Rodrigues, P. Cardoso, J. Monteiro, & M. Figueiredo (Eds.), *Handbook of Research on Human-Computer Interfaces, Developments, and Applications* (pp. 216–244). IGI-Global. doi:10.4018/978-1-5225-0435-1.ch009

Portalés, C., Viñals, M. J., Alonso-Monasterio, P., & Morant, M. (2010). AR-Immersive Cinema at the Aula Natura Visitors Center. *IEEE MultiMedia*, *17*(4), 8–15. doi:10.1109/MMUL.2010.72

Prabhakar, S., Mohanty, A. R., & Sekhar, A. S. (2002). Application of discrete wavelet transform for detection of ball bearing race faults. *Tribology International*, *35*(12), 793–800. doi:10.1016/S0301-679X(02)00063-4

Prakasam, K., & Ramesh, S. (2016). Testing and Analysis of Induction Motor Electrical Faults Using Current Signature Analysis. *Scientific Research-Circuits and Systems*, *7*(09), 2651–2662. doi:10.4236/cs.2016.79229

Qiao, Z., Shi, T., Wang, Y., Yan, Y., Xia, C., & He, X. (2013). New sliding-mode observer for position sensorless control of permanent-magnet synchronous motor. *IEEE Transactions on Industrial Electronics*, *60*(2), 710–719. doi:10.1109/TIE.2012.2206359

Rahman, S., Saeed, S. H., Siddiqui, M. J., & Rahman, S. (2013). Polymeric chemoresistor model use in design and simulation of sensor. *Multimedia, Signal Processing and Communication Technologies (IMPACT), 2013 International Conference on*, 273 – 277. 10.1109/MSPCT.2013.6782134

Rahman, S., Usmani, T., & Saeed, S.H. (2013). Review Of Electronic Nose and application. *International Journal of Computing and Corporate Research, 3*(2).

Rao, S. S. (2004). *Mechanical vibrations*. Delhi, India: Pearson Education Inc.

Rothenhagen, K., & Fuchs, F. W. (2006, November). Current sensor fault detection by bilinear observer for a doubly fed induction generator. In *IEEE Industrial Electronics, IECON 2006-32nd Annual Conference on* (pp. 1369-1374). IEEE. 10.1109/IECON.2006.347390

Rothenhagen, K., Thomsen, S., & Fuchs, F. W. (2007, September). Voltage sensor fault detection and reconfiguration for a doubly fed induction generator. In *Diagnostics for Electric Machines, Power Electronics and Drives, 2007. SDEMPED 2007. IEEE International Symposium on* (pp. 377-382). IEEE. 10.1109/DEMPED.2007.4393123

Runnemalm, A., Ahlberg, J., Appelgren, A., & Sjökvist, S. (2014). Automatic inspection of spot welds by thermography. *Journal of Nondestructive Evaluation, 33*(3), 398-406.

Sahraoui, M., Cardoso, A. J. M., & Ghoggal, A. (2015). The use of a modified prony method to track the broken rotor bar characteristic frequencies and amplitudes in three-phase induction motors. *IEEE Transactions on Industry Applications*, *51*(3), 2136–2147. doi:10.1109/TIA.2014.2375384

Salehifar, M., Arashloo, R. S., Moreno-Eguilaz, M., Sala, V., & Romeral, L. (2014). Observer-based open transistor fault diagnosis and fault-tolerant control of five-phase permanent magnet motor drive for application in electric vehicles. *IET Power Electronics*, *8*(1), 76–87. doi:10.1049/iet-pel.2013.0949

Sbartaï, Z. M., Breysse, D., Larget, M., & Balayssac, J. P. (2012). Combining NDT techniques for improved evaluation of concrete properties. *Cement and Concrete Composites*, *34*(6), 725–733. doi:10.1016/j.cemconcomp.2012.03.005

Sbartaï, Z. M., Laurens, S., Elachachi, S. M., & Payan, C. (2012). Concrete properties evaluation by statistical fusion of NDT techniques. *Construction & Building Materials*, *37*, 943–950. doi:10.1016/j.conbuildmat.2012.09.064

Seera, M., Lim, C. P., Nahavandi, S., & Loo, C. K. (2014). Condition monitoring of induction motors: A review and an application of an ensemble of hybrid intelligent models. *Expert Systems with Applications*, *41*(10), 4891–4903. doi:10.1016/j.eswa.2014.02.028

Sen, P. C. (1989). *Principles of Electric Machines and Power Electronics*. New York: Willey.

Seshadrinath, J., Singh, B., & Panigrahi, B. K. (2014). Vibration analysis based interturn fault diagnosis in induction machines. *IEEE Transactions on Industrial Informatics*, *10*(1), 340–350. doi:10.1109/TII.2013.2271979

Shao, S., Wheeler, P. W., Clare, J. C., & Watson, A. J. (2013). Fault detection for modular multilevel converters based on sliding mode observer. *IEEE Transactions on Power Electronics*, *28*(11), 4867–4872. doi:10.1109/TPEL.2013.2242093

Sharp, M. E. (2012). *Prognostic approaches using transient monitoring methods* (Ph.D. dissertation). University of Tennessee, Knoxville, TN.

Sheikh, M. A., Nor, N. M., Ibrahim, T., & Bakhsh, S. T. (2017). An Analytical and Experimental Approach to Diagnose Unbalanced Voltage Supply. *Arabian Journal for Science and Engineering*, 1–12.

Sheikh, M. A., Nor, N. M., Ibrahim, T., & Irfan, M. (2017). Unsupervised on-line method to diagnose unbalanced voltage in three-phase induction motor. *Neural Computing & Applications*, 1–16.

Shukla, S., Jha, M., & Qureshi, M. F. (2014). Motor Current Signature Analysis for Fault Diagnosis and Condition Monitoring of Induction Motors using Interval Type-2 Fuzzy logic. International Journal of Innovative Science. *Engineering & Technology*, *1*(5), 84–95.

Sick, B. (2002). On-line and indirect tool wear monitoring in turning with artificial neural networks: A review of more than a decade of research. *Mechanical Systems and Signal Processing*, *16*(4), 487–546. doi:10.1006/mssp.2001.1460

Siddiqui, K. M., Sahay, K., & Giri, V. (2014). Health Monitoring and Fault Diagnosis in Induction Motor-A Review," *International Journal of Advanced Research in Electrical. Electronics and Instrumentation Engineering*, *3*, 6549–6565.

Siddiqui, K. M., Sahay, K., Giri, V. K., & Gothwal, N. (2016). Diagnosis of airgap eccentricity fault in the inverter driven induction motor drives by transformative techniques. *Perspectives in Science, Elsevier*, *8*, 127–131. doi:10.1016/j.pisc.2016.04.014

Singh, S., Hines, E. L., & Garner, J. W. (1996). Fuzzy neural computing of coffee and tainted-water data from an electronic nose. *Sensors and Actuators. B, Chemical*, *6*(3), 185–190. doi:10.1016/0925-4005(96)80047-5

Si, X. S. (2015). An Adaptive Prognostic Approach via Nonlinear Degradation Modeling: Application to Battery Data. *IEEE Transactions on Industrial Electronics*, *62*(8), 5082–5096. doi:10.1109/TIE.2015.2393840

Si, X. S., Wang, W., Hu, C. H., Chen, M. Y., & Zhou, D. H. (2013). A Wiener-process-based degradation model with a recursive filter algorithm for remaining useful life estimation. *Mechanical Systems and Signal Processing*, *35*(1-2), 219–237. doi:10.1016/j.ymssp.2012.08.016

Si, X. S., Wang, W., Hu, C. H., Zhou, D. H., & Pecht, M. G. (2012). Remaining useful life estimation based on a nonlinear diffusion degradation process. *IEEE Transactions on Reliability*, *61*(1), 50–67. doi:10.1109/TR.2011.2182221

Sonje, D. M., & Monje, R. K. (2011). Rotor Cage Fault Detection in Induction Motors by Motor Current Signature Analysis. *International Conference in Computational Intelligence Proceedings*, 22-26.

Soomro, A. A., Kalwar, I. H., Kazi, K., Khoo, S. K., & Ansari, S. (2016). A Hybrid Monitoring Technique for Diagnosis of Mechanical Faults in Induction Motor. *Indian Journal of Science and Technology*, *9*(47), 1–9.

Soualhi, A., Medjaher, K., & Zerhouni, N. (2015). Bearing health monitoring based on Hilbert–Huang transform, support vector machine, and regression. *IEEE Transactions on Instrumentation and Measurement, 64*(1), 52–62. doi:10.1109/TIM.2014.2330494

Su, Z., Ye, L., & Lu, Y. (2006). Guided Lamb waves for identification of damage in composite structures: A review. *Journal of Sound and Vibration, 295*(3-5), 753–780. doi:10.1016/j.jsv.2006.01.020

Swann, S. A. (1963). Effect of rotor eccentricity on the magnetic fields of a non-salient pole machine. *Proceedings of IEE, 110*, 903–915.

Tallam, R. M., Habetler, T. G., Harley, R. G., Gritter, D. J., & Burton, B. H. (2000). *Neural Network Based On-Line Stator Winding Turn Fault Detection for Induction Motors. In Conference Record of the IEEE Industry Application Conference.* IAS.

Tandon, N., & Choudhury, A. (1999). A review of vibration and acoustic measurement methods for the detection of defects in rolling element bearings. *Tribology International, 32*(8), 469–480. doi:10.1016/S0301-679X(99)00077-8

Tandon, N., & Choudhury, A. (2000). A theoretical model to predict vibration response of rolling bearings in a rotor bearing system due to distributed defects under radial load. *ASME Journal of Tribology, 122*(3), 609–615. doi:10.1115/1.555409

Tandon, N., Yadava, G. S., & Ramakrishna, K. M. (2007). A comparison of some condition monitoring techniques for the detection of defects in induction motor ball bearings. *Mechanical Systems and Signal Processing, 21*(1), 244–256. doi:10.1016/j.ymssp.2005.08.005

Tang, S.-L., Kwoh, C.-K., Teo, M.-Y., Sing, N. W., & Ling, K.-V. (1998). Augmented reality systems for medical applications. *IEEE Engineering in Medicine and Biology Magazine, 17*(3), 49–58. doi:10.1109/51.677169 PMID:9604701

Tavner, P., Ran, L., Penman, J. & Sedding, H. (2008). *Condition monitoring of rotating electrical machines.* London: The Institution of Engineering and Technology.

Teti, R., Jemielniak, K., O'Donnell, G., & Dornfeld, D. (2010). Advanced monitoring of machining operations. *CIRP Annals-Manufacturing Technology, 59*(2), 717–739. doi:10.1016/j.cirp.2010.05.010

Thatoi, D. N., Das, H. C., & Parhi, D. R. (2012). Review of techniques for fault diagnosis in damaged structure and engineering system. *Advances in Mechanical Engineering, 4*, 327569. doi:10.1155/2012/327569

Thomson, W. T., & Culbert, I. (2017). Current Signature Analysis for Condition Monitoring of Cage Induction Motors. Wiley Publishers.

Thomson, W. T., & Dahleh, M. D. (2007). *Theory of vibration with applications.* Delhi, India: Pearson Education Inc.

Thomson, W. T., & Fenger, M. (2001). Current Signature Analysis to Detect Induction Motor Faults. *IEEE Industry Applications Magazine*, *7*(4), 26–34. doi:10.1109/2943.930988

Thomson, W. T., & Gilmore, R. J. (2003). Motor Current Signature Analysis to detect faults in Induction Motor Drives-Fundamentals, Data Interpretation and Industrial Case Histories. *Proceedings of the 32nd turbomachinery Symposium*, 145-156.

Thorsen, O. V., & Dalva, M. (1995). A survey of faults on induction motors in offshore oil industry, petrochemical industry, gas terminals, and oil refineries. *IEEE Transactions on Industry Applications*, *31*(5), 1186–1196. doi:10.1109/28.464536

Tian, J., Morillo, C., Azarian, M. H., & Pecht, M. (2016). Motor Bearing Fault Detection Using Spectral Kurtosis-Based Feature Extraction Coupled With K-Nearest Neighbor Distance Analysis. *IEEE Transactions on Industrial Electronics*, *63*(3), 1793–1803. doi:10.1109/TIE.2015.2509913

Tian, Y., Guan, T., & Wang, C. (2010). Real-time occlusion handling in augmented reality based on an object tracking approach. *Sensors (Basel)*, *10*(4), 2885–2900. doi:10.3390100402885 PMID:22319278

Tian, Z., Lin, D., & Wu, B. (2012). Condition based maintenance optimization considering multiple objectives. *Journal of Intelligent Manufacturing*, *23*(2), 333–340. doi:10.100710845-009-0358-7

Toliyat, H. A., Nandi, S., Choi, S., & Meshgin-Kelm, H. (2013). Electric machines: Modeling, condition monitoring and fault diagnosis. New York: Taylor & Francis Group.

Toloie-Eshlaghy, A., & Homayonfar, M. (2011). MCDM methodologies and applications: A literature review from 1999 to 2009. *Research Journal of International Studies*, *21*, 86–137.

Torri, L., Sinelli, N., & Limbo, S. (2010). Shelf life evaluation of fresh-cut pineapple by using an electronic nose. *Postharvest Biol. Technol.*, *56*, 239-245.

Trutt, F. C., Sottile, J., & Kohler, J. L. (2001). Detection of AC machine winding deterioration using electrically excited vibrations. *IEEE Transactions on Industry Applications*, *37*(1), 10–14. doi:10.1109/28.903120

Tse, P. W., Peng, Y. H., & Yam, R. (2001). Wavelet analysis and envelope detection for rolling element bearing fault diagnosis – their effectiveness and flexibilities. *ASME Journal of Vibration and Acoustics*, *123*(3), 303–310. doi:10.1115/1.1379745

Unity Technologies. (2018). *Unity*. Retrieved from https://unity3d.com/es

Vaimann, T., & Kallaste, A. (2012). Condition monitoring of electrical machines. In *11th International Symposium "Topical Problems in the Field of Electrical and Power Engineering" and "Doctoral School of Energy and Geotechnology II"* (pp. 209-212). Academic Press.

Vargas, H., Sánchez, J., Jara, C. A., Candelas, F., Reinoso, O., & Díez, J. L. (2010). Docencia en Automática: Aplicación de las TIC a la realización de actividades prácticas a través de Internet. *Revista Iberoamericana de Automática e Informática Industrial RIAI*, *7*(1), 35–45. doi:10.1016/S1697-7912(10)70006-6

Varmaziari, H., & Dehghani, M. (2017, May). Cyber-attack detection system of large-scale power systems using decentralized unknown input observer. In *Electrical Engineering (ICEE), 2017 Iranian Conference on* (pp. 621-626). IEEE. 10.1109/IranianCEE.2017.7985114

Vera, L., Gimeno, J., Coma, I., & Fernández, M. (2011). *Augmented mirror: interactive augmented reality system based on kinect.* Paper presented at the IFIP Conference on Human-Computer Interaction. 10.1007/978-3-642-23768-3_63

Vijay Kumar, E., Chaturvedi, S. K., & Deshpandé, A. W. (2009). Maintenance of industrial equipment: Degree of certainty with fuzzy modelling using predictive maintenance. *International Journal of Quality & Reliability Management*, *26*(2), 196–211. doi:10.1108/02656710910928824

Vilanova, X., Llobat, E., Brezmes, J., & Correig, X. (1998). *Influence of moisture and working temperature on the steady state and transient response of thick film tin oxide gas sensors, Internal Report GSM 98/01*. Universitat Rovira i Virgili.

Walker, D. N., Adams, S. L., & Placek, R. J. (1981). Torsional vibration and fatigue of turbine-generator shafts. *IEEE Transactions on Power Apparatus and Systems*, *PAS-100*(11), 4373–4380. doi:10.1109/TPAS.1981.316848

Wang, L. (2007). *Induction motor bearing fault detection using a sensorless approach.* Texas A&M University.

Wang, W. J., & McFadden, P. D. (1996). Applications of wavelets to gearbox vibration signals for fault detection. *Journal of Sound and Vibration*, *192*(5), 927–939. doi:10.1006/jsvi.1996.0226

Wang, W., Carr, M., Xu, W., & Kobbacy, A. K. H. (2011). A model for residual life prediction based on Brownian motion with an adaptive drift. *Microelectronics and Reliability*, *51*(2), 285–293. doi:10.1016/j.microrel.2010.09.013

Wang, Y., Peng, Y., Zi, Y., Jin, X., & Tsui, K. L. (2016). A Two-Stage Data-Driven-Based Prognostic Approach for Bearing Degradation Problem. *IEEE Transactions on Industrial Informatics*, *12*(3), 924–932. doi:10.1109/TII.2016.2535368

Wang, Y., Wang, J., Zhou, B., & Lu, Q. (2009). Monitoring storage time and quality attribute of egg based on electronic nose. *Analytica Chimica Acta*, *650*(2), 183–188. doi:10.1016/j.aca.2009.07.049 PMID:19720190

Wang, Z. Q., Hu, C. H., & Fan, H. D. (2018). Real-Time Remaining Useful Life Prediction for a Nonlinear Degrading System in Service: Application to Bearing Data. *IEEE/ASME Transactions on Mechatronics*, *23*(1), 211–222. doi:10.1109/TMECH.2017.2666199

Warren, V., & Stone, G. (n.d.). *Review of Electrical Tests to Assess Motor Windings*. I. P. Engineering.

Willcox, M., & Downes, G. (2003). *A brief description of NDT techniques*. Toronto: NDT Equipment Limited.

Wilson, J., & Instruments, B. (2003) Current state of surge testing induction machines. *Iris Rotating Machine Conference*, 1-13.

Wu, H.-K., Lee, S. W.-Y., Chang, H.-Y., & Liang, J.-C. (2013). Current status, opportunities and challenges of augmented reality in education. *Computers & Education*, *62*, 41–49. doi:10.1016/j.compedu.2012.10.024

Wu, L., Habetler, T., & Harley, R. (2005). Separating Load Torque Oscillation and Rotor Fault Effects in Stator Current-Based Motor Condition Monitoring. *International Electric Machines and Drives Conference*, 1889-1894.

Wu, L., Habetler, T., & Harley, R. (2007). A Review of Separating Mechanical Load Effects from Rotor Faults Detection in Induction Motors. *IEEE International Symposium on Diagnostics for Electric Machines, Power Electronics and Drives*. 10.1109/DEMPED.2007.4393098

Wu, Q., Yang, X., & Zhou, Q. (2012). Pattern recognition and its application in fault diagnosis of electromechanical system. *Journal of Information and Computational Science*, *9*(8), 2221–2228.

Xianghui, H., Habetler, T. G., Harley, R. G., & Wiedenbrug, E. J. (2007). Using a Surge Tester to Detect Rotor Eccentricity Faults in Induction Motors. *Industry Applications. IEEE Transactions on*, *43*, 1183–1190.

Xu, B., Sun, L., Xu, L., & Xu, G. (2013, March). Improvement of the hilbert method via esprit detecting rotor fault in induction motors at low sli. *IEEE Trans. Energ. Conv.*, *28*(1), 225–233. doi:10.1109/TEC.2012.2236557

Yang, F., Habibullah, M. S., Zhang, T., Xu, Z., Lim, P., & Nadarajan, S. (2016). Health Index-Based Prognostics for Remaining Useful Life Predictions in Electrical Machines. *IEEE Transactions on Industrial Electronics*, *63*(4), 2633–2644. doi:10.1109/TIE.2016.2515054

Yang, F., & Mao, K. Z. (2011). Robust feature selection for microarray data based on multicriterion fusion. *IEEE/ACM Transactions on Computational Biology and Bioinformatics*, *8*(4), 1080–1092. doi:10.1109/TCBB.2010.103 PMID:21566255

Yang, S. J. (1981). *Low noise electric motors*. Oxford, UK: Clarendon Press.

Yan, X. G., & Edwards, C. (2008). Adaptive sliding-mode-observer-based fault reconstruction for nonlinear systems with parametric uncertainties. *IEEE Transactions on Industrial Electronics*, *55*(11), 4029–4036. doi:10.1109/TIE.2008.2003367

Yin, S., & Zhu, X. (2015). Intelligent particle filter and its application to fault detection of nonlinear system. *IEEE Transactions on Industrial Electronics*, *62*(6), 3852–3861. doi:10.1109/TIE.2015.2399396

Yulianti, I., Addawiyah, A., & Setiawan, R. (2018, March). Optimization of exposure factors for X-ray radiography non-destructive testing of pearl oyster. *Journal of Physics: Conference Series*, *983*(1), 012004.

Zhang, P., Lu, B., & Habetler, T. G. (2008). A Remote and Sensorless Stator Winding Resistance Estimation Method for Thermal Protection of Soft- Starter- Connected Induction Machines. *IEEE Transactions on Industrial Electronics*, *55*(10), 3611–3618. doi:10.1109/TIE.2008.927241

About the Contributors

Muhammad Irfan received the B.Sc degree in Mechatronics & Control Engineering from University of Engineering & Technology Lahore, Pakistan in 2009. He joined Roshan Packages Pvt Ltd in 2009 as Electrical & Control Engineer. He had worked on the condition monitoring and preventive maintenance of plant equipment for two years and resigned in 2011. He has completed his Master and Ph.D Degrees in Electrical & Electronics Engineering from Universiti Teknologi PETRONAS, Malaysia in 2013 and 2016 respectively. Currently, He is working as an assistant professor in Electrical Engineering Department at Najran University, Saudi Arabia. His main research areas are intelligent control, Instrumentation, automation & process control, preventive maintenance, condition monitoring, fault analysis, Power Electronics and AC/DC Machines.

* * *

H. H. Alhelou received the B.Sc. (Hons.) degree in elec. eng. from Tishreen University, Lattakia, Syria, in 2011, the M.Sc. (Hons.) degree in elec. eng. from Isfahan University of Technology, Isfahan, Iran, in 2014, and the Ph.D. Candidate degree in elec. eng. from Isfahan University of Technology, Isfahan, Iran, in 2017. He was the recipient of the Outstanding Reviewer Award for Energy Conversion and Management Journal in 2016, ISA Transactions Journal in 2018 and many other Awards.

Abdullah Alwadie is Associate Professor and Dean Faculty of Engineering at Najran University.

Azeddine Bendiabdellah, MIEE Member, received his Bachelor Engineering degree (Honors class) and his Ph. D. degree in Electrical Engineering from the University of Sheffield, England, in 1980 and 1985 respectively. He is currently Lecturer/Researcher and Director of both the Laboratory of Electrical Drives Development (LDEE) and the Diagnostic Group at the University of Sciences and Technology of Oran Mohamed Boudiaf (USTO-MB), Algeria. His research interests include Electrical Machines and Drives Analysis and Modelling, Electrical Machines and Drives Control, Electrical Machines and Drives Diagnosis, Power Static Converters Fault Detection & Fault Tolerance.

Noureddine Benouzza was born in Oran, Algeria, on August 17, 1963. He received the electrical engineering, master's and doctorate degrees from the University of Sciences and Technology of Oran (USTO-MB), Oran, in 1989,1993, and 2006, respectively, all in electrical engineering. He is currently a Professor and Researcher at the University of Sciences and Technology of Oran, Algeria. His research interests include electrical machines and drives control, as well as electrical machines faults diagnosis.

Ahmed Hamida Boudinar was born in Oran, Algeria, on September 10, 1969. He received the diploma (Hons.), master's (Hons.), and doctorate (Hons.) degrees from the University of Sciences and Technology of Oran, (USTO-MB), Oran, in 1992, 1997, and 2007, respectively, all in electronic engineering. He is currently a Professor and Researcher with the Electrical Engineering Department, USTO-MB. His research interests include condition monitoring and diagnosis of electrical machines and drives using numerical methods for signal processing.

Sergio Casas-Yrurzum has a master's degree in Computer Engineering and also a bachelor's degree in Telecommunications Engineering - Telematics Specialty. He received the Spanish National Award on University Studies in 2008. He received his PhD in Computational Mathematics at the University of Valencia in 2014. He works as a senior researcher in the Robotics Institute (IRTIC) of the University of Valencia, where he is also a part-time professor at the School of Engineering (ETSE). His expertise is in the simulation field with special focus on Virtual Reality, Augmented Reality and motion cueing.

Achintya Choudhury graduated in Mechanical Engineering from NIT Silchar in India and obtained Masters' and Doctoral degrees from IIT Kharagpur and IIT Delhi respectively. His primary research interest is in the field of vibration monitoring of rolling bearings and gears. He is actively involved in dissemination of research outcomes and published research papers in journals of repute and attended conferences in India and abroad. He has 34 years' experience in teaching, research and academic administration in different Institutes of repute including NIT Silchar, Sikkim Manipal Institute of Technology, Asansol Engineering College and Sir Padampat Singhania University.

Muthiah Geethanjali received B.E., M.E. in Electrical and Electronics and Power Systems Engineering from Thiagarajar College of Engineering in 1989 and 1990 respectively. She received Ph.D degree in Electrical Engineering from Madurai Kamaraj University, India in the year 2008. Currently, she is working as an Assistant Professor in Department of Electrical and Electronics Engineering in Thiagarajar College of Engineering, India. She was a recipient of Young Scientist Fellowships-TNSCST, Chennai. She has published more than 50 articles in reputed journals and presented around 10 guest lectures. Her research interests include Power system protection, Smart Grid, Wide-area monitoring and Control and Renewable Energy Resources.

Jesús Gimeno Sancho received the Master degree in Computer Science from the University of Valencia (2008), with a dissertation about "Motion capture and augmented reality techniques for virtual prototyping". He is a PhD student and part-time professor at the University of Valencia (Spain). His research interests include augmented reality, motion capture, real time simulation, advanced user interfaces and mobile computing. In the last years he has served also as visiting scientist at the Augmented Reality Group at Bauhaus Universitat (Weimar, Germany, 2008) and at VRAC Center at Iowa State University (Ames IO, USA, 2010).

Vinoth Kumar K. was born on 30th April, 1984, in Vellore, Tamil Nadu, India. He received his Undergraduate degree in Electrical and Electronics Engineering from Anna University, Chennai, India in the year 2006 and Post Graduate degree specializing in Power Electronics and Drives from VIT University, Vellore, India in the year 2008. He is currently Assistant Professor in the Department of Electrical Sciences at Karunya University, Coimbatore, India. His research interests include Fault diagnosis in industrial drives, Neuro Fuzzy computing and Power Electronics & Drives. He has been actively involved in the Internal Quality Assurance Cell at Karunya University. His contribution in setting up the experiments for the Power Electronics Laboratory is also commendable. He published research papers in both international and national journals as well as in conferences. He is authored the Book titled "Basic Electrical and Electronics Engineering" published by John Wiley & Sons India in 2014 and Neural Networks, Soft Computing textbooks in 2012. He has delivered lectures on Energy Conservation in Industries, New techniques for Energy auditing and other related areas in various workshop and seminars as well as guided 14 UG and 14 PG Student projects and One International student project.

Ashish Khaira is a Mechanical Engineer, Researcher (PhD), MATLAB Programmer, Book Author (LAP Germany), Website Creator and Creative Doer. Ashish have multiple years industrial and academic experiences; along with published book, book chapter, journal articles, conference papers and magazine articles. He is also experienced in working with MATLAB and R programming language. Ashish's first book is available on 35+ international websites for ordering/purchasing that's by In Google search multiple pages are dedicated on him. He has 2600+ connections (First) on LinkedIn with 1000+ views per month on updates and 5000+ followers on twitter. Regarding Ashish's educational qualification he had picked the threads of profession during his graduation in Mechanical Engineering (BE-honours 79.38%) from SATI-India, post graduation in Maintenance Engineering (M. Tech GGPA 9.98) from MANIT-India and completed PhD (Full Time) in Mechanical Engineering from MANIT, Bhopal-India. He has also completed few certification coursers from John Hopkins University-USA through Coursera MOOC. Presently, he is working as an Assistant Professor (Project Based) @SATI, Vidisha-India. Apart from this he is avid reader & writer, amateur painter & photographer and also loves music & movies.

Souad Laribi was born in Tiaret, Algeria in 1972. She received engineering Diploma in Electromechanic from the Department of Electrical Engineering, University of Tiaret,Algeria and the Magister degree and PHD degree from the Department of Electrical Engineering, University of Sciences and Technology of Oran Mohamed Boudiaf, Algéria Currently, she is a conference Professor with the Department of Electrical Engineering Ibn Khaldoun University, Tiaret, Algéria, Her research interests include electrical machines associated with static converter, control, modeling-diagnosis by artificial intelligence and spectral analysis.

Maheswari M. is currently working as Professor and Head, Department of Electrical and Electronics Engineering in Malla Reddy Engineering College (Autonomous), Hyderabad. She completed her B.E. Electrical and Electronics Engineering degree in K.S. Rangasamy College of Technology Thiruchengode with first class with distinction, M.E. Power Systems in Government College of Technology, Coimbatore and with first class with distinction and Ph.D. at Anna University, Chennai in the year 2015. She has 17 years of teaching experience and having various administrative positions. She published her research work in ten various international journals.

Saifur Rahman completed his B.Tech from U.P. Technical University in the year 2005 from Electronics & Communication Engineering. He then pursued and completed his M.Tech and Ph.D in the year 2008 and year 2015 from Integral University, Lucknow, India. He has done his research in the field of sensors. He has done his research on the applications of sensors. He has number of research papers on national and international conference and journals.

Cristina Portalés Ricart (PhD in Surveying and Geoinformation, with specialization in Augmented Reality, 2008; IEEE Computer Society member) has been recently (2012-2015) a Juan de la Cierva post-doc fellow at the Institute of Robotics and Information and Communication Technology (IRTIC) at Universitat de València (Spain), where she currently works as full PhD senior researcher. She was formerly graduated with a double degree: Engineer in Geodesy and Cartography from the Universidad Politécnica de Valencia (Spain) and MSc in Surveying and Geoinformation from the Technische Universität Wien (Austria), with the specialization in photogrammetry/computer vision. She obtained her first diploma degree (Bachelor) with honours, for having the best academic record, being awarded with the San Isidoro prize. She was an ERASMUS, PROMOE and Leonardo da Vinci research fellow at the Institute of Photogrammetry and Remote Sensing (Vienna, 1999-2002), a PhD research fellow at the Mixed Reality Laboratory of the University of Nottingham (UK, 2005) and at the Interaction and Entertainment Research Centre of the Nanyang University of Singapore (2006). She received an outstanding PhD. Award by the UPV. First woman receiving the EH Thompson Award (best paper), given by the Remote Sensing and Photogrammetry Society (2010). During 2008-2010 she worked at the Photogrammetry and Laser Scanning Research Group (GIFLE) of the UPV, and during 2011-2012 she at the Technological Institute of Optics, Colour and Imaging (AIDO), being primarily involved in computer-vision related projects and in the project FP7-SYDDARTA, coordinating the technical work of the WP dedicated to software implementation and carrying out managerial tasks. She has been designed (since 2014) as the proposal coordinator for her research group (ARTEC). She is author of more than 60 scientific publications including international conferences, high impact journals, books and book chapters. She has been invited speaker by Univ. Granada, Aula Natura, UNITEC (Honduras), RUVID & Univ. Gjøvik (Norway). She is S&T program committee of diverse international conferences (e.g. ACM SIGCHI ACE, GECCO), highlighting her involvement in the IEEE ISMAR (CORE A*) for taking decisions on the selected papers. She is also reviewer of scientific journals with impact factors (e.g. MDPI Sensors, Springer Journal of Digital Imaging, Elsevier Computers in Industry). Cristina has co-organized the successful ACM Advances in Computer Entertainment Technology Conference 2005. Expert evaluator of FP7 and H2020 proposals. She is Deputy Editor of the scientific journal Multimodal Technologies and Interaction (MTI), and Editor in Chief of the International Journal of Virtual and Augmented Reality.

Hemavathi Ramadoss received B.E. EEE in 2009 from Kumaraguru College of Technology, Coimbatore. She received M.Tech. Control and Instrumentation Engineering from Thiagarajar College of Engineering in 2011.At present she is a full-time research scholar in Department of Electrical and Electronics Engineering in Thiagarajar College of Engineering, India. Her research interests include Power system protection and control, Smart Grid, Wide-area monitoring and Control and Renewable Energy Resources.

K. C. Ramya has research interests which include Fault diagnosis in industrial drives, Neuro Fuzzy computing and Power Electronics & Drives. She has been actively involved in the Internal Quality Assurance Cell at Karunya University. Her contribution in setting up the experiments for the Power Electronics Laboratory is also commendable. She published research papers in both international and national journals as well as in conferences.

Gunasekharan S. is currently working as Professor and Head, Department of Mehcanical Engineering in Malla Reddy Engineering College (Autonomous), Hyderabad. He completed his B.E. Mechanical Engineering degree in K.S. Rangasamy College of Technology, Thiruchengode, M.E. Industrial Engineering in Kumaraguru College of Technology, Coimbatore and Ph.D. at Karpagam University, Coimbatore in the area of Lean Manufacturing. He has 17 years of teaching experience and having various administrative positions. He published his research work in seven various international journals.

Mohamed Salahuddin received the Bachelor's degree (first class honors) in mechanical engineering and the Ph.D. degree in computational engineering from the University of Leicester, Leicester, U.K., in 1999 and 2004, respectively. Currently, he is a Senior Scientist and Capability Group Manager with the Department of Computing Science, Institute of High Performance Computing (IHPC), Agency for Science, Technology, and Research (A*STAR), Singapore. At IHPC, he is involved in many R&D projects with industry and public agencies in diverse computational engineering fields. His research interests include prognostics, optimization, safety and reliability, and system-level integration.

Muhammad Aman Sheikh has received the B.Sc degree in Electronics Engineering, in 2010. He has completed his Master degree and Ph.D degree in Electrical & Electronics Engineering from Universiti Teknologi PETRONAS, Malaysia in 2013 and 2018 respectively. His main research areas are intelligent control, automation & process control, preventive maintenance, condition monitoring, Power Electronics, AC/DC machines, vibration analysis, wireless sensor networks, wireless communication, and distributive scheduling algorithms.

Feng Yang (M'10) received the B.Eng. degree in information engineering and the M.Eng. degree in control engineering from Xi'an Jiaotong University, Xi'an, China, and the Ph.D. degree in machine learning and bioinformatics from Nanyang Technological University (NTU), Singapore, in 2012. From June 2011 to March 2012, he was a Research Associate at NTU. In 2012, he joined the Institute of High Performance Computing (IHPC), Agency for Science, Technology, and Research (A*STAR), Singapore. Currently, he is a Scientist in the Department of Computing Science, IHPC. His research interests include machine learning, signal processing, information retrieval, big data, and data-based diagnosis and prognosis.

Index

3D registration 100, 103, 105-107, 109, 114, 120

A

air gap eccentricity 141, 168
ANFIS 208-209, 214-216, 219
artificial intelligence 8-9, 190-193, 198, 206
Augmented Reality 98-100, 102, 104-106, 108, 113, 120
authoring tool 98, 106, 109-113, 116-117

B

bearing defects 122-124, 166, 184
Bearing Failure 20, 170
Bearing Fault 5, 19, 122-125, 127, 131, 144, 146
broken rotor bars 18, 59, 75, 77, 83, 85, 139, 167, 173, 176, 205, 207, 210, 218-219

C

cage induction motor 3, 62, 76, 91, 94, 135, 139-140, 151, 154
condition monitoring 1-2, 8, 12-13, 23, 30, 33, 122-123, 131, 152-157, 159, 163-164, 184, 189, 191, 193, 196, 198, 222, 227, 229
correlation matrix 67, 226
cracks 140, 166-168, 244
crawling 5, 131, 151-152

D

decision making 223, 229-230, 232-233
diagnosis 1-4, 8-11, 13, 16, 18, 22-24, 28-33, 39, 44, 46, 54, 59-60, 62, 75-76, 89, 91, 123-126, 163, 169, 172, 193, 205
dynamic smoothing 247-249

E

eccentricity 5, 18-19, 24-25, 29-30, 90, 141, 156-157, 166, 168
Eigen decomposition 70
electrical machine 98-99, 108-109, 113, 116, 120, 173, 176, 184
electronic nose 261

F

fault detection 1-2, 12-13, 15, 24, 29, 31-32, 38-39, 41-47, 49, 52, 54, 60, 90, 95, 153, 155, 157, 163-164, 189-192, 222-223, 233
fault diagnosis 1-3, 8-11, 16, 18, 23-24, 28-30, 32-33, 39, 46, 54, 89, 123, 193
fault isolation 45
feature extraction 30, 39, 193, 242-243, 262
feature selection 30, 192, 242-244
frequency band 59, 65, 74, 77, 82, 84

H

health degradation 245, 252, 255

health index (HI) 237
High Resolution 61, 67, 80

I

Induction Motor 1-8, 13-14, 23, 29-33, 43-45, 59-62, 76-77, 88-91, 93-96, 122-123, 125, 127, 131-135, 137-141, 144, 146-147, 150-154, 159, 166, 237, 239, 241, 251-252, 254-255
interaction 90, 103-106, 120, 125, 157, 168, 170, 174, 182, 263
invasive methods 8, 122-123

M

Mean Time Between Failure (MTBF) 190, 223
Mean Time to Repair (MTTR) 190, 223
misalignment 141, 156, 165, 195
Mobile AR 98, 100, 109-110, 117, 121
Motor Current Signature Analysis 1, 10, 92, 172

N

neuro-fuzzy 205-206, 219
Non-destructive testing (NDT) 223

O

optimization 1, 54, 189-190, 193, 195-196, 198, 200, 246
overall vibration 170, 183

P

PCA 30, 261, 265
Periodogram 59-60, 62, 65-67, 78
power system 42-43, 54
Prognostics and health management (PHM) 237
PSD 59-60, 62, 65, 67, 78-79, 82, 84-85

R

Remaining Useful Life (RUL) 237-238, 256
Rotor Fault 14, 24, 29, 62, 75, 85, 94, 96, 131, 139-141, 158-159
rotor mass unbalance 5, 131, 141, 144

S

SCADA 98, 100-102, 106-117, 121
spectral analysis 3, 10, 13-14, 19, 60-62, 67, 90, 173-174, 176-177, 184, 205-206, 210-211, 213, 215
Squirrel Cage Rotor 78, 135, 157
stator current 8, 10-11, 13-14, 17-22, 24, 29-30, 33, 42, 44, 60, 62-63, 65-69, 74-75, 77-85, 140, 154, 156-158, 205-207, 210-211, 213-215, 218-219
Stator Fault 13, 96, 131, 146, 154, 168, 205
stator winding faults 7, 17, 124-126, 147
supervision 98-107, 113-114, 116-117, 121

T

thermal aging 241-242, 252, 255
three phase induction motor 3, 90, 95, 137, 150

U

unbalance 5, 21-22, 131, 141-142, 144, 165

V

virtual panels 98, 109-113, 116-117

W

wavelets 14, 178-179, 193

www.ingramcontent.com/pod-product-compliance
Lightning Source LLC
LaVergne TN
LVHW061545290926
849162LV00004B/905